F. Hall, H.N.C., C.Ed, C.G.F.T.C, M.N.A.P.T, R.P.

Building services and equipment

Volume 1
Second edition

Longman
Scientific &
Technical

Longman Scientific & Technical,
Longman Group UK Limited,
Longman House, Burnt Mill, Harlow,
Essex CM20 2JE, England
and Associated Companies throughout the world.

First published 1976
Seventh impression 1985
Second edition 1987
Fifth impression 1991

British Library Cataloguing in Publication Data

Hall, F.
  Building services and equipment. – 2nd ed. –
(Longman technician series. Construction and
civil engineering)
Vol. 1
1. Buildings-Environmental engineering
I. Title
696   TH6021

ISBN 0-582-41394-X

LONGMAN TECHNICIAN SERIES
General Editor – Construction and Civil Engineering
C. R. Bassett, B.Sc. F.C.I.O.B.

Formerly Principal Lecturer in the Department of Building
and Surveying, Guildford Coonty College of Technology

Produced by Longman Singapore Publishers (Pte) Ltd.
Printed in Singapore.

Contents

Acknowledgements

We are grateful to the following for permission to reproduce copyright material: British Standards Institution for extracts from the following Codes of Practice: CP 2005 – Sewerage, BS 6700 – Water Supply, CP 342 – Centralised Hot Water Supply, CP 352 – Mechanical Ventilation, CP 8301 – Building Drainage, BS 5572 – Soil & Waste Systems, CP 3 Ch. VII – Engineering & Utility Services, CP 302 – Small Sewerage Treatment Works. Reproduced by permission of BSI, 2 Park Street, London W1A 2BS from whom complete copies can be obtained; The Controller of Her Majesty's Stationery Office for extracts from *Building Regulations and Model Bylaws;* Building Research Establishment for two tables from Digests 115 and 151; The Chartered Institution of Building Services for tables from *CIBS Guide*.

We would appreciate any information that would enable us to trace the copyright holder of a table entitled 'Drainage and Sewerage' from the *Clay Pipe Development Association Journal*.

We are also grateful to the following for permission to reproduce diagrams:

Building Research Establishment; Elsan Sewage Systems Limited; Klargester Environmental Engineering Limited; Marley Plumbing; Marscar Limited; Spirax Sarco Limited; Twyfords Limited; Wavin Plastics Limited.

Chapter 1

Cold-water supply

Water supply

A plentiful supply of wholesome water is essential for the occupants of the buildings intended for human habitation. Most buildings can obtain their supply from the Water Authorities' main, but in rural areas it is sometimes necessary to obtain water from private sources, such as streams, rivers, lakes, wells, springs or by catchment areas from roofs and paved surfaces. The water used must be colourless, free from small suspended matter and harmful bacteria, pleasant to taste and for health reasons moderately 'hard'. The River Pollution Commissioners classify water from the various sources as follows:

Wholesome	1. Spring water 2. Deep well water	Very palatable
	3. Uplands surface water	Moderately palatable
Suspicious	4. Stored rainwater	
	5. Surface water from cultivated lands	Palatable
Dangerous	6. River water to which sewage gains access 7. Shallow well water	

Figures 1.1 and 1.2 show various sources of water supply.

Hardness of water

Generally, surface waters are 'soft' and subterranean waters 'hard'; a great deal however depends upon the type of earth strata with which the water comes into contact.

Note: The term 'hardness' means it is difficult to obtain a lather with soap. Water is an excellent solvent and the presence of dissolved carbon dioxide increases this solvent power.

Two types of hardness

1. *Temporary:* If the water passes through strata containing a carbonate of calcium, or magnesium, a certain amount of these salts will be taken into the solution, depending upon the amount of carbon dioxide present in the water. Upon being dissolved the carbonate becomes bicarbonate due to the presence of carbon dioxide. This type of hardness can be removed by boiling the water, hence the term 'temporary'. It causes scaling or furring of hot-water pipes and boilers unless an indirect system is used (see Chapter 2, Hot-water supply).

2. *Permanent:* If the water passes through the strata containing calcium sulphate, calcium chloride or magnesium chloride, the salts are readily dissolved in the water without the presence of carbon dioxide. This type of hardness cannot be removed by boiling the water and hence the term 'permanent'. It will not cause scaling or furring, unless the water is brought up to high temperatures and pressures, but it may cause corrosion. Most waters contain both temporary and permanent hardness, usually more temporary; for example, Guildford in Surrey uses water containing 200 and 50 parts per million (p.p.m.) of temporary and permanent hardness respectively. The generally accepted classification of hardness is given in Table 1.1.

Table 1.1

Type	Hardness — p.p.m.
Soft	0–50
Moderately soft	50–100
Slightly hard	100–150
Moderately hard	150–200
Hard	200–300
Very hard	over 300

Removal of temporary hardness

Temporary hardness, as already stated, can be removed by boiling which drives off the carbon dioxide, allowing the carbonate of calcium or magnesium to precipitate. Whilst small quantities of water may be softened in this way it is, however, impractical for large-scale treatment.

Clark's process: This consists of adding small quantities of lime water or cream of lime to the supply and this takes up the carbon dioxide from the bicarbonate present, resulting in the precipitation of the insoluble carbonate and the removal

2

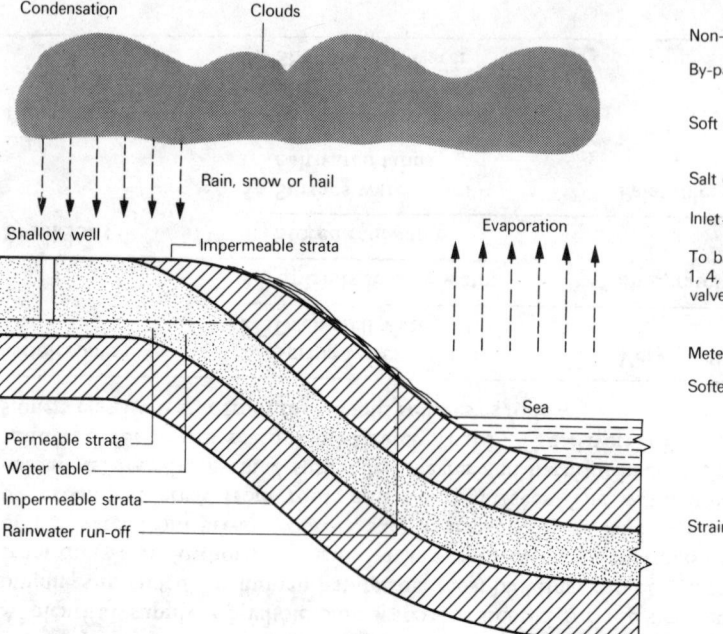

Fig. 1.1 The rain cycle

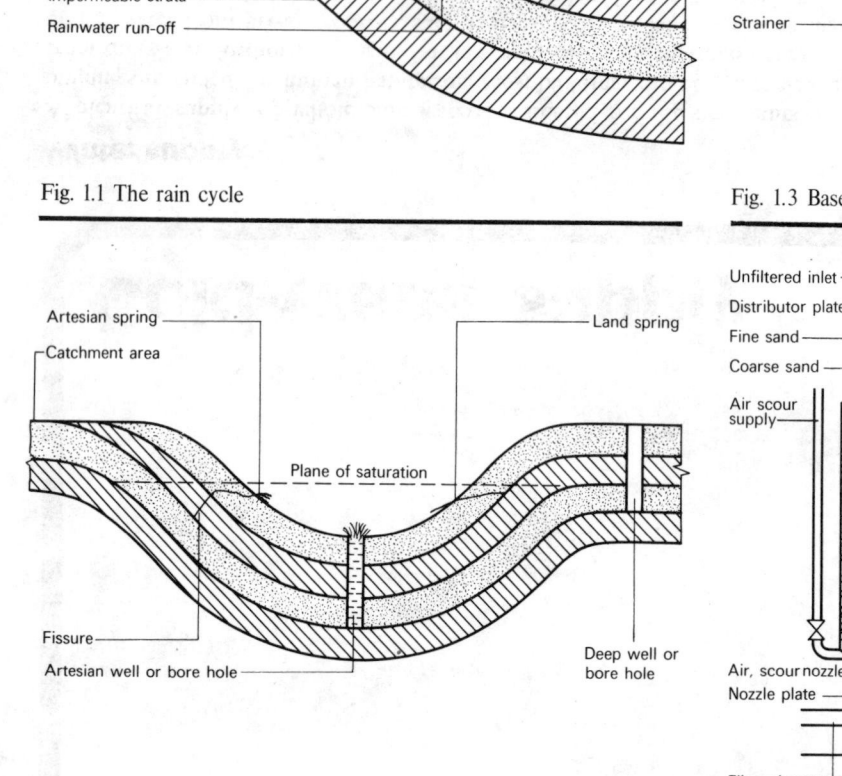

Fig. 1.2 Wells and springs

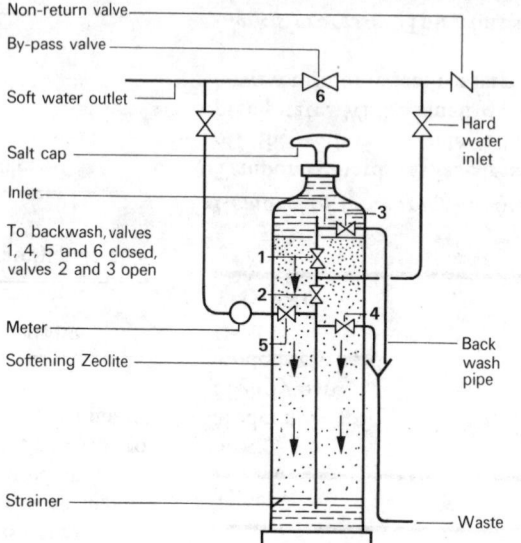

Fig. 1.3 Base-exchange water softener

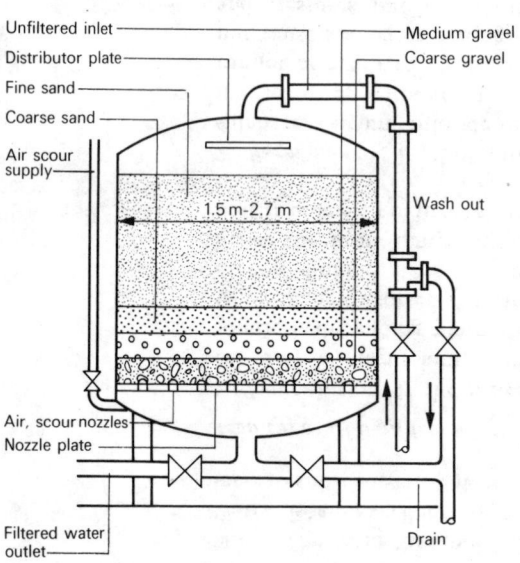

Fig. 1.5 Pressure filter

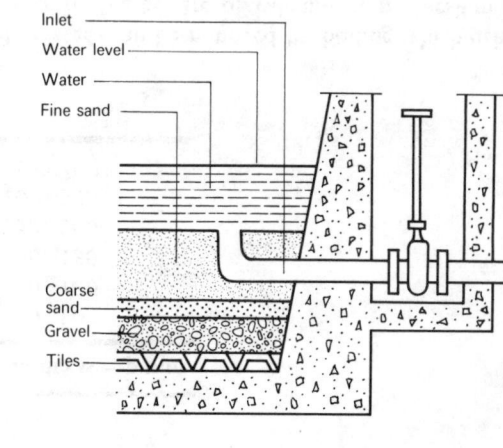

Fig. 1.4 Slow sand filter

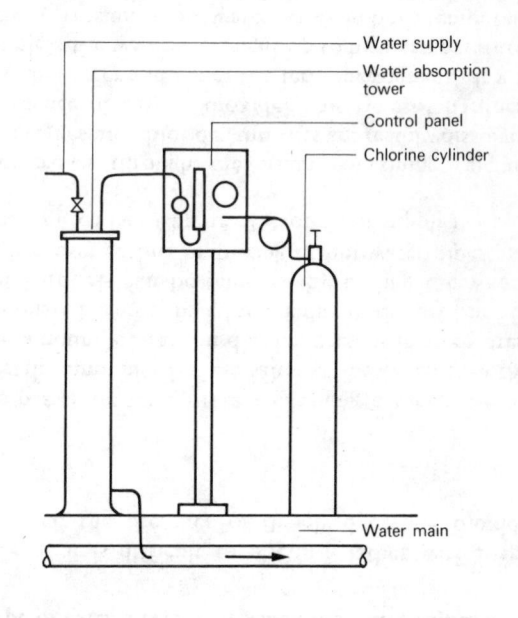

Fig. 1.6 Chlorinating plant

of most of the temporary hardness. The precipitated carbonate is either allowed to settle out in a tank or is arrested in fine screens.

Removal of permanent hardness

The term 'permanent' is a misnomer, and has been used because this type of hardness cannot be removed by boiling the water. The hardness, however, may be removed by the addition of washing soda (sodium carbonate). The sodium carbonate becomes sodium sulphate which remains in solution in the water and is not harmful.

Base exchange process (Fig. 1.3)

This process removes both temporary and permanent hardness very efficiently by passing the water through zeolites contained in a steel cylinder. Zeolites have the property of exchanging their sodium base for the magnesium or calcium base, hence the term 'base exchange'.

The process is as follows:

Sodium Zeolite + Calcium Sulphate or carbonate
  (in softener)   (in water)

becomes

Calcium Zeolite + Sodium Sulphate or carbonate
(held in softener)   (in solution with the water,
                                but harmless)

After a period of use, the sodium zeolite is converted into calcium and magnesium zeolite, thus losing its softening power. It is then regenerated by the addition of a strong solution of common salt (sodium chloride). The salt is kept in contact with the calcium zeolite for about half an hour, in which time the calcium zeolite is converted back into sodium zeolite.

The process is as follows:

Calcium Zeolite + Sodium Chloride
  (exhausted   (common salt)
    sodium)

becomes

Sodium Zeolite + Calcium Chloride
  (regenerated)   (flushed to drain)

Lime soda process

For industrial use, water which contains both temporary and permanent hardness may be softened by the addition of lime and soda in the correct proportions and the precipitated salts again settled out.

Scale reducers

In order to reduce scale formation and corrosion in hot-water systems, low concentration of metaphosphates may be used. Technically, the water remains hard and therefore these methods of water treatment should not be regarded as water softeners. A special dispenser may be connected to the cold-water main or suspended in the cold-water cistern.

Soft water

Although soft water readily lathers it is not as palatable for drinking as is water containing some degree of hardness. Soft water may cause corrosion of iron, steel, zinc, copper and lead. Copper is a safe metal to use because a small amount is non-poisonous. If lead, however, is taken into solution (Plumbo solvency) lead poisoning would result. The corrosion of metals is accelerated if the water passes through organic soils and it should be neutralised by passing it through limestone or chalk.

Filtration

Slow sand filters (Fig. 1.4)

These consist of rectangular tanks constructed of brickwork, or concrete, with porous slabs or perforated tiles over collecting channels at its base. The filter bed consists of a layer of fine sand about 1 m in depth, the top of which is kept flooded with water to a depth of about 1.2 m. When the filter is first put into operation it only removes suspended matter by straining. Clean fine sand will not keep back bacteria, but the surface of the sand soon becomes covered with a thin layer of colloidal matter, deposited from the water. This gelatinous film constitutes the actual filter and is a barrier to the passage of bacteria. The production of this film is sometimes done artificially, by the use of a chemical coagulant, such as aluminium sulphate.

Pressure filters (Fig. 1.5)

These are now much used instead of the slow sand filter because they take less space for their installation and it is easier to clean the sand. The principle of operation is the same, but filtration is much quicker. purification of the water takes place in closed steel cylinders working under pressure. The efficiency of the filter is increased by adding a small dose of aluminium sulphate to the inlet water which forms a gelatinous film on the top of the sand, as described for the slow sand filter. This film is known as the 'vital layer'.

Sterilisation (Fig. 1.6)

All the water used for human consumption must be free from harmful bacteria. Chlorine is added to the water, which oxidises the organic compounds in it. The dosage of chlorine is strictly regulated so that there is enough to destroy any bacteria present, but not too much to give an unpleasant taste to the water. The chlorine is stored as a liquid under pressure in special steel cylinders similar to oxygen cylinders, but painted yellow. On being released from the cylinder the liquid converts into a gas, which is then injected into the water main or storage tank. The flow of chlorine is controlled automatically to give the correct dosage.

Note: It is advisable to maintain a residual of free chlorine in the water to deal with any subsequent contamination and chlorination should always follow the filtration process.

Sampling

Where any doubt exists regarding the purity of the water, or for a new source of supply, samples of the water should be examined by a qualified analyst. The

analyst will supply the necessary bottles and instructions for taking the samples, which are both chemical and biological.

Chemical symbols

In order to make the descriptions easier to follow, the chemical symbols have been omitted from the text. For those students wishing to know the symbols, the following list should be useful:

Aluminium sulphate	$Al_2(SO_4)_3$	Magnesia	MgO
Calcium bicarbonate	$Ca(HCO_3)_2$	Magnesium carbonate	$MgCO_3$
Calcium carbonate	$CaCO_3$	Magnesium bicarbonate	$Mg(HCO_3)_2$
Calcium sulphate	$CaSO_4$	Magnesium sulphate	$MgSO_4$
Calcium zeolite	CaZ	Sodium carbonate	Na_2CO_3
Calcium chloride	$CaCl_2$	Sodium sulphate	Na_2SO_4
Hydrated lime	$Ca(OH)_2$	Sodium zeolite	Na_2Z
		Sodium chloride	$NaCl$

Cold-water system

Before designing a cold-water system for a building it is essential to know the local Water Authority requirements. There are two distinct systems, namely: direct and indirect, but some Water Authorities will allow some modifications to these systems.

Direct (Fig. 1.7)

This is a system used extensively in northern districts, where large high-level reservoirs provide a good mains supply and pressure. It is however permitted by several Water Authorities in other districts. In it all sanitary fittings are supplied with cold water direct from the main, and a cold-water feed cistern is required only to 'feed' the hot-water storage cylinder. The capacity in litres of the feed cistern is required to be at least equal to the capacity in litres of the hot-water cylinder. The Water Regulations require a cistern of 114 litres (minimum) capacity and is therefore small enough to be accommodated in the top of an airing cupboard, thus saving lagging of the cistern and pipework.

Indirect (Fig. 1.8)

In this system all the sanitary fittings, except drinking water draw-offs at sinks and fountains, are supplied indirectly from a cold-water storage cistern. Since the cistern supplies cold water to baths, basins, showers, etc., and also feeds the hot-water cylinder, its capacity in litres will be approximately double that required for the direct system. The Water Regulations require a cistern of 227 litres minimum capacity, and therefore it will have to be accommodated in the roof space and will require lagging (see Fig. 1.9). For larger buildings, the capacity of the storage cistern will have to be estimated and Table 1.2 gives the storage requirements for various types of buildings.

It is not always possible at the early design stage to know the exact number of people that will occupy the building but the number and type of sanitary fittings are known. Table 1.3 may be used as a guide to finding the storage of cold water, by estimating the possible maximum use of the fittings per day.

Table 1.2 Provision of cold water to cover 24-hour interruption of supply C.P. Water Supply

Type of building	Storage in litres
Dwelling houses and flats *per resident*	91
Hostels *per resident*	91
Hotels *per resident*	136
Offices without canteens *per head*	37
Offices with canteens *per head*	45
Restaurants *per head/per meal*	7
Day schools *per head*	27
Boarding schools *per head*	91
Nurses' homes and medical quarters *per resident*	114

Table 1.3 Volumes of water, hot and cold, added together required for single use of appliances

Appliance	Volume required in litres
Wash basin	
hand wash	5
hand and face wash	10
hair wash	20
Shower	40
Bath	110
W.C.	10
Washing machine	150
Sink	
wash up	15
cleaning	10

Back siphonage

This is the back flow of water into the drinking water supply main. In order for back siphonage to occur a partial vacuum must be created in the pipe connected to a valve or tap, with its outlet submerged in water, which may be contaminated. This is possible when the demand on the water main is sufficient to draw back the water, thus leaving behind a partial vacuum.

In order to prevent back siphonage, the following points must be observed.

1. Ball valves in cisterns should be fitted above the overflow pipe, and if a silencer pipe is fitted its outlet must be above the valve (see Fig. 1.30).
2. The outlets of taps connected to sinks, baths and basins should be well above the flooding level of the fitting.
3. Fittings having low-level water inlets, for example, bidets, should be supplied with cold water from a storage cistern and never direct from the water main.

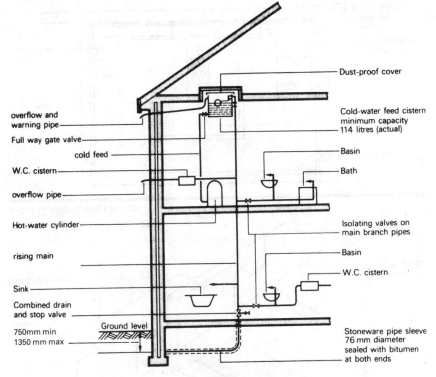

overflow and warning pipe
Full way gate valve
cold feed
W.C. cistern
overflow pipe
Hot-water cylinder
rising main
Sink
Combined drain and stop valve
750mm min 1350 mm max
Ground level

Dust-proof cover
Cold-water feed cistern minimum capacity 114 litres (actual)
Basin
Bath
Isolating valves on main branch pipes
Basin
W.C. cistern
Stoneware pipe sleeve 76 mm diameter sealed with bitumen at both ends

Fig. 1.7 Direct system of cold water supply

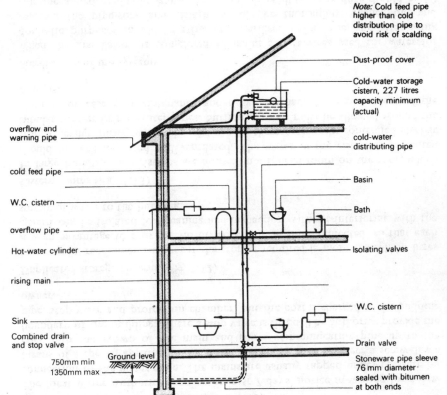

Note: Cold feed pipe higher than cold distribution pipe to avoid risk of scalding

overflow and warning pipe
cold feed pipe
W.C. cistern
overflow pipe
Hot-water cylinder
rising main
Sink
Combined drain and stop valve
750mm min 1350mm max
Ground level

Dust-proof cover
Cold-water storage cistern, 227 litres capacity minimum (actual)
cold-water distributing pipe
Basin
Bath
Isolating valves
W.C. cistern
Drain valve
Stoneware pipe sleeve 76 mm diameter sealed with bitumen at both ends

Fig. 1.8 Indirect system of cold water supply

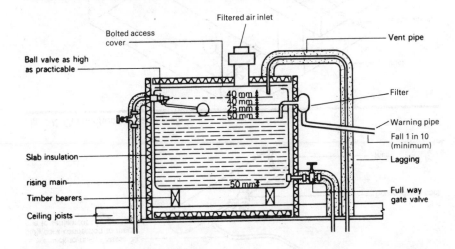

Bolted access cover
Filtered air inlet
Vent pipe
Ball valve as high as practicable
40 mm
40 mm
25 mm
50 mm
Filter
Warning pipe
Fall 1 in 10 (minimum)
Slab insulation
50 mm
Lagging
rising main
Timber bearers
Ceiling joists
Full way gate valve

Fig. 1.9 Detail of cold water cistern

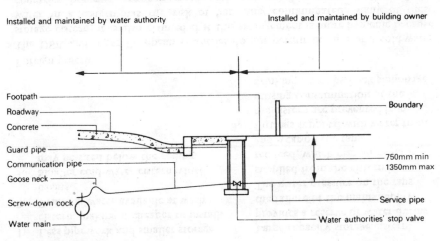

Installed and maintained by water authority
Installed and maintained by building owner

Footpath
Roadway
Concrete
Guard pipe
Communication pipe
Goose neck
Screw-down cock
Water main
Boundary
750mm min 1350mm max
Service pipe
Water authorities' stop valve

Fig. 1.10 Connection to water main

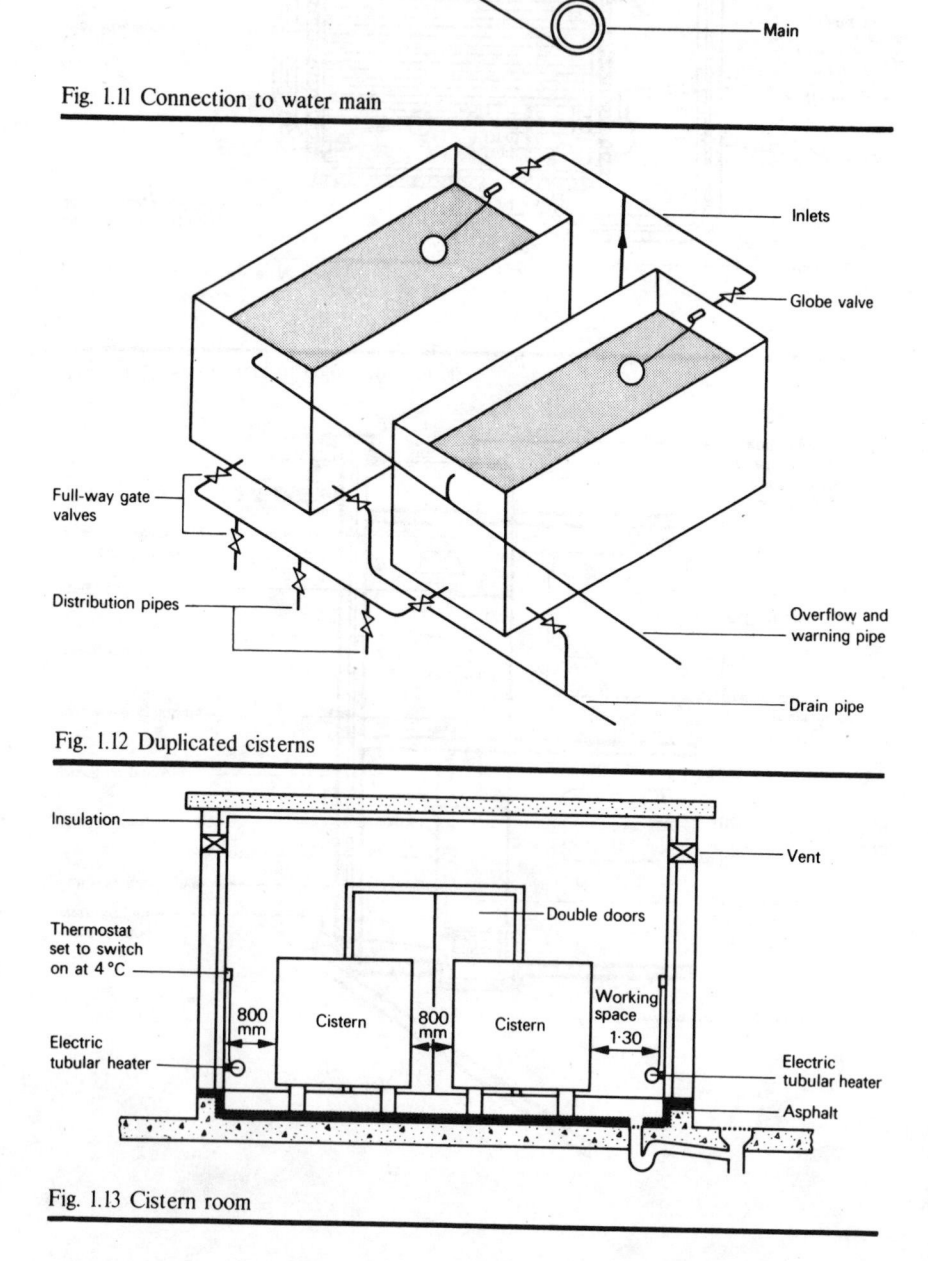

Goose neck to relieve stress on plug cock connection to main due to settlement of pipe

Screw-down cock

Main

Fig. 1.11 Connection to water main

Inlets

Globe valve

Full-way gate valves

Distribution pipes

Overflow and warning pipe

Drain pipe

Fig. 1.12 Duplicated cisterns

Insulation

Vent

Double doors

Thermostat set to switch on at 4 °C

800 mm | Cistern | 800 mm | Cistern | Working space 1·30

Electric tubular heater

Electric tubular heater

Asphalt

Fig. 1.13 Cistern room

Advantages of direct and indirect systems

Direct

1. Less pipework and smaller storage cistern, making it cheaper to install.
2. Drinking water available at wash basins.
3. Smaller cold-water cistern which may be sited below the ceiling.

Indirect

1. Large capacity storage cistern, provides a reserve of water during the failure of the mains supply.
2. The water pressure on the taps supplied from the cistern is reduced, which minimises noise and wear on the taps.
3. Fittings supplied with water from the cistern are prevented from causing contamination of the drinking water by back siphonage.

Foreign practice

The USA and most European countries do not permit the use of a cold-water storage cistern; it is also claimed that this arrangement is more hygienic because water in a cistern has the risk of becoming contaminated. Many of these countries however are concerned with the possible contamination of the drinking water by back siphonage.

Connection to water main (Figs 1.10 and 1.11)

The local Water Authority requires at least 7 days' notice in writing for a new connection to the water main. The mains are usually tapped under pressure by means of a special apparatus. The Water Authority then connect a screw-down cock into the crown of the main and run a communication pipe up to the boundary of the building, where a stop valve is fitted. The authority provide the pipe, stop valve and protection chamber, but the cost is charged to the building owner.

Duplicated storage cisterns (Fig. 1.12)

It is usually recommended that if the storage required is more than 4500 litres it is an advantage to provide two or more cisterns interconnected, so that each cistern can be isolated for cleaning and inspection without interfering with the supply of water to the building.

Cistern rooms (Fig. 1.13)

In large buildings the cisterns are housed in a special room on the roof, which should be well insulated and ventilated. The water in the cistern is prevented from freezing during winter by the use of thermostatically controlled electric tubular heaters or hot-water pipes. Sufficient room must be allowed around the cisterns for ease of maintenance and provision made for replacement of the cisterns.

Sterilisation of the systems

When the installation is completed all mains and services used for water for domestic purposes should be efficiently sterilised before use, and after being repaired the pipework and cisterns should be thoroughly flushed out and chlorine added gradually whilst the cistern is refilling. The dose should be 50

parts of chlorine to 1 million parts of water. When the cistern is full all taps on the distributing pipes should be opened successively, working progressively away from the cistern. Each tap should be closed when the water discharged begins to smell of chlorine and the cistern should then be refilled and more chlorine added as before. The cistern and pipes should then remain charged for at least 3 hours and then tested for residual chlorine; if none is found sterilisation will have to be carried out again. Finally, the cistern and pipes should be thoroughly flushed out.

Water supplies for buildings where the mains pressure is insufficient

Principles

For high-rise buildings or for buildings constructed on high ground where the pressure is low, it is necessary to pump the water from inside the building. Before designing the water system it is necessary to ascertain the pressure on the water main during the peak demand period. If for example the water mains pressure during the peak demand is 300 kPa this pressure will supply water inside the building up to a height of approximately 30 m. In order to give a good supply of water, a residual head is required above the highest fittings and it is therefore usual to deduct 6 m from the mains pressure head. Therefore, in this example, the water main would supply water up to 24 m without pumping.

Direct pumping from the main (Fig. 1.14)

Where the water authority will allow this method of boosting, the pressure of water required to be developed by the pump will be reduced by the amount equal to the pressure of water on the main. The disadvantage of this method is that if the building is large the amount of water pumped from the main may cause a serious drop in pressure in other buildings supplied from the same main. The method does however save the cost of a break cistern, space inside the building and costs.

Use of drinking water header-pipe (Fig. 1.14)

In order to supply the upper floors above the water mains pressure with drinking water, a horizontal or vertical header-pipe may be installed and to prevent stagnation of the water inside the pipe its size should not be greater than will supply up to 4.5 litres per dwelling. The pipe greatly reduces the frequency with which the pump would be switched on and off. When the pump is off drinking water is supplied by gravity from the drinking water heater, and if this is eventually emptied the pipeline switch will switch on the pump.

Indirect pumping from the main (Fig. 1.15)

Many water authorities require a break cistern installed between the main and the pumping unit; this cistern will serve as a pumping reservoir and prevent lowering the pressure on the main. A low-level switch fitted on the cistern will protect the pump if the water level drops too low and at low water level the switch cuts out the pump.

Use of drinking water cistern (Figs 1.16 and 1.17)

In buildings of twenty storeys or more in height the use of one header-pipe will give too much pressure on the drinking water taps on the lower floors; to prevent this, the floors are zoned by use of a drinking water cistern, so that the head of water above the lowest fitting does not exceed 30 m.

Use of pneumatic cylinder (Figs 1.18, 1.19 and 1.20)

As an alternative to a drinking water header-pipe, a steel cylinder placed either vertically or horizontally may be used for boosting the drinking water to the upper floors. The cylinder contains a cushion of air under pressure, in contact with the water to be boosted; this cushion of air forces water up to the drinking water taps and as the water is drawn off the air expands and its pressure falls. At a predetermined low water level a low-level switch cuts in the pump, the pump will then satisfy the demand taking place and gradually increase the air pressure in the cylinder, until at a predetermined high water level a high-level switch cuts out the pump. The pressure switch settings are such that the minimum water pressures at the roof level cistern ball valve are 70 kPa and 100 kPa respectively.

Purpose of the air compressor

After a period of use some of the air is absorbed into the water and a gauge glass is usually fitted to give a visual indication of the water level. As more air becomes absorbed a smaller quantity is available to boost the water and the frequency of the pump operation is increased. To prevent this, a float switch is fitted in the vessel and is arranged to start the air compressor at a predetermined high water level. The compressor will force air into the cylinder until the correct water level is achieved, thus allowing the float switch to lower and thus cutting out the compressor. The ball valves in the water cisterns served from a pneumatic cylinder should preferably be the delayed action type to conserve the air pressure in the cylinder and minimise the number of pump operations. This type of ball-valve arrangement does not allow the valve to open until the cistern is approximately two-thirds empty. It will also allow the ball valves to stay fully open until the cistern is almost full, whereas a normal ball valve closes very gradually and takes longer to fill the cistern (see Fig. 1.21).

Supply to baths, showers, W.C.s (Figs 1.14, 1.17, 1.19 and 1.20)

So far the problem of supplying drinking water to sinks or fountains has been considered. In high-rise buildings the water supply to baths, showers, W.C.s, etc., must also be considered. As already stated for drinking water fittings, the maximum head of water is 30 m and therefore the floors must be zoned, by use of break pressure cisterns or by pressure-reducing valves.

Taps and valves

The term 'taps', 'valves' and 'cocks' are used indiscriminately to name fittings required to control the flow of fluids, either along or at the end of a pipeline. Valves are usually used to control the flow along a pipeline, whilst taps are usually used at the end of a pipeline for draw-off purposes. Cocks consist of a

8

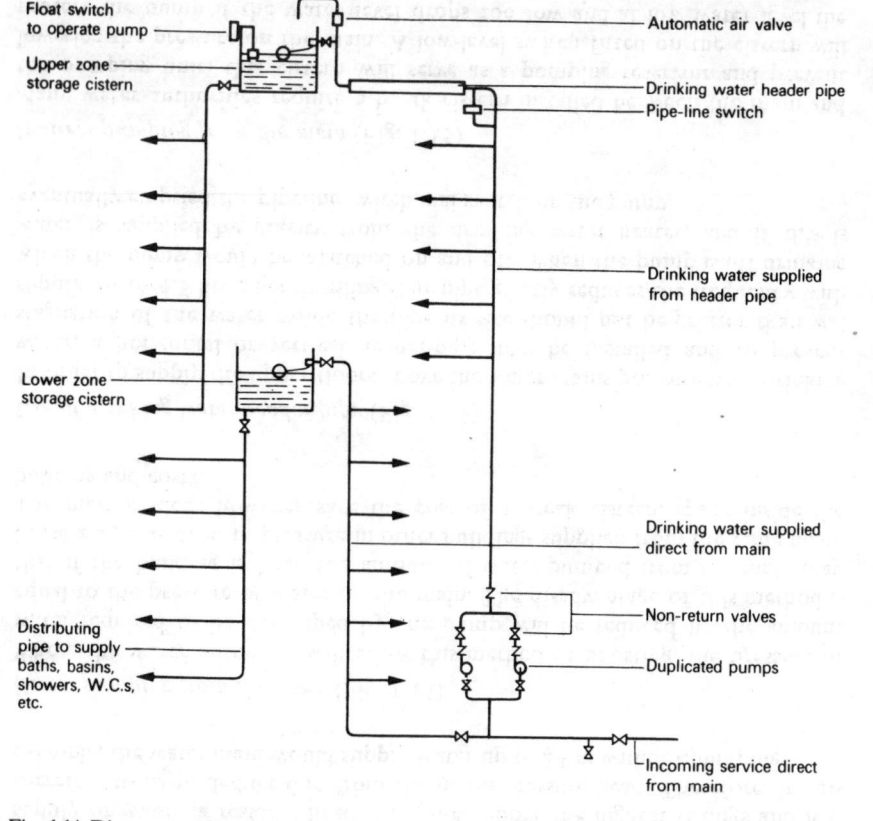

Float switch to operate pump

Upper zone storage cistern

Automatic air valve

Drinking water header pipe

Pipe-line switch

Drinking water supplied from header pipe

Lower zone storage cistern

Drinking water supplied direct from main

Non-return valves

Duplicated pumps

Distributing pipe to supply baths, basins, showers, W.C.s, etc.

Incoming service direct from main

Fig. 1.14 Direct pumping from main

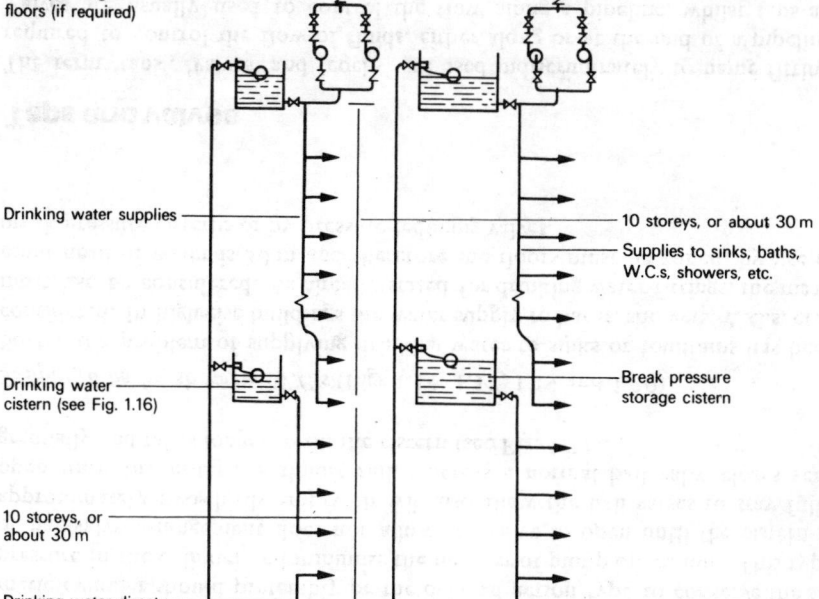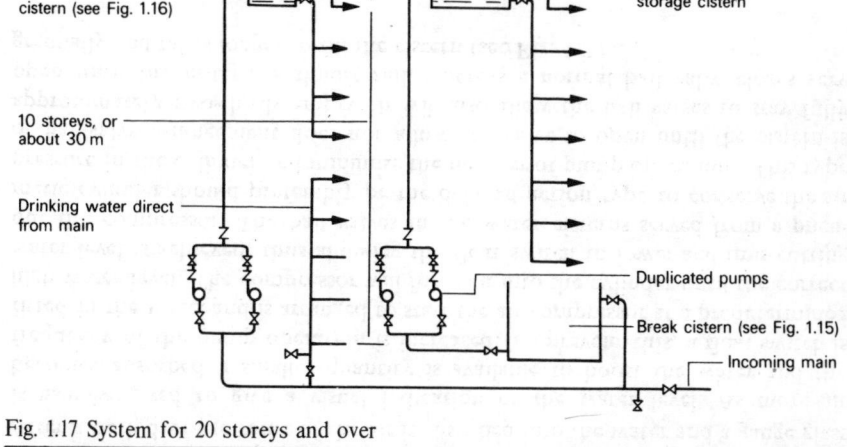

Supply to higher floors (if required)

Drinking water supplies

10 storeys, or about 30 m

Supplies to sinks, baths, W.C.s, showers, etc.

Drinking water cistern (see Fig. 1.16)

Break pressure storage cistern

10 storeys, or about 30 m

Drinking water direct from main

Duplicated pumps

Break cistern (see Fig. 1.15)

Incoming main

Fig. 1.17 System for 20 storeys and over

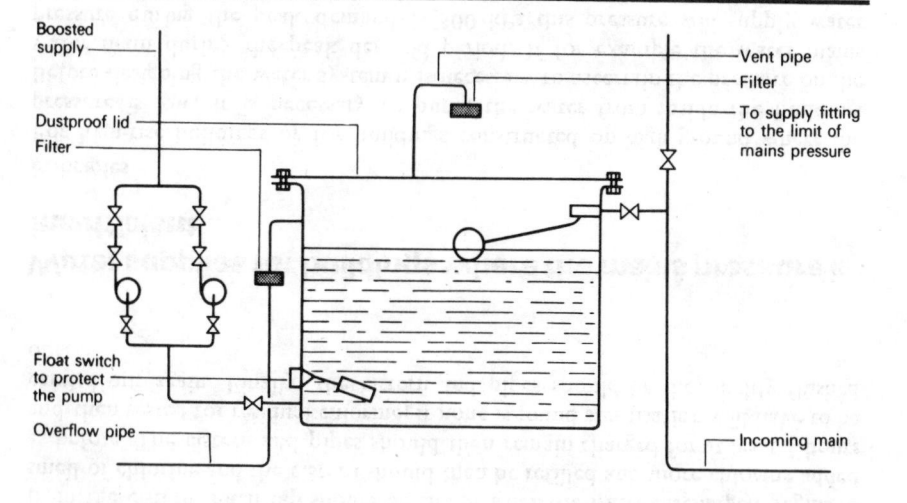

Boosted supply

Vent pipe

Filter

To supply fitting to the limit of mains pressure

Dustproof lid

Filter

Float switch to protect the pump

Overflow pipe

Incoming main

Fig. 1.15 Indirect pumping from main

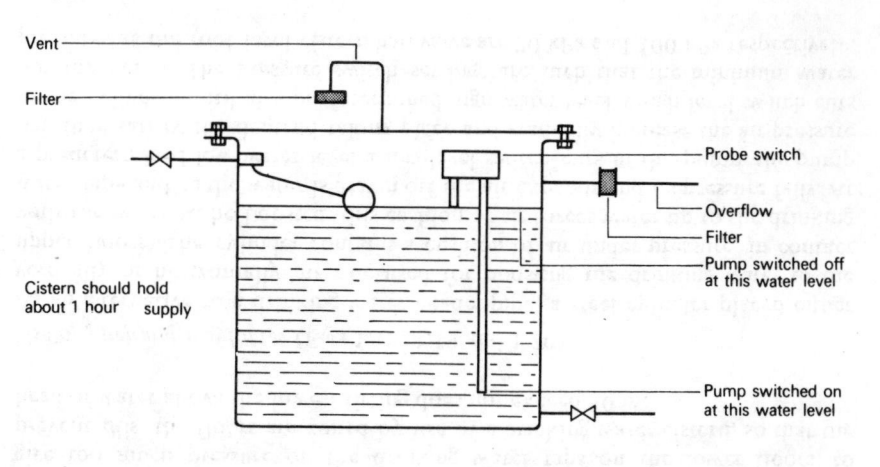

Vent

Filter

Probe switch

Overflow

Filter

Pump switched off at this water level

Cistern should hold about 1 hour supply

Pump switched on at this water level

Fig. 1.16 Drinking water break cistern

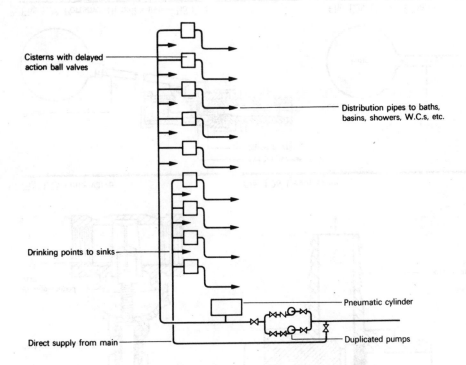

Cisterns with delayed action ball valves

Distribution pipes to baths, basins, showers, W.C.s, etc.

Drinking points to sinks

Pneumatic cylinder

Direct supply from main

Duplicated pumps

Fig. 1.19 Pneumatic cylinder with individual cisterns for a tall block of flats

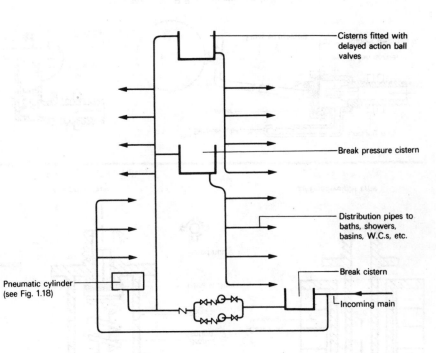

Cisterns fitted with delayed action ball valves

Break pressure cistern

Distribution pipes to baths, showers, basins, W.C.s, etc.

Break cistern

Pneumatic cylinder (see Fig. 1.18)

Incoming main

Fig. 1.20 Pneumatic cylinder for an office block

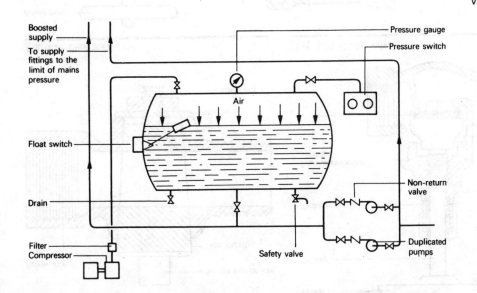

Boosted supply

To supply fittings to the limit of mains pressure

Pressure gauge

Pressure switch

Air

Float switch

Drain

Non-return valve

Filter
Compressor

Safety valve

Duplicated pumps

Fig. 1.18 Automatic pneumatic cylinder

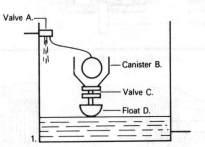
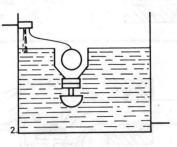

Valve A.

Canister B.

Valve C.

Float D.

1.

2.

3.

1. Water level below float D, valves A and C open

2. Cistern full of water and overflowing into canister and valve A closing and valve C closed

3. Cistern being emptied whilst valve C is closed and valve A is delayed from opening

Fig. 1.21 Delayed action ball valve

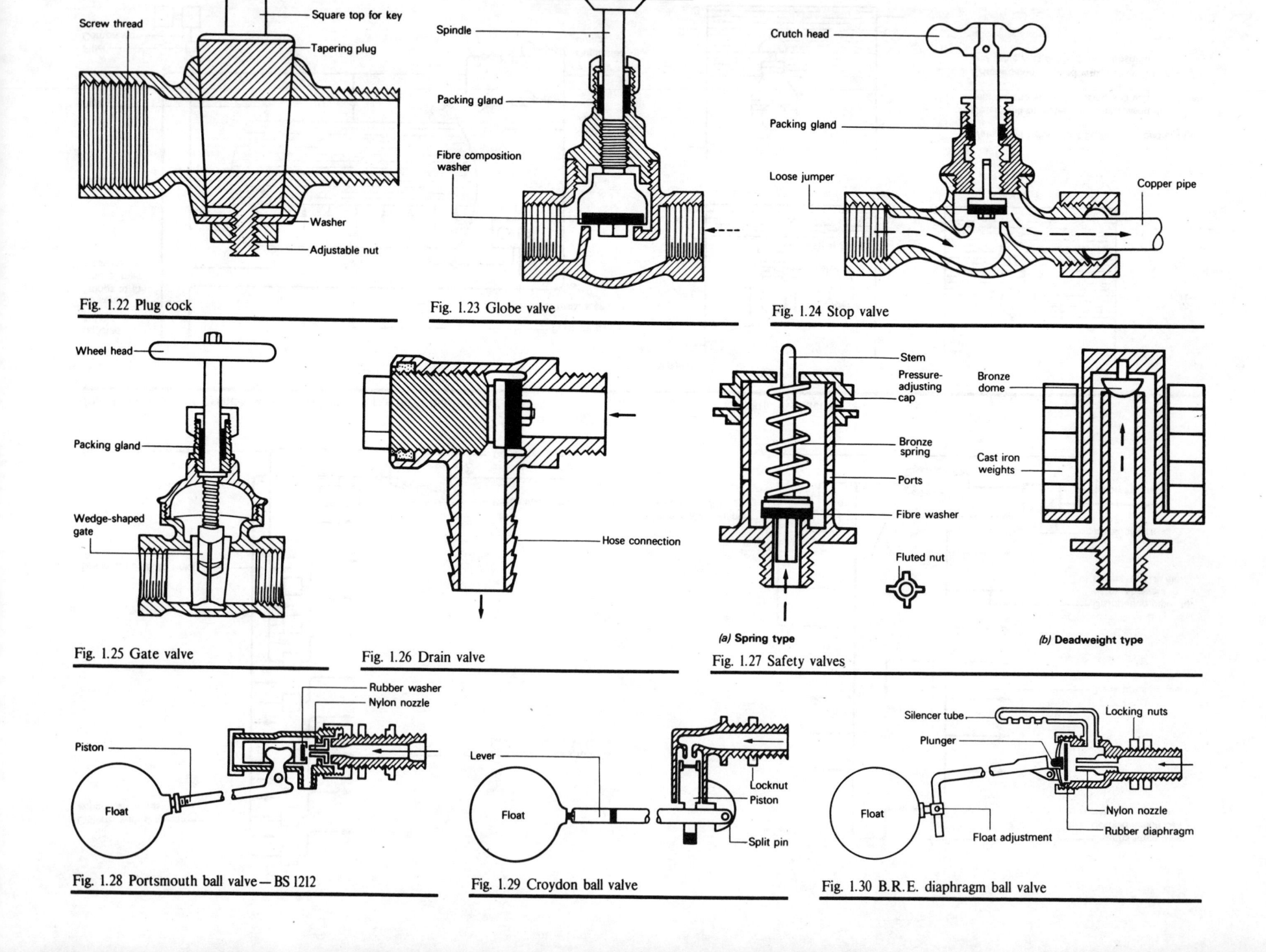

10

Fig. 1.22 Plug cock

Screw thread
Square top for key
Tapering plug
Washer
Adjustable nut

Fig. 1.23 Globe valve

Wheel
Spindle
Packing gland
Fibre composition washer

Fig. 1.24 Stop valve

Crutch head
Packing gland
Loose jumper
Copper pipe

Fig. 1.25 Gate valve

Wheel head
Packing gland
Wedge-shaped gate

Fig. 1.26 Drain valve

Hose connection

Fig. 1.27 Safety valves

Stem
Pressure-adjusting cap
Bronze spring
Ports
Fibre washer
Fluted nut

(a) **Spring type**

Bronze dome
Cast iron weights

(b) **Deadweight type**

Fig. 1.28 Portsmouth ball valve — BS 1212

Rubber washer
Nylon nozzle
Piston
Float

Fig. 1.29 Croydon ball valve

Lever
Float
Locknut
Piston
Split pin

Fig. 1.30 B.R.E. diaphragm ball valve

Silencer tube
Locking nuts
Plunger
Float
Nylon nozzle
Rubber diaphragm
Float adjustment

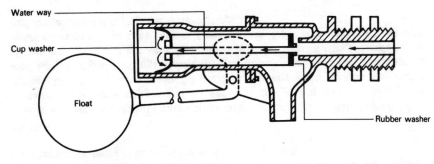

Fig. 1.31 Equilibrium ball valve — Portsmouth type

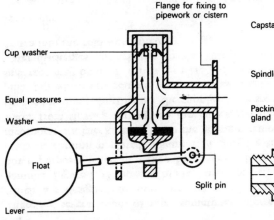

Fig. 1.32 Equilibrium ball valve — Croydon type

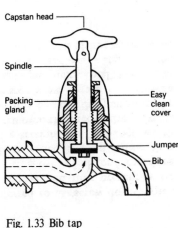

Fig. 1.33 Bib tap

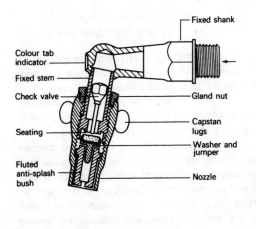

Fig. 1.34 'Supatap' bib tap

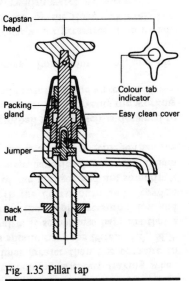

Fig. 1.35 Pillar tap

body holding a tapered plug (see Fig. 1.22). The plug has a hole cast or drilled in the middle through which the fluid can flow. The cock can be fully opened or fully closed by turning the plug through an angle of 45° (quarter turn). It can be closed very quickly and may cause water hammer; it is therefore not usually permitted in water supply systems, but they are used extensively in gas installations.

Valves

Valves used to control the flow along a pipeline are known as the globe or gate types. Both types close slowly and therefore do not usually give rise to problems of water hammer.

Globe valves

These are used on high-pressure systems. Figure 1.23 shows a section of one type of globe valve; the metal-to-metal seating type is often used for heating systems and the composition valve for very high-pressure systems where a complete shut off is required. Figure 1.24 shows a section of a stopvalve used for domestic water installations. When the valve is used on cold water service pipework the jumper should be loose which tends to act as a non-return valve and prevent backflow into the main.

Gate valves (see Fig. 1.25)

These are used for the control of fluids in low-pressure systems, such as on distribution pipework from storage cisterns, or on low-pressure heating systems. They offer much less resistance to the flow of fluids than the globe valve.

Drain valves (see Fig. 1.26)

These are used to drain boilers, cylinders and sections of pipework.

Safety valves

These are used to relieve excess pressure on boilers, tanks and pipework. Figure 1.27A, B shows spring and deadweight types of safety valves.

Ball valves

These are used to supply water to storage and flushing cisterns and to automatically shut off the supply when the correct water level has been reached. The valve is operated by a float which allows the valve to be fully open when it is in the lower position. As the water level rises, the float also rises which gradually closes the valve and shuts off the supply of water.

Types: There are four types, namely; (1) Portsmouth, (2) Croydon, (3) Diaphragm, (4) Equilibrium.

Portsmouth and Croydon are similar in construction except in the former type the plunger moves horizontally and in the latter the plunger moves vertically. Figure 1.28 shows the Portsmouth type, which complies with BS 1212, and Fig. 1.29 shows the Croydon type. The diameter of the orifice is a very important factor and governs the type of valve; for example, whether high, medium or low pressure. An orifice of smaller diameter allows the valve to close at a higher

water pressure. The 1986 Water Supply Byelaws state that on installation every float-operated valve shall be capable of withstanding without leaking when closed, an internal hydraulic pressure 1.5 times greater than the pressure to which it will ordinarily be subject. The valves should have the letters H.P., M.P. or L.P. cast or stamped on the body of the valve. It sometimes happens that a high-pressure ball valve is fitted to a cistern where the supply pressure is low and this results in slow filling of the cistern. Both the Portsmouth and Croydon pattern ball valves give trouble with sticking of the plunger and are sometimes noisy. The Building Research Station Establishment have developed a diaphragm type of ball valve that is quieter in operation and is also more reliable than the Portsmouth or Croydon types.

Figure 1.20 shows a section of a diaphragm ball valve which is designed to have the working parts out of contact with the water. This prevents the working parts from sticking, due to the deposit of salts and rust from the water.

Equilibrium ball valve (see Fig 1.31, which shows a Portsmouth type, and Fig. 1.32, which shows a Croydon type).

The principle of the valve is to transmit equal water pressures at both ends of the piston and thus reduce the force produced by the float and level. The valve is used for large diameter supplies, or for very high water pressures.

Taps

There are several types of taps, sometimes referred to as 'screw down' types, which are designed to shut off the supply slowly and thus prevent water hammer. Figure 1.33 shows a 'bib tap' used for fitting over a sink, or for washing down purposes when it is then fitted with a hose outlet. The tap can be plain brass or chromium plated. Figure 1.34 shows a section of a 'Supatap' which incorporates a check valve. When the nozzle is removed the check valve prevents water from flowing through to the outlet and this permits the jumper to be changed without shutting off the supply at the stop valve. Figure 1.35 shows a pillar tap which can be used for baths, wash basins and sinks. Dual flow swivel sink taps may be used, which separate the flow of hot and cold water, until the water discharges through the nozzle. This prevents the risk of hot water being drawn into the cold-water main.

Spray taps

These are used for hand washing in factories, schools and offices and result in the saving of approximately 50 per cent of the water that would be used with ordinary taps; because less hot water is used there is also a saving in fuel. Hot and cold water supplies are connected to the same valve and are blended together before being discharged through a spray outlet.

Materials

Taps and valves are manufactured from either brass or gunmetal. English brass contains 64 per cent copper and 36 per cent zinc and is used for high grade fittings. Common brass contains 50 per cent copper and 50 per cent zinc and is used for lower grade fittings. Admiralty gunmetal contains 88 per cent copper, 10 per cent tin and 2 per cent zinc and is used for steam fittings. Leaded gunmetal contains 85 per cent copper, 5 per cent tin, 5 per cent zinc and 5 per cent lead and is used for high grade water fittings.

Materials for pipes

Lead and lead alloy

Lead is very resistant to corrosion by atmospheric gases or acid soils, but is affected by contact with cement mortar or concrete, and if there is a danger of corrosion the pipes should be wrapped with 'Denso' tape or coated with bitumen. The pipes are attacked internally by soft water which may cause minute particles to pass into the water system and endanger the health of the user. Lead pipe is easily bent when cold, cut and jointed and for this reason is very useful for short connections to sanitary fittings and cisterns. Figures 1.36 and 1.37 show wiped soldered joints on lead pipes. The pipes may also be jointed by means of lead welding.

The 1986 Water Supply Byelaws prohibit the use of lead pipe for the repair or replacement of similar pipe. Copper pipe may be used providing suitable means are employed to prevent, so far as is reasonably practicable, corrosion through galvanic action.

To prevent galvanic action the Byelaws state that copper pipe must not be used upstream of lead pipe.

Copper

Copper pipes are used for hot and cold water, heating, gas and drainage installations. They have high tensile strength and may therefore be of thin walls, which make them comparatively cheap and light in weight. Some acid soils may attack copper, but if there is a danger of corrosion the pipes may be obtained with a plastic covering. The pipes have smooth internal surfaces and offer little resistance to flow of fluids and therefore smaller bore pipes may be used. Copper pipes are easily bent by a machine or spring, they are neat in appearance and if required may be chromium plated. Jointing can be achieved by non-manipulative compression joints (Fig. 1.38), manipulative compression joints (Fig. 1.39), or soldered capillary joints (Fig. 1.40). n.b. Lead-free solder joints must be used for potable water systems.

The pipes may also be jointed by bronze welding or silver soldering. Where there is a risk of the joint being pulled out due to fibration or settlement, manipulative compression or soldered capillary joints should be specified and these types of joints are therefore used for underground service pipes.

Specification: BS 2871, Part 1, Table X applies to light gauge, half-hand temper tube, supplied in straight lengths of 6 m for use above ground. BS 2871, Part 1, Table Y applies to light gauge annealed tube, supplied in coils, in lengths of up to 60 m. The tube is used for underground water and gas services and for panel heating installations. BS 2871, Part 1, Table Z applies to hand-tempered, thin-walled tube supplied in straight lengths for use above ground. The tube is not suitable for bending, but has the same outside diameter of the other tubes and standard non-manipulative and capillary fittings may therefore be used.

Note: All tube sizes refer to the outside diameter and sizes from 6 mm to 159 mm O.D. may be obtained.

Mild steel

Mild steel pipes may be obtained galvanised or ungalvanised; the former type are used for hot- and cold-water installations and the latter for heating installations. The pipes are comparatively cheap and have a very high tensile strength, which

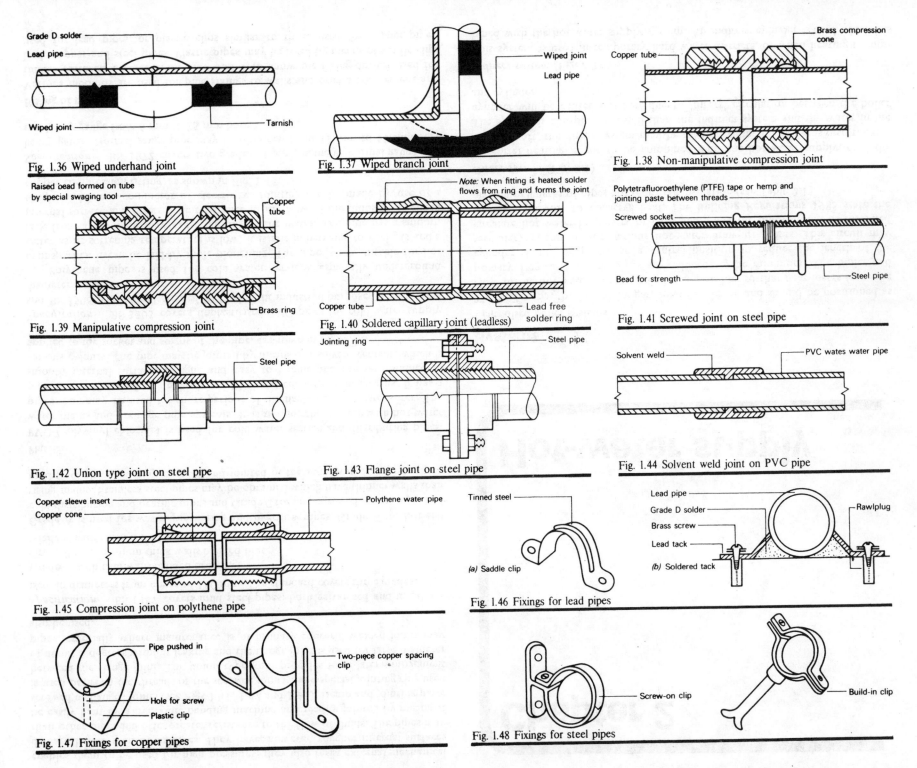

Fig. 1.36 Wiped underhand joint

Fig. 1.37 Wiped branch joint

Fig. 1.38 Non-manipulative compression joint

Fig. 1.39 Manipulative compression joint

Fig. 1.40 Soldered capillary joint (leadless)

Fig. 1.41 Screwed joint on steel pipe

Fig. 1.42 Union type joint on steel pipe

Fig. 1.43 Flange joint on steel pipe

Fig. 1.44 Solvent weld joint on PVC pipe

Fig. 1.45 Compression joint on polythene pipe

Fig. 1.46 Fixings for lead pipes

Fig. 1.47 Fixings for copper pipes

Fig. 1.48 Fixings for steel pipes

enables them to be used for high pressures; they also resist external mechanical damage better than other pipes. They have, however, rougher internal surfaces than other pipes and offer greater resistance to the flow of fluids. The pipes may be easily bent by a hydraulic bending machine and readily jointed by means of screwed or flanged joints (see Figs 1.41, 1.42 and 1.43). Hemp and jointing paste is used between the threads of the screwed joints and a special jointing ring used between the flange joint. The union type joint provides for a lateral movement of about 5 degrees without leaking and the flange joint is used for large diameter pipes, especially where maintenance is of primary concern. Welded joints may also be used.

Specification: BS 1387 covers mild steel pipes, both galvanised and ungalvanised, in diameters from 6 mm to 150 mm. The standard covers three grades:

Grade A with thin walls banded brown;
Grade B with medium thick walls banded blue;
Grade C with thick walls banded red.

Grade A is used for waste, ventilating and overflow pipes, Grade B for hot and cold distribution and heating pipes and Grade C for underground service pipes and rising mains. Stainless steel pipes may be obtained which have thinner walls than mild steel pipes and may be bent and jointed in the same way as copper pipes.

Plastic

PVC (Polyvinyl chloride) is used for cold water service and distributing pipes, water mains and drainage pipe systems. It is not suitable for use with hot-water pipes, but may be used for short periods of discharge of hot waste water. The pipe is manufactured as unplasticised PVC and cut into 10 m lengths. It has a smooth internal surface, is light and easy to handle and can be obtained in various colours. The pipe may be jointed by means of a solvent cement, which is applied to the socket and spigot of the pipe, as shown in Fig. 1.44.

Specification: BS 3505 covers unplasticised PVC pipes for cold water supply and BS 3506 covers unplasticised PVC pipe for industrial purposes. The internal diameters range from 9 mm to 150 mm.

Polythene pipe is used for cold water services, especially underground mains, when it can be laid by mole ploughing. The pipe is not suitable for hot water as its softening temperature is low, it is not impervious to coal gas and a leak from a gas main may cause tainting of the water. The pipe has a smooth internal surface and offers little resistance to the flow of the fluids, it is light in weight and a non-conductor of electricity. Jointing can be made by use of a copper compression fitting, as shown in Fig. 1.45.

Specification: BS 1972 covers two grades of pipe, namely: normal gauge and heavy gauge. Normal gauge pipe may be obtained from 13 mm to 50 mm bore and heavy gauge from 6 mm to 25 mm bore.

Fixing of pipes

Various methods are used for pipe fixings to walls and brickwork, cement and timber surfaces. Figures 1.46, 1.47 and 1.48 show the fixing devices used for lead, copper and steel pipes. Plastic pipes may be fixed by means of saddle clips used for lead pipes, or plastic clips similar to those used for copper pipes.

Chapter 2

Hot-water supply

Systems

Centralised boiler systems

The boiler may be heated by gas, solid fuel or oil and should be positioned as close to the hot-water storage cylinder as possible, so that heat losses from the primary flow and return pipes are reduced to the minimum. The boiler and cylinder should be placed in a central position to reduce the length of the secondary circuit to the various hot-water draw-off points. This circuit may circulate hot water by gravity or by a pump and the pipes should be insulated wherever possible to conserve heat. The Building Regulations 1985 state the maximum lengths of pipes not requiring insulation (see Table 2.1).

Direct system (Fig. 2.1)

If central heating is not to be combined with the hot-water supply, or if the water is soft, the direct system may be used providing the boiler is rustprooved. The system is cheaper to install than the indirect system and the water in the cylinder will be heated quicker, due to 'direct' circulation between the boiler and cylinder.

Indirect system (Fig. 2.2)

This system is used in temporary hard water districts, or when heating is combined with the hot water supply system. An indirect cylinder is used which has

Table 2.1 Water Authorities Regulations for the maximum distance between the hot water apparatus and the draw off points

Outside diameter of pipe (mm)	Maximum length of uninsulated pipe (m)
Not more than 12	20
More than 12 but not More than 22	12
More than 22 but not More than 28	8
More than 28	3

an inner heat exchanger. The water from the boiler circulates through this heat exchanger and heats the water in the cylinder indirectly. Since the water in this heat exchanger and boiler is not drawn off through the hot-water taps, lime is precipitated only after the initial heating of the water, and afterwards (unless the system is drained) there is no further occurrence and therefore no scaling. This same water also circulates through the steel or cast iron radiators, and after heating the water is freed from carbon dioxide which then reduces corrosion of the radiators.

Primatic cylinder (Fig. 2.3)

For small installations. This type of self-venting indirect cylinder may be used. When the water is heated the heat exchanger has two air locks, which prevent the secondary water in the cylinder from mixing with the primary water in the heat exchanger and boiler. The cylinder is connected with pipework similar to the direct system and therefore there is a saving in cost, due to the absence of an expansion and feed cistern, primary cold feed and primary vent pipes (see Fig. 2.4).

Combined system

Figure 2.5 shows a system having a 'closed' primary circuit, which, like the primatic cylinder, saves an expansion and feed cistern, primary cold feed and primary vent pipes. The closed expansion vessel contains air or nitrogen which takes up the expansion of the water when heated. A 'micro bore' heating system is shown, with the radiators served with 6, 8, 10 or 12 mm outside-diameter soft copper pipes, from a 22 mm outside-diameter copper manifold. The system saves about 15 per cent of costs when compared with a conventional system of heating with larger pipes. When the pump is working the injector tee draws water down from the return pipe and promotes better circulation through the heat exchanger in the cylinder.

Systems for larger buildings

Figure 2.6 shows a system for a three-storey hotel with bathrooms and towel rails on each floor. Each floor is zoned with valves, so that a repair on one floor may be carried out without draining the whole of the secondary pipework.

Figure 2.7 shows a duplicated plant which allows the repair or renewal of one of the boilers or calorifiers without shutting down and draining the others. This will provide a supply of hot water at all times, and in order to accomplish

this every pipe connection to the boilers and calorifiers must be provided with valves for isolating purposes. The three-way vent valves will ensure that the boilers are open to atmosphere at all times and thus avoid the risk of an explosion. The plant is essential in hospitals, large factories, colleges, hotels and offices.

Figure 2.8 shows a detail of a large horizontal steam-heated calorifier, which is an ideal method of providing hospitals and factories with hot water when steam is used for space heating. The calorifiers may be sited at various strategic points in the building; for example, close to the various hot water draw-off points.

Figure 2.9 shows a hot-water system for a tall building. The floors are zoned, so that the maximum head of water on the lowest draw-off taps does not exceed 30 m. This will reduce noise and wear of valves on the lower floors. The head tank improves the flow of hot water to the taps on the upper floors of each zone.

Estimation of hot-water storage

The Code of Practice gives the storage requirements shown in Table 2.2.

Table 2.2

Type of building	Storage per person (litres)	Type of building	Storage per person (litres)
Colleges and schools		Hospitals	
Boarding	23	General	27
Day	4.5	Infectious	45.5
Dwelling houses	45.5	Maternity	32
Factories	4.5	Nurses' homes	45.5
Flats	32	Hostels	32
Hotels (average)	35	Offices	4.5
		Sports pavilions	36

By use of Table 2.2, a dwelling house with three occupants will require 45.5 × 3 = 136.5 litres of hot-water storage. If the number of occupants of the building is not known Table 2.3 may be used.

Table 2.3 Volumes of water used at each appliance

Appliance	Volumes of hot water (litres)
Wash basin	
Hand wash	1.5
Wash	3
Hair wash	6
Shower	13
Bath	70
Washing machine	70
Sink	
Wash-up	15
Cleaning	5

16

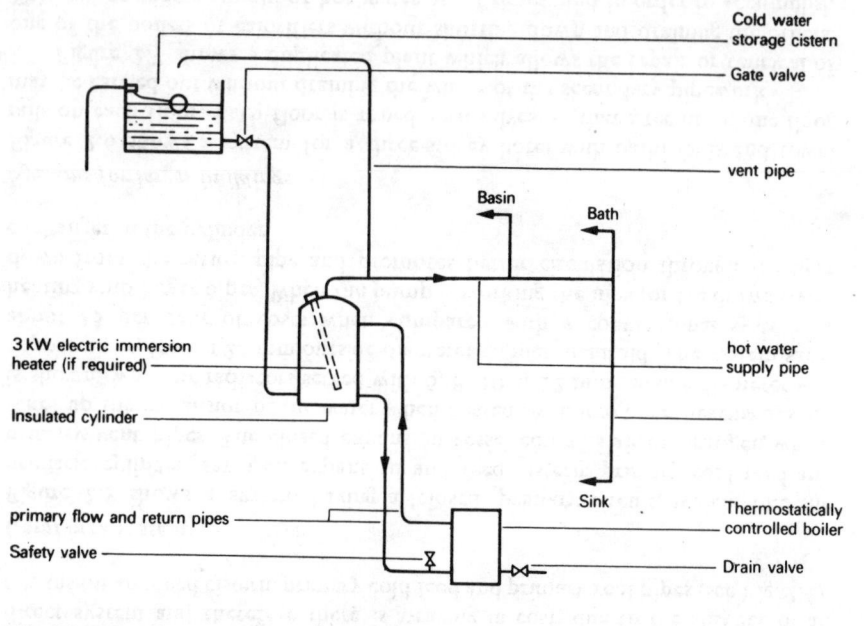

Fig. 2.1 'Direct' system of hot water supply

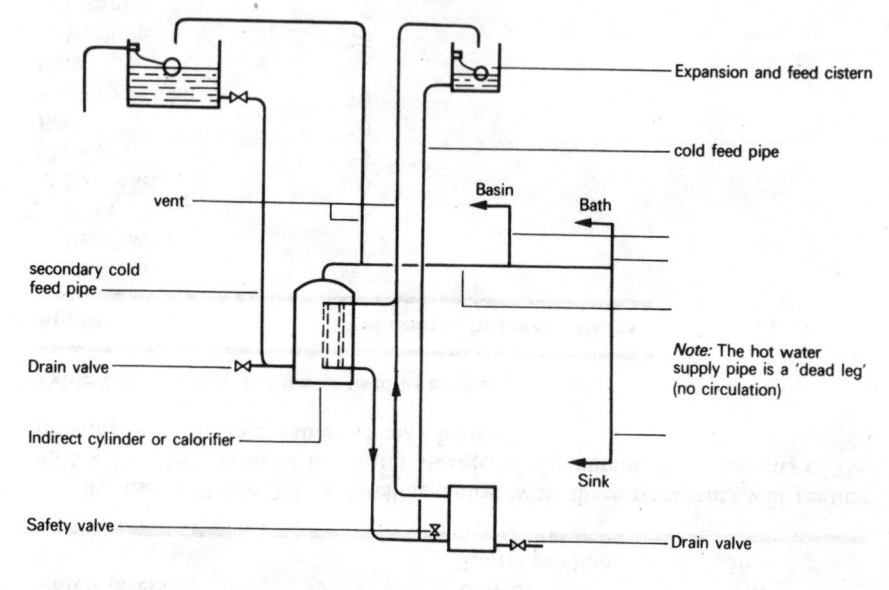

Fig. 2.2 'Indirect' system of hot water supply

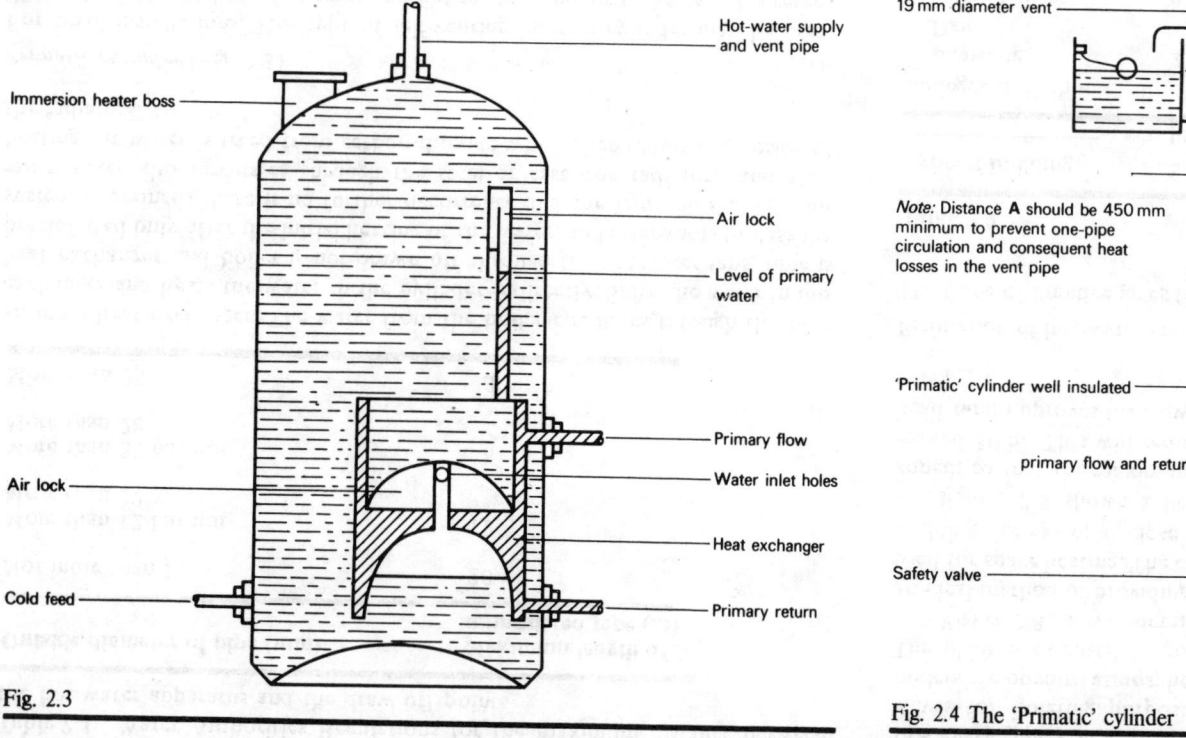

Fig. 2.3

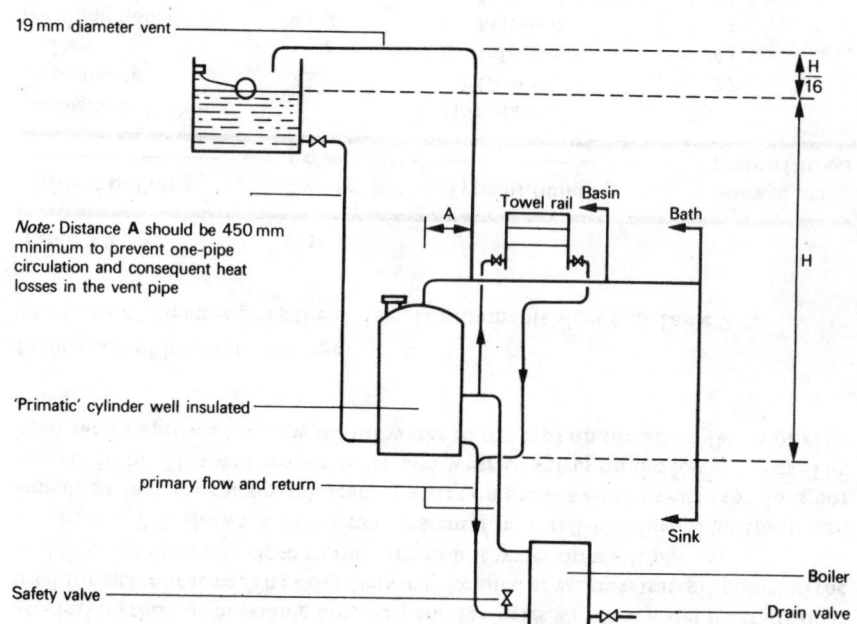

Fig. 2.4 The 'Primatic' cylinder

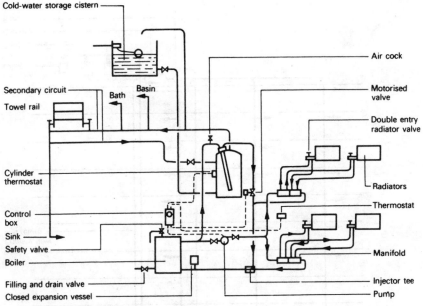

- Cold-water storage cistern
- Air cock
- Secondary circuit
- Motorised valve
- Towel rail
- Bath
- Basin
- Double entry radiator valve
- Cylinder thermostat
- Radiators
- Thermostat
- Control box
- Sink
- Safety valve
- Boiler
- Manifold
- Filling and drain valve
- Injector tee
- Closed expansion vessel
- Pump

Fig. 2.5 Combined heating and hot water system

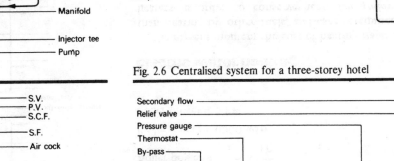

- Cold-water storage cistern
- Expansion and feed cistern
- Towel rail
- Shower
- Basin
- Zoning valves
- Secondary circuit
- Pump

Fig. 2.6 Centralised system for a three-storey hotel

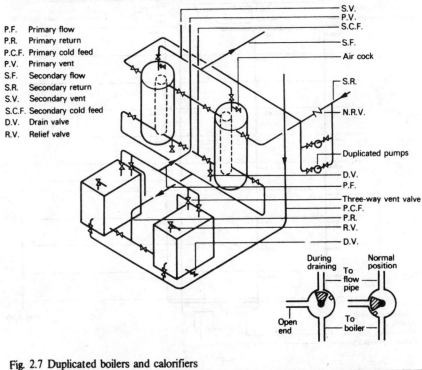

P.F.   Primary flow
P.R.   Primary return
P.C.F. Primary cold feed
P.V.   Primary vent
S.F.   Secondary flow
S.R.   Secondary return
S.V.   Secondary vent
S.C.F. Secondary cold feed
D.V.   Drain valve
R.V.   Relief valve

- S.V.
- P.V.
- S.C.F.
- S.F.
- Air cock
- S.R.
- N.R.V.
- Duplicated pumps
- D.V.
- P.F.
- Three-way vent valve
- P.C.F.
- P.R.
- R.V.
- D.V.

During draining
To flow pipe
Open end
Normal position
To boiler

Fig. 2.7 Duplicated boilers and calorifiers

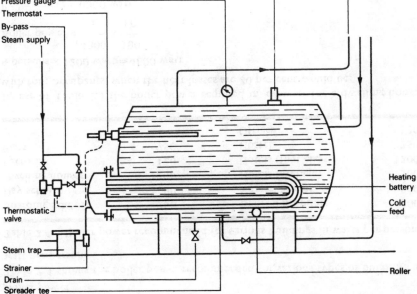

- Secondary flow
- Relief valve
- Pressure gauge
- Thermostat
- By-pass
- Steam supply
- Thermostatic valve
- Steam trap
- Strainer
- Drain
- Spreader tee
- Heating battery
- Cold feed
- Roller

Fig. 2.8 Large horizontal steam heated calorifier

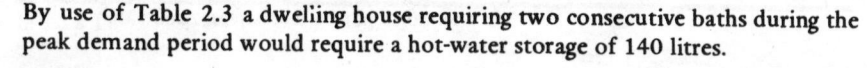

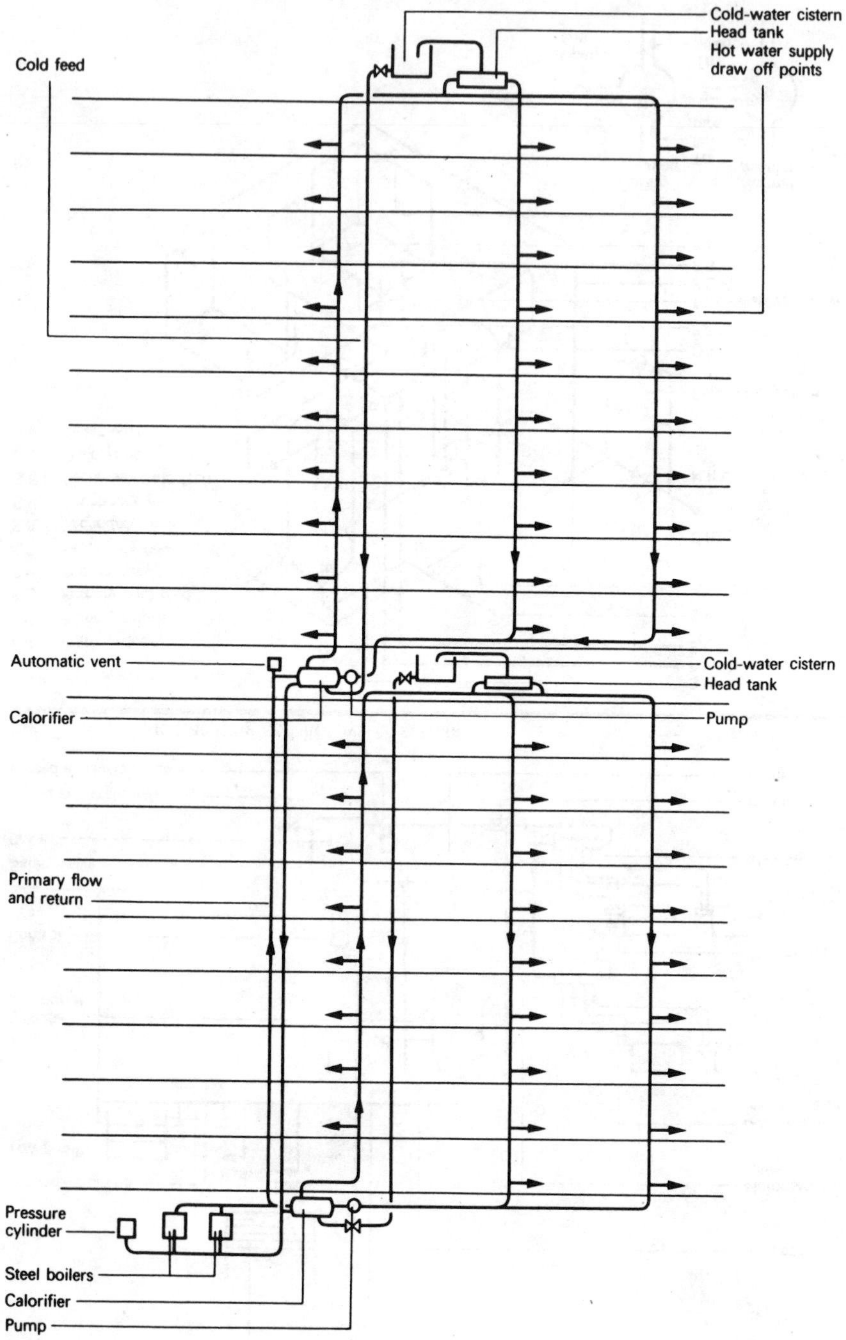

Cold feed

Cold-water cistern
Head tank
Hot water supply
draw off points

Automatic vent

Calorifier

Cold-water cistern
Head tank

Pump

Primary flow
and return

Pressure
cylinder

Steel boilers
Calorifier
Pump

Fig. 2.9 Hot water system for a tall building

By use of Table 2.3 a dwelling house requiring two consecutive baths during the peak demand period would require a hot-water storage of 140 litres.

Boiler power

Table 2.4 shows the boiler power recommended for various types of buildings in watt per person.

Table 2.4 Boiler power recommended for various buildings in watts per person

Boarding schools	750	General hospitals	1500
Day schools	90	Nurses' homes	900
Dwelling houses	1200	Hotels (average)	900
Factories	120	Hotels (first class)	1200
Flats	750	Hostels	750
		Offices	120

By use of Table 2.4 the boiler power required in kilowatt for a dwelling house with four occupants, when the heat losses are 20 per cent would be:

4 persons x 1200 watt = 4800 watt

$$\text{Boiler power} = \frac{4800}{1} \times \frac{100}{80}$$

Boiler power = 6000 watt

Boiler power = 6 kW

Electric water heating

At the present moment the cost of heating water by electricity is more expensive than heating by other fuels, and care is required when installing electric water heaters in order to conserve heat. *The following points therefore must be observed:*

1. The hot-water storage vessel must be well insulated, with a minimum thickness of 50 mm — preferably 75 mm — of good insulation.
2. The heated water must not circulate through towel rails or radiators.
3. The length of hot water draw-off points, particularly to the sink, should be reduced to the minimum.
4. Single-pipe circulation in the hot-water pipes, or vent pipes, must be avoided.
5. Airing cupboards should not be heated by leaving part of the hot-water storage vessel uninsulated.
6. An effective thermostat must control the temperature of the water at a maximum of 60 °C for temporary hard water and at a maximum of 71 °C for soft water. The lower temperature for temporary hard water greatly reduces the deposit of lime.

Pressure type heaters (Fig. 2.10)

These must always be supplied from a cold-water storage cistern. They are available in capacities from 50 to 450 litres. One of the most useful type is called

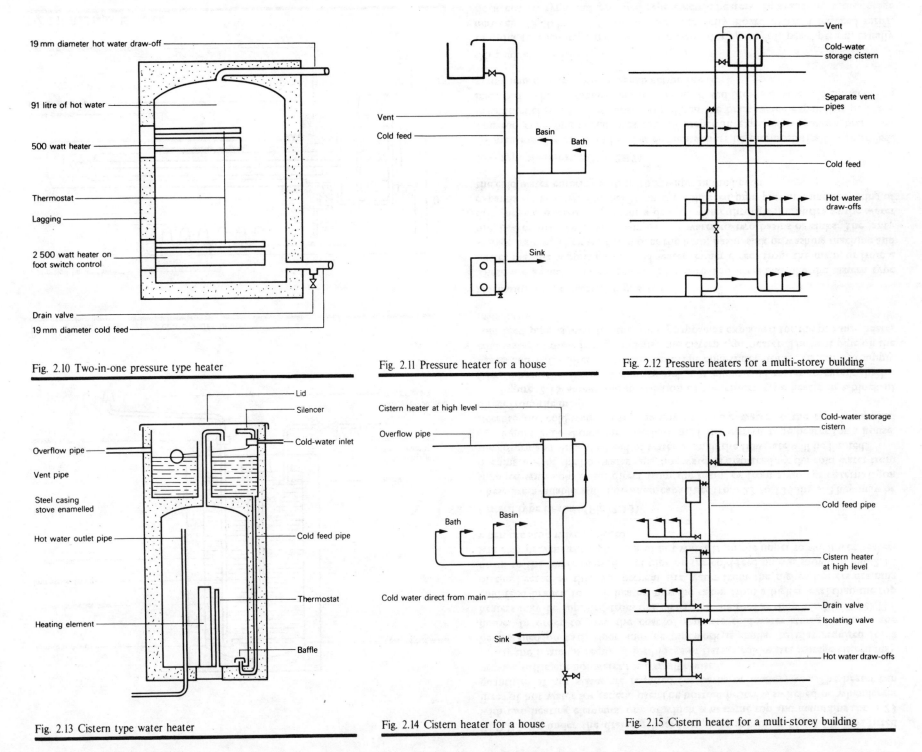

19 mm diameter hot water draw-off

91 litre of hot water

500 watt heater

Thermostat

Lagging

2 500 watt heater on foot switch control

Drain valve

19 mm diameter cold feed

Fig. 2.10 Two-in-one pressure type heater

Vent

Cold feed

Basin

Bath

Sink

Fig. 2.11 Pressure heater for a house

Vent

Cold-water storage cistern

Separate vent pipes

Cold feed

Hot water draw-offs

Fig. 2.12 Pressure heaters for a multi-storey building

Lid

Silencer

Cold-water inlet

Overflow pipe

Vent pipe

Steel casing stove enamelled

Hot water outlet pipe

Cold feed pipe

Thermostat

Heating element

Baffle

Fig. 2.13 Cistern type water heater

Cistern heater at high level

Overflow pipe

Bath

Basin

Cold water direct from main

Sink

Fig. 2.14 Cistern heater for a house

Cold-water storage cistern

Cold feed pipe

Cistern heater at high level

Drain valve

Isolating valve

Hot water draw-offs

Fig. 2.15 Cistern heater for a multi-storey building

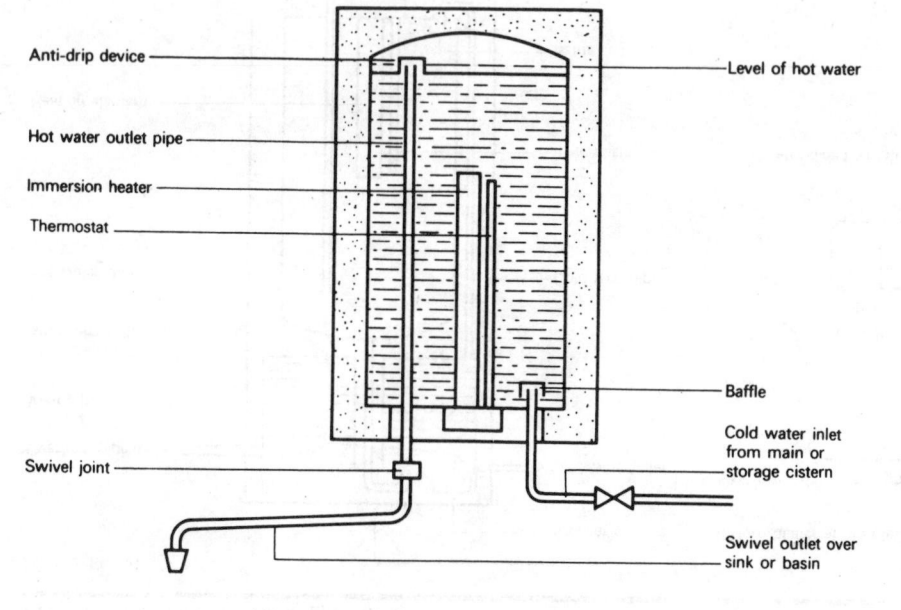

Anti-drip device —

Hot water outlet pipe —

Immersion heater —

Thermostat —

Level of hot water

Baffle

Cold water inlet from main or storage cistern

Swivel joint —

Swivel outlet over sink or basin

Fig. 2.16 Open outlet type electric water heater

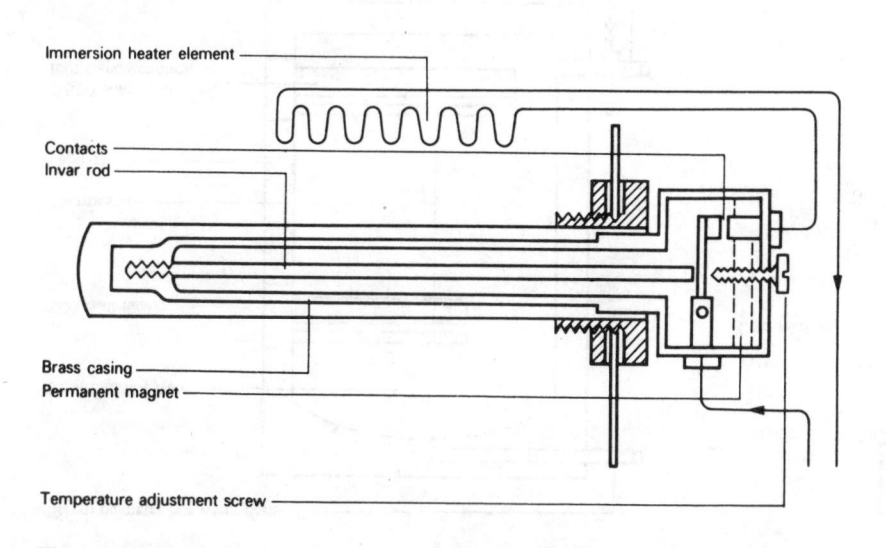

Immersion heater element —

Contacts —
Invar rod —

Brass casing —
Permanent magnet —

Temperature adjustment screw —

Fig. 2.17 Rod type thermostat

'U.D.B.', or under the draining board, type as shown in Fig. 2.11. It is fitted with two heating elements, one of which is near the top and maintains about 23 litres of hot water for general use. The bottom heater is switched on when larger quantities of hot water are required for baths or laundry use. The heater can provide sufficient hot water for a small house.

If the heater is required for blocks of flats, a cold-water storage cistern may be installed on each floor and the pipework is similar to that required for a house. In order to save the cost of separate cold-water storage cisterns, the heaters may be supplied from one cistern on the roof as shown in Fig. 2.12. The cold feed branch to each heater must be taken from a higher level than the top of the heater, as this will prevent the water from the higher heaters draining down to the lower ones. A vent pipe on the cold-feed pipe as shown in Fig. 2.12 will also prevent the siphoning of hot water from the upper to the lower heaters when the stop valve is closed.

Cistern type heaters (Fig. 2.13)

These are available with hot-water capacities from 23 to 136 litres. They may be supplied with cold water direct from the main, or from a storage cistern. Upon opening one of the hot-water taps, hot water is displaced by the cold water from the cistern and since the head of water is small the flow rate will be limited.

Figure 2.14 shows the installation of the cistern type heater for a house. Because the cold-water cistern is small the cold water to the taps is supplied direct from the main.

Figure 2.15 shows the installation of the cistern type heater in a block of flats, where the Water Authority will not allow cold water to supply baths, basins and W.C.s direct from the main. The cold-water storage cistern will supply cold water to these fittings and also the cistern type heater. The vent pipe on the cold feed pipe is used for the same purpose as explained for the pressure heater installation.

Open outlet type heaters (Fig. 2.16)

These are available in capacities from 6 to 136 litres and like the cistern type heater may be supplied with cold water, either direct from the main, or from a storage cistern. They are fitted over the bath, basin, sink or washing machine and one heater may be used to supply hot water to two basins or sinks. The 'anti-drip' device is used to prevent a drip of water through the outlet as the water expands on heating. The baffle on the cold-feed pipe prevents undue mixing of the cold water entering with the hot water in the heater.

Rod type thermostat (Fig. 2.17)

As mentioned earlier, all heating elements must be thermostatically controlled. The rod type thermostat uses the difference in expansion between brass and invar (nickel steel) for its operation. When the brass casing expands it pulls the invar rod (which expands very little) with it and breaks the electric contact. The permanent magnet ensures a snap action break.

Off peak electricity

In order to encourage the use of electricity during the 'off peak' period, usually between 23.00 hr and 07.00 hr, the Electricity Board offers a reduced tariff. Both cistern type and pressure type electric heaters are available, with storage

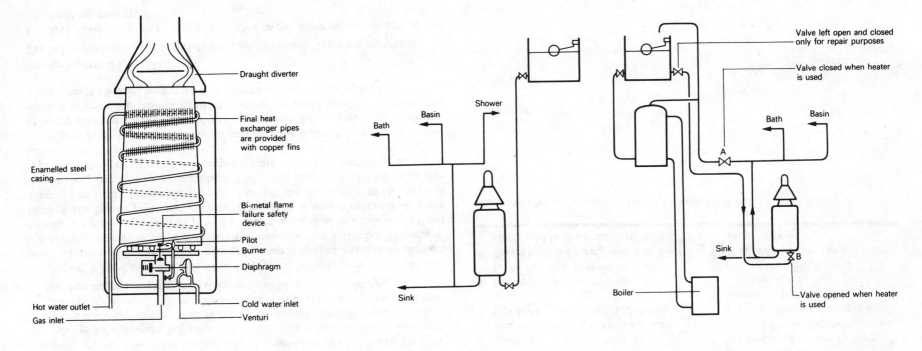

Draught diverter

Final heat exchanger pipes are provided with copper fins

Enamelled steel casing

Bi-metal flame failure safety device

Pilot

Burner

Diaphragm

Hot water outlet

Gas inlet

Cold water inlet

Venturi

Fig. 2.18

Basin

Shower

Bath

Sink

Fig. 2.19 Instantaneous gas water heater

Valve left open and closed only for repair purposes

Valve closed when heater is used

Bath

Basin

A

Sink

Boiler

B

Valve opened when heater is used

Fig. 2.20

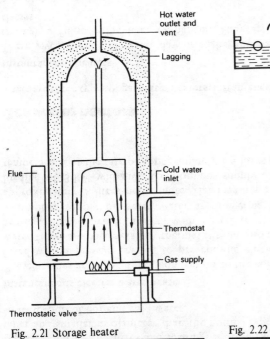

Hot water outlet and vent

Lagging

Flue

Cold water inlet

Thermostat

Gas supply

Thermostatic valve

Fig. 2.21 Storage heater

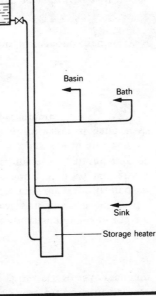

Basin

Bath

Sink

Storage heater

Fig. 2.22

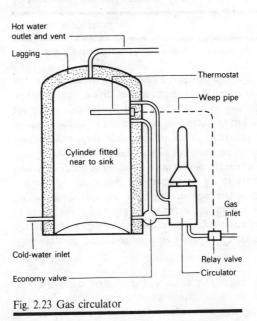

Hot water outlet and vent

Lagging

Thermostat

Weep pipe

Cylinder fitted near to sink

Gas inlet

Cold-water inlet

Economy valve

Relay valve

Circulator

Fig. 2.23 Gas circulator

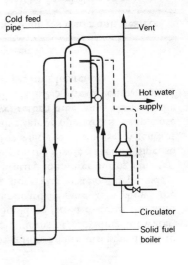

Cold feed pipe

Vent

Hot water supply

Circulator

Solid fuel boiler

Fig. 2.24 Auxiliary circulator

capacities up to 225 litres and loadings up to 6 kW. These 'off peak' heaters are taller and smaller in diameter than the normal type of heaters, so as to maintain stratification of hot and cold water.

Instantaneous electric water heaters

These are designed for direct connection to the cold-water supply main, and are fitted with a pressure switch to prevent the element switching on before the water flows and vice versa. A thermal cut-out also prevents the water from over-heating. Shower water heaters with an electrical loading of 6 kW will give a continuous supply of warm water at showering temperature, at the rate of approximately 3 litres per minute. Hand-wash water heaters with an electrical loading of 3 kW will give a continuous supply of warm water at hand wash temperature, at the rate of approximately 1.4 litres per minute.

Gas water heaters

There are three types of gas water heaters; instantaneous, storage and circulatory.

Instantaneous heaters

These give instant hot water day or night, the water being heated as it passes through a finned pipe heat exchanger. They are obtainable in the following classes;

1. *Sink heaters*, single or multipoint, to serve a sink, or basin, or a sink and basin where these are close together. They are fitted with a swivel spout outlet.
2. *Bath heaters*, a swivel spout can serve a wash basin in addition to the bath.
3. *Boiling water heaters*, will provide hot or boiling water for making beverages and a humming device fitted to the heater sounds when the water is boiling.
4. *Multipoint heaters*, will provide sufficient hot water to bath, basin and sink, or they may be used for several basins or sinks in larger buildings.

Figure 2.18 shows a detail of a multipoint instantaneous water heater, which operates as follows: When a hot-water tap is opened water flows through the venturi, which causes a differential pressure across a flexible diaphragm, causing the main gas valve to open, the burners are ignited and cold water flows through the finned pipe heat exchanger, where it is heated before leaving the heater. The flame failure device will shut off the gas supply to the burners should the pilot light be extinguished. The heater may be of the balanced flue type and will deliver about 5.6 litres per minute, with the water raised through a temperature of 38 °C.

Figure 2.19 shows the method of installing the heater for a house; the head of water above the heater should not be less than 3 m. Figure 2.20 shows the method of installing the heater to an existing hot water system. To obtain hot water from the heater, stop valve A is closed and stop valve B opened.

Storage heaters (Fig. 2.21)

The hot water is stored in an insulated copper cylinder. There are two types:

1. *Sink heaters*, which are provided with a swivel spout outlet, usually of 9 to 23 litres capacity.

2. *Multipoint storage heaters*, which like the instantaneous multi[...] may supply bath, basin and sink, as shown in Fig. 2.22. [...] hot-water storage is 114 litres, but other capacities are obtainab[...]

Circulators

This type of heater should be fitted with its own flow and return pip[...] the storage cylinder. The heater may be the sole means of providing [...], as shown in Fig. 2.23. If possible the heater should be fitted close to the sink, but it may also be installed in the airing cupboard. Figure 2.24 shows the method of installing the heater to an existing hot-water system, to provide a boost to the boiler or an alternative means of heating the water during summer. The three-way economy valve permits the closing of the low-level return pipe, so that only a small quantity of water is heated for sink and basin use. When hot water is required for bath or laundry use, the valve can be operated so as to close the high-level return pipe and open the low-level return pipe, thus allowing the whole of the water in the cylinder to be heated. Table 2.5 gives a comparison of heating water by centralised plant and electric or gas unit heaters.

Table 2.5　Summary of the hot-water heating systems

Centralised systems	Electric or gas water unit heaters
1. Provides large bulk storage for hospitals, hotels and factories	1. May be fitted close to the fittings being supplied, thus saving a great deal of pipework
2. One central boiler plant requires less maintenance than several unit heaters	2. Saving in boiler house space and possible fuel store
3. Cheaper fuel may be used	3. A number of heaters in large buildings require either separate gas or electrical connections
4. Requires long lengths of secondary pipework, which can lead to large heat losses	4. Greater risk of fire inside the building, due to more gas or electrical connections
5. Reduction in flue construction (unless electric heaters are used)	5. No need to pump the hot water supply circuit
6. Towel rails and airing coils may be connected to the secondary circuit	

See Appendix A for a discussion of unvented hot water storage systems.

Chapter 3

Low-pressure hot-water heating

Systems

When low-pressure hot water is used for heating systems the temperature of the water is below boiling point, usually about 80 °C on the flow pipe and between 60 °C and 70 °C on the return pipe.

Whater has the high specific heat capacity of 4.2 kJ/kg K and although it is more difficult to heat than other media, more heat can be transferred from the boiler to the various heat emitters with pipes of a relatively small diameter. The higher the temperature of the water the greater the amount of heat transferred, but some of this heat is lost by the higher heat losses on the pipework. This higher temperature may also cause injury to persons coming into contact with the radiators, etc., and special precautions must be taken (see Chapter 7, Medium and high-pressure hot-water systems).

Water circulation

Water may be circulated in the system either by natural convection (thermo-siphonage) or by means of a centrifugal pump. Thermo-siphonage is produced by the difference in temperature between the flow and return pipes. The denser cooler water in the return pipe forces the less dense hotter water in the flow pipe through the circuit. Pump circulation has now replaced thermo-siphonage circulation for all but the smallest house installation. It has the advantage of reducing the heating-up period and also smaller pipes may be used.

Circuit arrangement

Various circuits may be used depending upon the type and layout of the building. In some buildings two or more circuits may be used supplied from the same boiler plant. One-pipe or two-pipe circuits may be used and although the one-pipe is simpler than the two-pipe it has two disadvantages:

1. The cooler water passing out of each heat emitter flows to the next one, resulting in the emitters at the end of the circuit being cooler than those at the beginning. This can be reduced by shutting down the lock shield valves at the start of the circuit thus allowing more water to pass to the emitters at the end of the circuit.
2. Even if the circuit is pumped, the pump pressure does not force water through the heat emitters. Water is induced to circulate through the emitters by thermo-siphonage created by the difference in density between the water entering and leaving the emitters. A one-pipe system therefore cannot be used to supply heat to various types of emitters that have a comparatively high resistance to the flow of water.

Injector tees can be used to induce water through emitters on a one-pipe circuit, but these increase the resistance to the flow of water.

The two-pipe system requires more pipework, but this can be reduced in diameter as it passes around the circuit. The pump pressure acts throughout the circuit and therefore any type of heat emitter can be used. There is much less cooling down of the emitters at the end of the circuit, due to the cooler water passing out of the emitters being returned to the boiler via the return pipe.

One-pipe ring circuit (Fig. 3.1): suitable for small single-storey buildings and if the circuit is pumped the heat emitters may be on the same level as the boiler. All the other circuits shown may be used for multi-storey buildings.

One-pipe ladder (Fig. 3.2): suitable when it is possible to use horizontal runs either exposed in a room or in a floor duct.

One-pipe drop (Fig. 3.3): this circuit requires provision at the top of the building, for a horizontal main distributor pipe and at the bottom of the building, for a main horizontal collector pipe. It is suitable for offices, etc., having separate rooms on each floor where the vertical pipe can be housed in a corner duct. It has the advantage of making the heat emitters self venting.

One-pipe parallel (Fig. 3.4): similar to the ladder system and is used when it is impractical to install a vertical main return pipe.

Two-pipe parallel (Fig. 3.5): again similar to the ladder system, but possessing the advantages of a two-pipe circuit.

Two-pipe upfeed (Fig. 3.6): suitable for buildings having floors of various heights, or for groups of buildings. An automatic vent valve can be used at the top of each flow pipe. The system is also used for embedded panel heating, using a panel of pipes instead of radiators.

Two-pipe high level return (Fig. 3.7): used when it is impractical to install a main horizontal collector pipe. The system is most useful when installing heating in existing buildings having a solid ground floor.

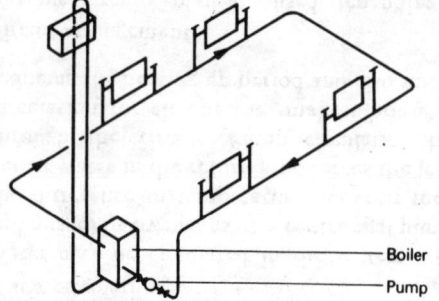

Fig. 3.1 One-pipe ring

Boiler
Pump

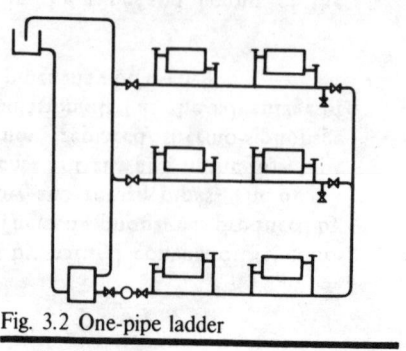

Fig. 3.2 One-pipe ladder

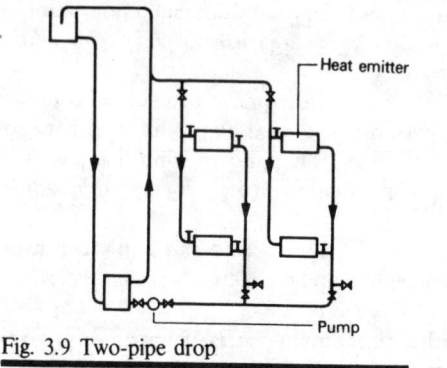

Heat emitter

Pump

Fig. 3.9 Two-pipe drop

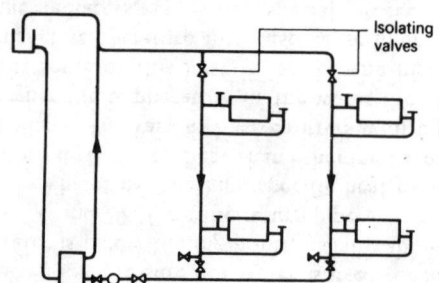

Isolating valves

Fig. 3.3 One-pipe drop

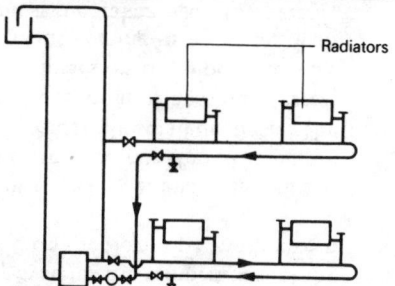

Radiators

Fig. 3.4 One-pipe parallel

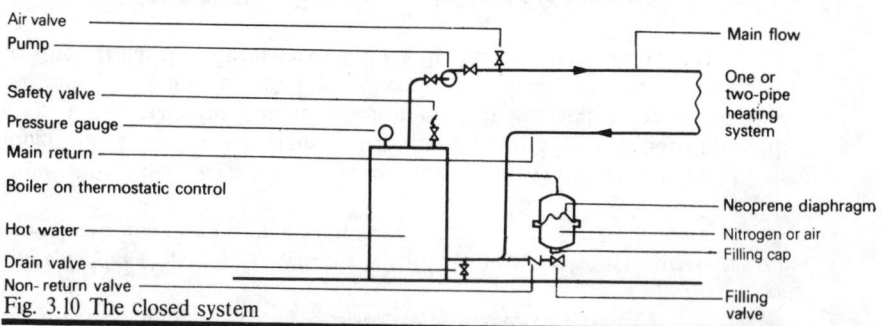

Air valve — Main flow
Pump
One or two-pipe heating system
Safety valve
Pressure gauge
Main return
Neoprene diaphragm
Boiler on thermostatic control
Hot water
Nitrogen or air
Filling cap
Drain valve
Non-return valve
Filling valve

Fig. 3.10 The closed system

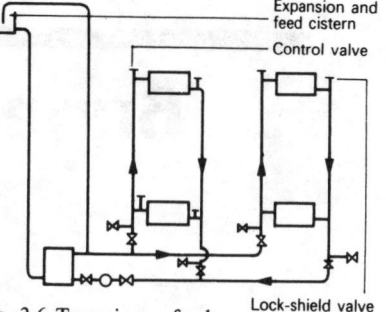

Drain valve

Fig. 3.5 Two-pipe parallel

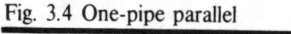

Expansion and feed cistern
Control valve

Fig. 3.6 Two-pipe upfeed

Lock-shield valve

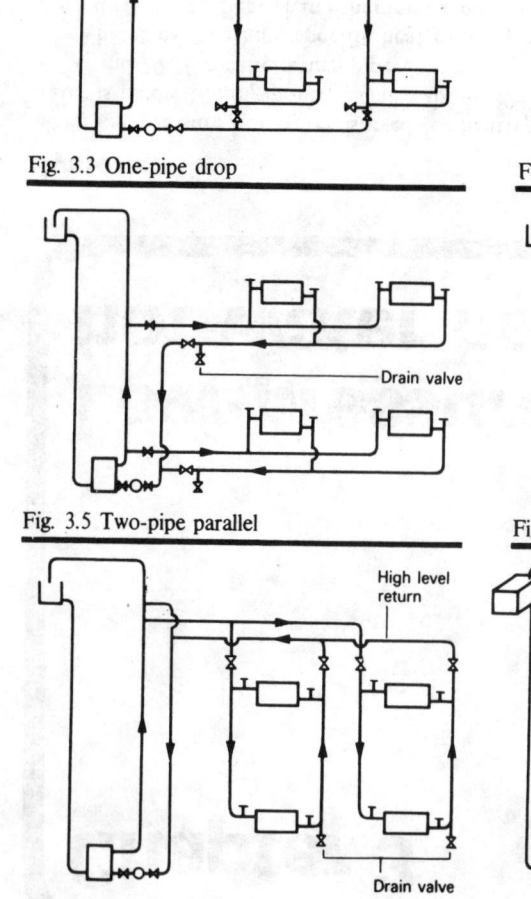

Feed and expansion cistern

13 mm cold feed pipe

Radiators in bedrooms
19 mm vent pipe
13 mm

13 mm

Calorifier

Lounge
13 mm

Hall
25 mm primary return
25 mm primary flow
19 mm
Pump

Dining room
13 mm
Study
Thermostatic zoning valves

Boiler

19 mm

Fig. 3.11 Small-bore heating system

High level return

Drain valve

Fig. 3.7 Two-pipe, high-level return

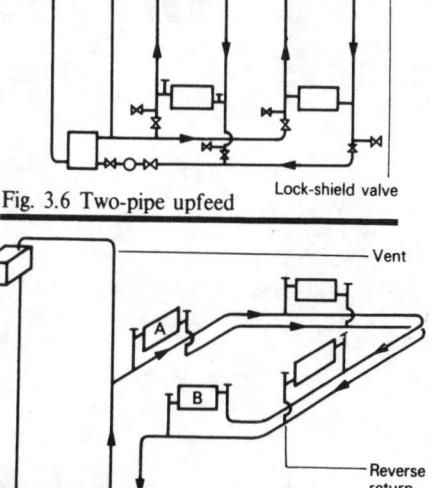

Vent

A

B

Reverse return

Fig. 3.8 Two-pipe reverse return

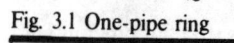
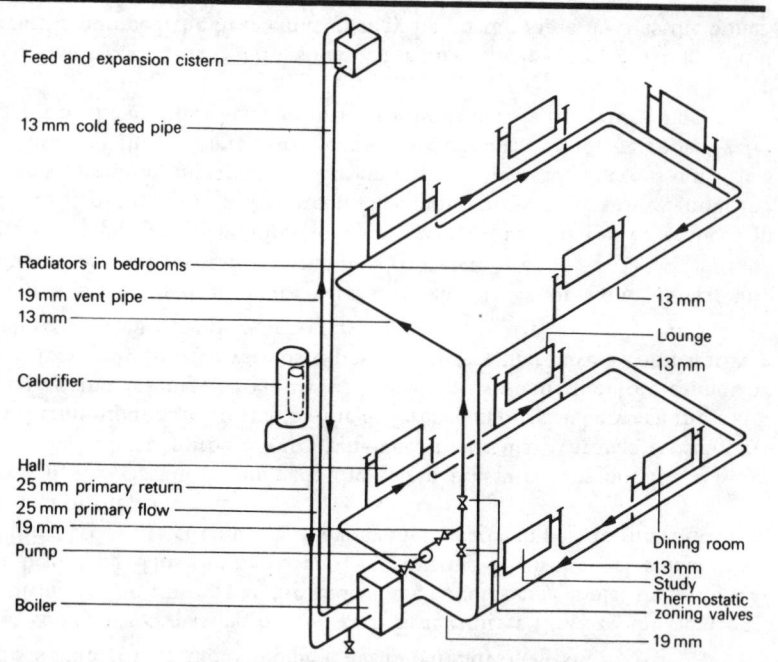
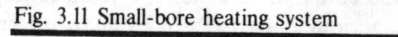

Two-pipe reverse return (Fig. 3.8): the predominant feature of this system is the equal travel to each heat emitter which provides a well-balanced circuit. Note that the length of the circuit for emitter A is the same as for emitter B.

Two-pipe drop (Fig. 3.9): suitable for buildings where it is possible to install a main horizontal distributor pipe and a main horizontal collector pipe in ducts. The system is self venting and therefore periodic venting of the radiators is unnecessary.

Isolating valves

Each heat emitter should be provided with a control valve on the inlet and a lock shield valve on the outlet. The lock shield valve is used by the heating engineer to balance the circuit as described earlier.

Isolating valves will also be required on the boiler, pumps and on all main branches. Drain valves must also be provided, to allow the various sections to be drained down.

Expansion and feed cistern

When the system is cold the ball valve should be set so that there is not more than 100 mm of water in the cistern and the capacity of the cistern should be such that when the water is raised to its working temperature the water level does not rise within 50 mm from the bottom of the warning pipe.

Feed and vent pipes

The feed pipe should be of adequate diameter and taken from the bottom of the expansion cistern, without any branch pipes, direct to the bottom of the boiler or return pipe. If a valve is fitted to the feed pipe, it should be of the lock shield type to prevent unauthorised use.

The vent pipe should be taken from the boiler and turned over the expansion cistern. A valve should never be fitted on the vent pipe, unless a three-way type is used when two or more boilers are to be installed (see centralised hot water supply).

Position of pump

This may be fitted on either the main flow or main return pipe. If fitted on the flow pipe, the heating system is under positive pressure and there is less risk of air being drawn into the circuit.

If fitted on the return pipe, the pump is normally easier to install and is at a lower temperature. There is however a greater risk of drawing air into the circuit, due to negative pump pressure.

Closed systems (Fig. 3.10)

In place of an expansion and feed cistern, the expansion of the heated water may be accommodated in a closed expansion vessel fitted to the boiler. The vessel contains nitrogen or air above a neoprene diaphragm and as the water expands the gas is compressed and its pressure rises. The vessel should accommodate the expansion of the heated water in the system.

Small-bore heating systems (Fig. 3.11)

The system is used extensively for houses and other types of small buildings. Use is made of small-bore copper pipes, usually 13 mm or 19 mm bore, depending upon the heating load to be carried in the circuit.

The boiler also heats the hot-water calorifier by natural convection. Thermostatic control may be by zoning valves shown, separate thermostatic radiator valves, a three-way mixing valve or by switching on and off the pump by a wall thermostat.

Thermal comfort, hot water and steam heat emitters

Thermal comfort

The thermal comfort of a human body is governed by the following factors:

1. The heat lost in radiation from the body, through clothing and exposed skins surfaces to the cooler surroundings.
2. The heat lost by convection from the body through clothing and exposed skin surfaces due to contact with the surrounding air, the temperature of which is considerably lower than that of the body.
3. The heat lost from the body by evaporation from the skin, due to perspiring:

The normal losses from the body, from the above sources, approximate to the following:

Radiation	45 per cent
Convection	30 per cent
Evaporation	25 per cent

In order to preserve the normal temperature of the body, these heat losses must be balanced by the heat gained. In the absence of sufficient warmth from the sun, or heat gains inside the building from lighting, people or machines, the necessary heat must be provided by heat emitters. It follows therefore that for various types of human activity there must be the correct proportions of radiant and convection heat, which will provide the most comfortable artificial warmth.

The rate of heat losses from the body can be controlled by the following:

1. *Radiation* — by the mean radiant temperature of the surrounding surfaces.
2. *Convection* — by the air temperature and air velocity.
3. *Evaporation* — by the relative humidity of the air and air velocity.

The purpose of a heat emitter is to maintain, at economic cost, the conditions of mean radiant temperature, air temperature and velocity that will give a suitable balance between the three ways in which heat is lost from the body. Although a heat emitter will provide radiant and convection heating, the correct control of relative humidity may require a system of air conditioning, which is described on p. 54.

Figure 3.12 shows how the heat losses and heat gains from and to a person in a room are balanced, also the external heat gains and losses.

Hot water and steam heat emitters

A centralised hot-water heating system has three basic elements:

1. Boiler plant for heat generation.
2. Heat distribution circuit.
3. Heat emitter.

The main type of heat emitters used for centralised hot water heating systems are:

1. Radiators

These may be column, hospital or panel types, made from either steel or cast iron. Steel radiators are made from light gauge steel pressings welded together, they are modern in appearance and are used extensively for heating systems in houses and flats. Cast iron radiators are bulkier and heavier, but will stand up to rough use in schools, hospitals and factories.

If a radiator is fitted against a wall, staining of the wall above the radiator will occur due to convection currents picking up dust from the floor. To prevent this, a shelf should be fitted about 76 mm above the radiator, and the jointing of the shelf to the wall must be well made or otherwise black stains will appear above the shelf. End shields must be fitted to the shelf to prevent black stains at the sides.

Painting of radiators: The use of metallic paints reduce the heat emitted from a radiator and the best radiating surface is dead black. Any colour of non-metallic paints may be used, as these do not affect appreciably the amount of radiant heat emitted by the radiator. The heat emitted by convection is not affected by the painting of radiators. The name 'radiator' is misleading, for although heat is transmitted by radiation, a greater proportion of heat is transmitted by convection, depending upon the type of radiator used.

Position of radiators: The best position is under a window, so that the heat emitted mixes with the incoming cold air from the window and this prevents cold air passing along the floor, which would cause discomfort to the occupants of the room. Figures 3.13, 3.14 and 3.15 show column, hospital and panel types of radiators. Figure 3.16 shows a radiator shelf.

2. Radiant panels

These have flat faced metal fronts and are similar to panel radiators, but transmit a greater proportion of heat by radiation. The panels are particularly suited to the heating of workshops, where they may be suspended at heights from 3 to 4 m above the floor level and arranged so that the heat is radiated downwards. They have the advantage of giving comfortable conditions to the occupants of the room by providing radiant heat at a lower air temperature. There is also a lower temperature gradient between the floor and the ceiling, which added to the lower air temperature reduces the cost of heating by about 15 per cent.

Figure 3.17 shows a radiant panel made from steel and Fig. 3.18 shows the various positions of the panels in a workshop.

3. Natural convection

These can be cabinet or skirting types: the cabinet type comprises a finned tubular heating element fitted near to the bottom of the casing, so that a stack effect is created inside the cabinet. The column of warm air above the heating element is displaced through the top of the cabinet by the cool air entering at the bottom. The greater the height of the cabinet the greater will be the air flow through it. Skirting types provide a good distribution of heat in a room and are very neat in appearance. If the floor is to be carpeted care should be taken to ensure that a gap is left at the bottom of the heater casing or convection through

the heater will be prevented. Figure 3.19 shows a cabinet type convector and Fig. 3.20 shows a skirting type convector.

4. Fan convectors

These have a finned tubular heating element, usually fitted near the top of the casing. The fan or fans fitted below the element draw air in from the bottom of the casing, the air is then forced through the heating element, where it is heated, before being discharged through the top of the cabinet into the room. The fans may have two- or three-speed control and thermostatic control of the heat output is often by switching off, or a change of speed of the fans. If required, a clock can also control the switching on and off of the fans. The convector may be fitted with an air filter below the heating element, which is not possible with the natural convector. They also have the advantage of quickly heating the air in the room and give a good distribution of heat. Figure 3.21 shows a fan convector incorporating an air filter.

5. Overhead unit heaters

These are similar to the fan convector and have a finned tubular heating element with a fan to improve the circulation of warm air. The method of fixing the fan and the heating element however differs from the fan convector to permit the overhead installation of the heater in factories, garages and warehouses, so that the warm air is blown down on to the working area. They usually operate on high temperature hot water or steam and the louvres can be adjusted so as to alter the direction of the warm air. Figure 3.22 shows an overhead unit heater and Fig. 3.23 shows how several heaters may be used to heat a workshop.

Note: The heating efficiency of both types of convectors and unit heaters is lowered more quickly than that of radiators, due to a drop in temperature of the heating element. For this reason, it is better to use a two-piped pumped distribution system.

6. Overhead radiant strips

These overcome the difficulty and cost of connecting separate radiant panels at high level. They consist of heating pipes, up to 30 m long, fixed to an insulated metal plate which also becomes heated by conduction from the pipe. The minimum mounting height of the strips is governed by the heating system temperature and ranges from about 3 m for low temperature hot water, to about 5 m for high temperature hot water and steam. Figure 3.24 shows a radiant strip having two heating pipes; one-to-four heating pipes may also be used. Figure 3.25 shows how radiant strips may be installed in a workshop.

7. Embedded pipe panels

Continuous coils of copper or steel pipes of 13 or 19 mm bore at 225 to 300 mm centres are embedded inside the building fabric, usually in the floor or ceiling, although wall panels may also be used. The following recommend panel surface temperatures may be used, except in special cases, such as may be required for entrance halls to public buildings or areas bordering on exposed outside walls.

Floors	26.7 °C
Ceilings	49 °C
Walls	43 °C

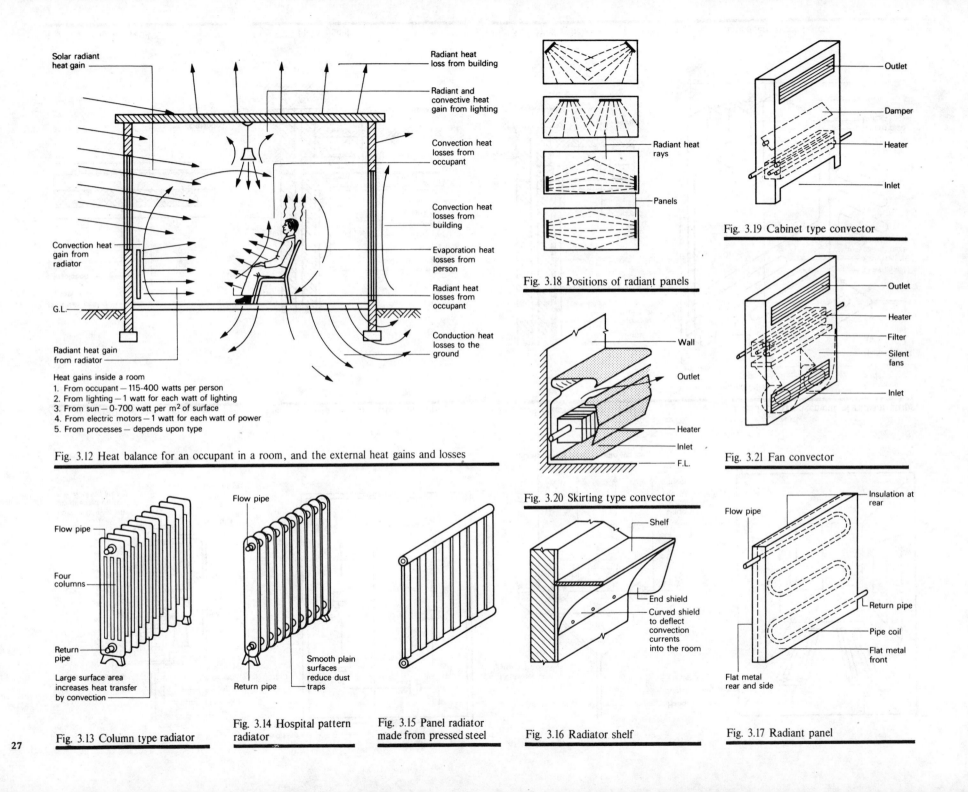

Solar radiant
heat gain

Radiant heat
loss from building

Radiant and
convective heat
gain from lighting

Convection heat
losses from
occupant

Convection heat
losses from
building

Convection heat
gain from
radiator

Evaporation heat
losses from
person

Radiant heat
losses from
occupant

G.L.

Radiant heat gain
from radiator

Conduction heat
losses to the
ground

Heat gains inside a room
1. From occupant — 115-400 watts per person
2. From lighting — 1 watt for each watt of lighting
3. From sun — 0-700 watt per m^2 of surface
4. From electric motors — 1 watt for each watt of power
5. From processes — depends upon type

Fig. 3.12 Heat balance for an occupant in a room, and the external heat gains and losses

Radiant heat
rays

Panels

Fig. 3.18 Positions of radiant panels

Wall

Outlet

Heater

Inlet

F.L.

Fig. 3.20 Skirting type convector

Outlet

Damper

Heater

Inlet

Fig. 3.19 Cabinet type convector

Outlet

Heater

Filter

Silent
fans

Inlet

Fig. 3.21 Fan convector

Flow pipe

Four
columns

Return
pipe

Large surface area
increases heat transfer
by convection

Fig. 3.13 Column type radiator

Flow pipe

Return pipe

Smooth plain
surfaces
reduce dust
traps

Fig. 3.14 Hospital pattern
radiator

Fig. 3.15 Panel radiator
made from pressed steel

Shelf

End shield

Curved shield
to deflect
convection
currents
into the room

Fig. 3.16 Radiator shelf

Insulation at
rear

Flow pipe

Return pipe

Pipe coil

Flat metal
front

Flat metal
rear and side

Fig. 3.17 Radiant panel

27

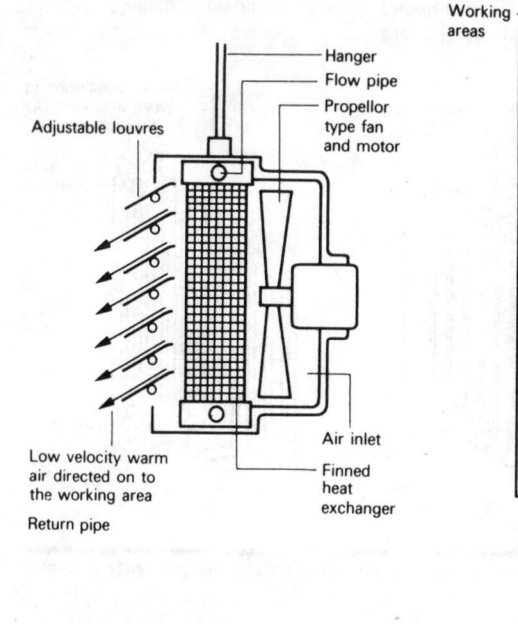

Hanger
Flow pipe
Propellor type fan and motor
Adjustable louvres
Low velocity warm air directed on to the working area
Return pipe
Air inlet
Finned heat exchanger

Fig. 3.22 Overhead unit heater

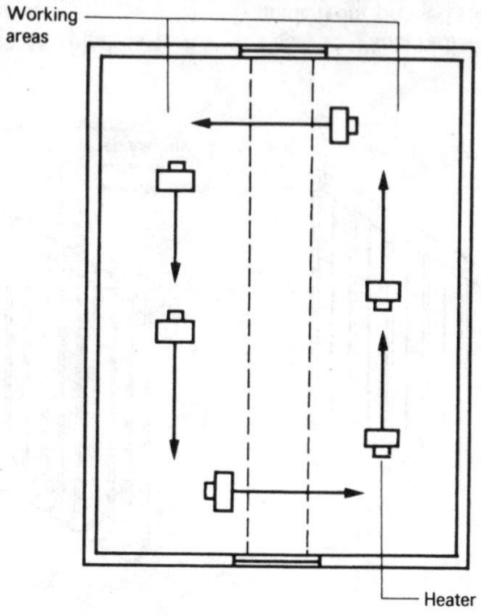

Working areas
Heater

Fig. 3.23 Arrangement of unit heaters

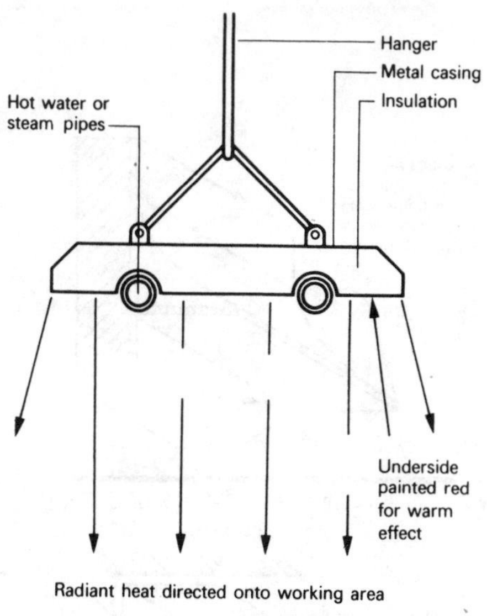

Hot water or steam pipes
Hanger
Metal casing
Insulation
Underside painted red for warm effect
Radiant heat directed onto working area

Fig. 3.24 Overhead radiant strip

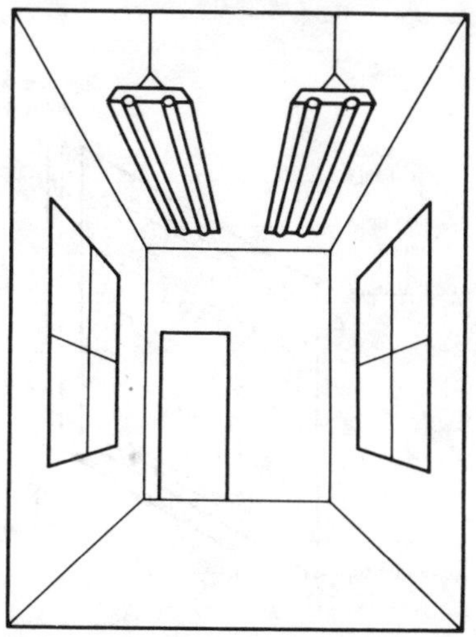

Fig. 3.25 Arrangement of radiant strips

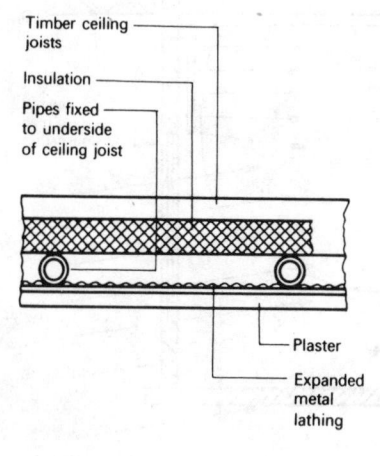

Timber ceiling joists
Insulation
Pipes fixed to underside of ceiling joist
Plaster
Expanded metal lathing

Fig. 3.26 Ceiling panel

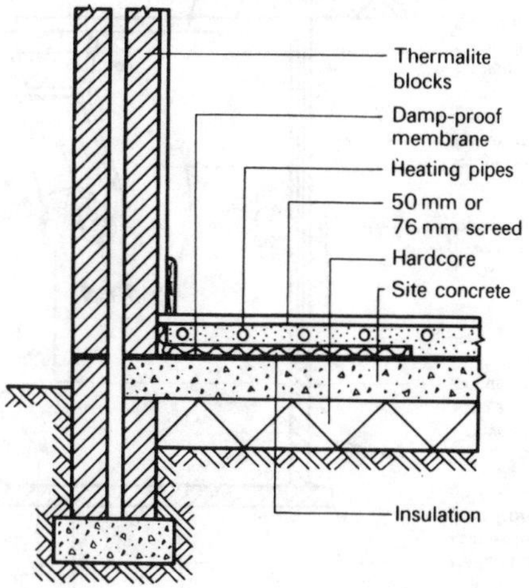

Thermalite blocks
Damp-proof membrane
Heating pipes
50 mm or 76 mm screed
Hardcore
Site concrete
Insulation

Fig. 3.27 Floor panel

Concrete insulation
Plaster
Pipes
Screed
D.P.C.
Hardcore

Fig. 3.28 Wall panels

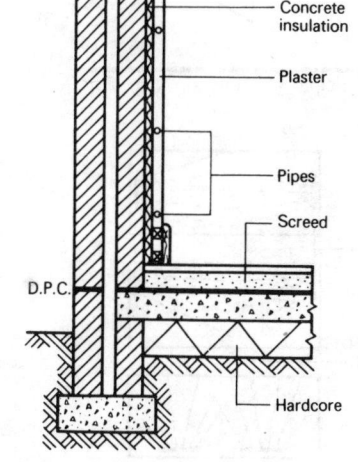

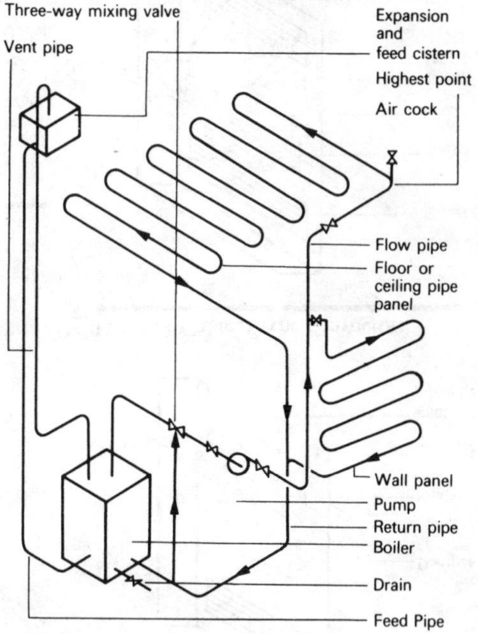

Three-way mixing valve
Vent pipe
Expansion and feed cistern
Highest point
Air cock
Flow pipe
Floor or ceiling pipe panel
Wall panel
Pump
Return pipe
Boiler
Drain
Feed Pipe

Fig. 3.29 Embedded panel system

With regards to these panel temperatures, the required water temperatures in the flow and return mains for the different panel locations are:

Panel location	Flow pipe	Return pipe
Ceilings	54 °C	43 °C
Floor	43 °C	35 °C
Wall	43 °C	35 °C

In order to control the temperature of the pipe panel, a modulating three-way mixing valve is used which allows the cooler water in the return pipe to mix with the hotter water in the flow pipe. The amount of radiant heat given off to the room, from different panel locations, is as follows:

Ceilings	65 per cent
Floors	50 per cent
Walls	50 per cent

Since radiant heat provides a feeling of warmth at a lower temperature than heat provided by convection, ceiling panels are to be preferred.

Figures 3.26, 3.27 and 3.28 show the methods of embedding the pipes in ceilings, floors and walls. Figure 3.29 shows an embedded panel system, including a three-way mixing valve.

Pipework

Copper pipes are usually jointed by means of soft soldered capillary fittings, but silver soldered or bronze welded joints are also suitable; steel pipes are welded. Compression joints for copper and screwed joints for steel are not to be recommended, due to risks of leaks caused by vibration of the building fabric. Before the pipes are embedded, they should be hydraulically tested at a pressure of 1400 kPa and this pressure should be maintained for 24 hours. The floor screed should be allowed to cure naturally before the pipes are heated and then heated gradually over a period of 10 days, before being put on full load.

Merits of heating by radiation

Radiant heating

1. Since about 45 per cent of the heat lost by the human body is due to radiation, the feeling of warmth derived from radiant heating is greater than by convection.
2. The radiant heat gives a greater feeling of warmth with a lower air temperature and this achieves about a 15 per cent saving in fuel costs.
3. In factories the lower air temperature gives a greater feeling of freshness, and production is known to increase.
4. The draughts are reduced to a minimum and dust is also kept down to a minimum.
5. Radiant heat does not heat the air through which it passes, but heats solid objects on which it falls, and so floors and walls derive warmth from the radiant heat rays. These warm surfaces set up convection currents, which reduce the heat lost from the human body by convection.

Chapter 4

Heat losses

Heat loss calculations

Definition of terms

1. *Thermal conductivity* (*k*)

The thermal transmission in unit time through unit area of a slab, or a uniform homogeneous material of unit thickness, when the difference of temperature is established between its surfaces.

The unit is W/m K

2. *Thermal resistivity* (*r*)

The reciprocal of the thermal conductivity.

The unit is m K/W

3. *Thermal conductance* (*c*)

The thermal transmission is unit time, through a unit area of a uniform structural component of thickness L per unit of temperature difference, between the hot and cold surfaces.

The unit is W/m² K

4. *Thermal resistance* (*R*)

The reciprocal of the thermal conductance.

$$R = \frac{L}{k}$$

where:

R = Thermal resistance of a material (m² K/W)

L = Thickness of the material in metres

k = Thermal conductivity of the material (W/m K)

Note: If the thickness L is given in millimetres, it must be converted to metres.

5. *Thermal transmission (U)*

The thermal transmission in unit time through unit area of a given structure, divided by the difference between the environmental temperature on either side of the structure.

The unit is W/m² K

6. *Standard thermal transmission (Standard U)*

The value of the thermal transmission of a building element related to standard conditions.

7. *Design thermal transmission (Design U)*

The value of the thermal transmission of a building element for prevailing design conditions.

8. *Emissivity (e)*

The ratio of the thermal radiation from unit area of a surface, to the radiation from unit area of a black body at the same temperature.

9. *Environmental temperature (t_e)*

A balanced mean, between the mean radiant temperature and the air temperature. It may be evaluated approximately from the following formula:

$$t_{ei} = \frac{2}{3} t_{ri} + \frac{1}{3} t_{ai}$$

where

t_{ri} = mean radiant temperature of all the room surfaces in °C

t_{ai} = inside air temperature in °C

The concept of environmental temperature for heat loss calculations provides a more accurate assessment of steady state heat loss than the conventional procedure that uses air temperature as a basis. The internal environmental temperature is also better than the internal air temperature as an index of the thermal comfort of the internal environment. This places the designer in a more favourable position for assessing the internal thermal comfort than is possible employing the air temperature. Conventional methods however may be followed when the difference between the mean radiant temperature and the air temperature is quite small. This occurs when rooms have little exposure to the outside and the standard of building thermal insulation is very high.

10. *Sol-air temperature (t_{eo})*

The outside air temperature which in the absence of solar radiation would give the same temperature distribution and rate of heat transfer, through the wall or roof, as exists with the actual outdoor temperature and the incidence of solar radiation.

11. *Mean radiant temperature (t_{rm})*

The temperature of a uniform block enclosure, in which a solid body or occupant would exchange the same amount of radiant heat as in existing non-uniform environment.

12. *Degrees Kelvin (K)*

This is defined in terms of the triple point of water as the fundamental fixed point, attributing to it the temperature of 273.16 K. Absolute zero is defined as O K. The unit may also be used for temperature interval.

13. *Degrees Celsius (°C)*

Degrees Celsius 0 °C = 273.16 K and the intervals on the Celsius and Kelvin scales are identical.

14. *Heat*

Heat is a form of energy, its quantity is measured in joules (J).

15. *Power*

Power is measured in watts (W). One watt is equal to one joule per second.

16. *Heat transfer*

There are three ways in which heat may be transferred from a material: (*a*) conduction, (*b*) convection and (*c*) radiation.

Conduction. The molecules of a material at a higher temperature will vibrate more than the molecules of a material at a lower temperature and this vibrational heat energy is transferred from the higher temperature to the lower temperature. This heat transfer takes place without movement of the hotter molecules to the cooler molecules, and the greater the temperature difference the greater the transfer of heat by conduction.

Conduction is greater in solids than in gases. Still air conducts heat very slowly and an unventilated cavity provides a good insulator. Still air pockets in insulating materials provide good heat-insulating characteristics.

Convection: this is the transfer of heat in a fluid. The hotter less dense liquid or gas is displaced by the more dense liquid or gas surrounding it, thus creating circulation. The natural circulation of air in a room, or water in a heating system, is brought about by convection.

Radiation: this is the transfer of heat in the form of electro-magnetic radiation from one body to another, without the need of a conducting medium. All bodies emit radiant heat and receive radiant heat back from other bodies. The higher the temperature of the body, the greater the radiant heat emitted. Matt black surfaces generally radiate or receive more heat than white, or bright shiny surfaces. White chippings on roofs will help to prevent a roof slab receiving radiant heat from the sun and aluminium foil will reflect radiant heat back from a heated room and this will act as a heat insulator. Figure 4.1 shows the transfer of heat through a wall by conduction, convection and radiation.

Temperature distribution

Figure 4.2 shows the temperature distribution for a solid wall. The wall with a high thermal conductivity will have a higher outside surface temperature that the wall with a low termal conductivity and therefore the temperature distribution will be greater. Figure 4.3 shows the temperature distribution for a cavity wall with a plastered aerated inner leaf, with a low thermal conductivity and a brick external leaf, and a fairly high thermal conductivity. In Fig. 4.4 the cavity is filled with urea formaldehyde foam, which will improve the thermal insulation and the temperature distribution through the cavity will be greater than the air cavity.

Standard U values

Steady state conditions

The rate of transfer of heat through a material may be affected by:

(a) Amount of moisture in the material.
(b) The variations in the composition of the material.
(c) Jointing of the component parts.

In heat loss calculations, however, it is assumed that a steady state exists which would result if the material is homogeneous; for example, it has the same composition throughout and the material is also dry, due to receiving heat from inside the building.

Computation of U values

The thermal transmission through the structure is obtained by combining the thermal resistance of its components and the adjacent air layers. The thermal transmission is found by adding the thermal resistances and taking the reciprocal.

$$U = \frac{1}{R_{si} + R_{so} + R_1 + R_2 + R_3 + R_a}$$

where

U = thermal transmission W/m^2 K

R_{si} = inside surface resistance m^2 K/W

R_{so} = outside surface resistance m^2 K/W

R_1 = thermal resistance of the structural components m^2 K/W

R_2 = thermal resistance of the structural components m^2 K/W

R_3 = thermal resistance of the structural components m^2 K/W

R_a = resistance of air space m^2 K/W

In computation of U values, the thermal resistances L/k are used. Where

k = thermal conductivity W/m K

L = thickness in (m) of a uniform homogeneous material

$$U = \frac{1}{R_{si} + R_{so} + \dfrac{L_1}{k_1} + \dfrac{L_2}{k_2} + \dfrac{L_3}{k_3} + R_a}$$

Table 4.1 gives the thermal conductivity for commonly used building materials.

Table 4.1 Thermal conductivities of common building materials

Material	k value W/m K
Asbestos cement sheet	0.40
Asbestos insulating board	0.12
Asphalt	1.20
Brickwork (commons)	
light	0.80
average	1.20
dense	1.47
Brickwork (engineering)	1.15
Concrete	
structural	1.40
aerated	0.14
Cork slab	0.40
Clinker block	0.05
Glass	1.02
Glass wool	0.034
Gypsum plasterboard	0.15—0.58
Linoleum (inlaid)	0.20
Plastering	
gypsum	0.40
vermiculite	0.20
Plywood	0.138
Polystyrene foam slab	0.04
Polyurethane foam	0.020—0.025
Foamed urea formaldehyde	0.030—0.036
Hair felt	0.43
Rendering (cement and sand)	0.53
Roofing felt	0.20
Rubber flooring	0.40
Slates	1.50
Soil	1.00—1.15
Stone	
granite	2.90
limestone	1.50
sandstone	1.30
Strawboard	0.09
Tiles	
burnt clay	0.83
plastic	0.50
Timber	
softwood	0.14
hardwood	0.16
Vermiculite	0.36—0.58
Wood	
chipboard	0.108
wool slabs	0.09

Surface resistances

The transfer of heat by convection to or from a homogeneous material depends upon the velocity at which the air passes over the surface of the material and the roughness of the surface. An internal wall may have a smooth plastered surface with very little air movement and an external wall may have a rough exterior surface and a high air movement.

The smooth internal surface will have a small amount of heat transfer by convection. The air forms a stagnant film, which tends to insulate the wall surface from the warmer air in the room. On the external surface of the material, the wind forces acting on a rough surface will cause eddy currents and heat will be transferred at a higher rate. Table 4.2 gives the inside surface resistances.

Table 4.2 Inside surface resistances R_{si} in m² K/W

Building element	Heat flow	Surface resistance in m² K/W	
		High emissivity	Low emissivity
Walls	Horizontal	0.123	0.304
Ceilings, flat or pitched roofs, floors	Upward	0.106	0.218
Ceilings and floors	Downward	0.150	0.562

Table 4.3 gives the external surface resistance.

Table 4.3 External surface resistance R_{so} in m² K/W for various exposures and surfaces

Building element	Emissivity of surface	Surface resistance for stated exposure (m² K/W)		
		Sheltered	Normal (standard)	Severe
Wall	High	0.08	0.055	0.03
	Low	0.11	0.067	0.03
Roof	High	0.07	0.045	0.02
	Low	0.09	0.053	0.02

Sheltered: up to third-floor buildings in city centres.
Normal: most suburban and country premises and the fourth to eighth floors, of buildings in city centres.
Severe: buildings on the coast, or exposed hill sites and above the fifth floor of buildings in suburban, or country districts, or above the ninth floor of buildings in city centres.

Thermal resistances of air spaces

Air spaces can be regarded as a further thermal resistance and an unventilated space offers more resistance than a ventilated one. Table 4.4 gives the standard unventilated resistances of unventilated air spaces.

Table 4.4

Air space thickness	Surface emissivity	Thermal resistance (m² K/W)	
		Heat flow horizontal or upwards	Heat flow downwards
5 mm	High	0.11	0.11
	Low	0.18	0.18
20 mm	High	0.18	0.21
	Low	0.35	1.06

Example 4.1 (Fig. 4.5). *A flat roof consists of 150 mm thick concrete covered by 20 mm of asphalt and sheltered conditions exists. Calculate the thermal transmittance (U) for the roof.*

Thermal conductivities
Concrete   1.4 W/m K
Asphalt   1.2 W/m K

Thermal resistances
Inside surface   0.11 m² K/W
Outside surface   0.07 m² K/W

$$U = \cfrac{1}{R_{si} + R_{so} + \cfrac{L_1}{k_1} + \cfrac{L_2}{k_2}}$$

$$U = \cfrac{1}{0.11 + 0.07 + \cfrac{0.15}{1.4} + \cfrac{0.02}{1.2}}$$

$$U = \frac{1}{0.11 + 0.07 + 0.11 + 0.02}$$

$$U = \frac{1}{0.31}$$

$$U = \underline{3.23 \text{ W/m}^2 \text{ K}}$$

Example 4.2 (Fig. 4.6). *A cavity wall consists of 13 mm thick plaster, 80 mm thick aerated concrete, 50 mm wide unventilated air space and 115 mm of external facing brick. Calculate the thermal transmission (U) for the wall.*

Thermal conductivities

Plaster	0.4 W/m K
Cellular concrete	0.25 W/m K
Brick	1.15 W/m K

Thermal resistances

Inside surface	0.123 m² K/W
Air space	0.18 m² K/W
Outside surface	0.055 m² K/W

$$U = \frac{1}{R_{si} + R_{so} + \frac{L_1}{k_1} + \frac{L_2}{k_2} + \frac{L_3}{k_3} + R_a}$$

$$U = \frac{1}{0.123 + 0.055 + \frac{0.013}{0.4} + \frac{0.08}{0.25} + \frac{0.115}{1.15} + 0.18}$$

$$U = \frac{1}{0.123 + 0.055 + 0.0325 + 0.32 + 0.1 + 0.18}$$

$$U = \frac{1}{0.8105}$$

$U = \underline{1.234\ \text{W/m}^2\ \text{K}}$

Example 4.3 (Fig. 4.7). *One wall of a building consists of 25 mm thickness of cedar boarding, 76 mm thickness of glass wool insulation and 13 mm thickness of plasterboard. Compare the overall thermal transmission of the wall, with the wall given in Example 4.2.*

Thermal conductivity

Cedar wood	0.14 W/m K
Glass wool	0.042 W/m K
Plasterboard	0.58 W/m K

Thermal resistances

Inside surface	0.123 m² K/W
Outside surface	0.055 m² K/W

$$U = \frac{1}{R_{si} + R_{so} + \frac{L_1}{k_1} + \frac{L_2}{k_2} + \frac{L_3}{k_3}}$$

$$U = \frac{1}{0.123 + 0.055 + \frac{0.013}{0.58} + \frac{0.076}{0.042} + \frac{0.025}{0.14}}$$

$$U = \frac{1}{0.123 + 0.055 + 0.0224 + 1.809 + 0.178}$$

$$U = \frac{1}{2.188}$$

$U = 0.457$

$U = \underline{0.457\ \text{W/m}^2\ \text{K}}$

If the inside and outside environmental temperatures are 20 °C and −2 °C respectively, the rate of heat loss per metre squared would be:

For solid wall

Heat loss = 1.234 x 1.0 x [20 − (−2)]

Heat loss = 27.148 W

For lightweight wall

Heat loss = 0.457 x 1.0 x [20 − (−2)]

Heat loss = 10.054 W

The heat loss through the solid wall is approximately 2.7 times greater than through the lightweight wall.

Example 4.4. *Figure 4.8 shows a section of a wall. Using the following data, calculate the thermal transmission* (U) *for the wall.*

Thermal conductivities

Brick	1.0 W/m K
Aerated concrete	0.14 W/m K
Plaster	0.7 W/m K
Expanded polystyrene	0.04 W/m K

Surface resistances

R_{si} External surface layer	0.12 m² K/W
R_{so} Internal surface layer	0.08 m² K/W
R_a Air space	0.18 m² K/W

$$U = \frac{1}{R_{si} + R_{so} + R_a + \frac{L_1}{k_1} + \frac{L_2}{k_2} + \frac{L_3}{k_3} + \frac{L_4}{k_4}}$$

$$U = \frac{1}{0.12 + 0.08 + 0.18 + \frac{0.11}{1} + \frac{0.11}{0.14} + \frac{0.013}{0.7} + \frac{0.003}{0.04}}$$

$$U = \frac{1}{0.12 + 0.08 + 0.18 + 0.11 + 0.78 + 0.02 + 0.08}$$

$$U = \frac{1}{1.37}$$

$U = \underline{0.73\ \text{W/m}^2\ \text{K}}$

The Chartered Institution of Building Services Guide Book give a comprehensive table of U values for various types of structures. Table 4.5 gives U values for common types of structures.

Table 4.5 U values for common types of structures

Construction	U value (W/m^2 K)		
	Sheltered	Normal	Severe
Walls (brickwork)			
220 mm solid brick wall unplastered	2.2	2.3	2.4
222 mm solid brick wall with 16 mm plaster	2.0	2.1	2.2
220 mm solid brick wall with 10 mm plaster-board on inside face	1.9	2.0	2.1
260 mm cavity wall (unventilated with 105 mm outer and inner leaves, with 16 mm plaster on inside face	1.4	1.5	1.6
Walls (brickwork)			
lightweight concrete block 260 mm cavity unventilated with 105 mm brick outer leaf, 100 mm lightweight concrete block inner leaf and 16 mm dense plaster on inside face	0.93	0.96	0.98
Walls, but with 13 mm expanded polystyrene board in cavity	0.69	0.70	0.71
Walls (lightweight concrete block)			
150 mm solid wall with 150 mm aerated concrete block, with tile hanging externally and with 16 mm plaster on inside face	0.95	0.97	1.0
Cavity wall (unventilated)			
with 76 mm aerated concrete block outer leaf, rendered externally 100 mm aerated concrete inner leaf and with 16 mm plaster on inside face	0.82	0.84	0.86
Concrete			
150 mm thick cast	3.2	3.5	3.9
200 mm thick cast	2.9	3.1	3.4
150 mm thick cast, with 50 mm woodwool slab permanent shuttering on inside face and 16 mm plaster	1.1	1.1	1.1

Table 4.5 – *continued*

Construction	U value (W/m^2 K)		
	Sheltered	Normal	Severe
Roofs (flat or pitched)			
19 mm asphalt on 150 mm solid concrete	3.1	3.4	3.7
19 mm asphalt on 150 mm hollow tiles	2.1	2.2	2.3
19 mm asphalt on 13 mm cement and sand screed on 50 mm metal edge reinforced woodwool slabs on steel framing, with vapour barrier at inside	0.88	0.90	0.92
19 mm asphalt on 13 mm cement and sand screed, 50 mm woodwool slabs on timber joists and aluminium foil – backed 10 mm plasterboard ceiling, sealed to prevent moisture penetration	0.88	0.90	0.92
Roofs, but with 25 mm glass fibre insulation laid between joists	0.59	0.60	0.61
Tiles on battens, roofing felt and rafters, with roof space and aluminium foil-backed 10 mm plasterboard on ceiling joists	1.4	1.5	1.6
Tiles, but with boarding on rafters	1.4	1.5	1.6
Corrugated asbestos cement sheeting	5.3	6.1	7.2
Floors			
Suspended timber floor above ground:			
150 mm x 60 mm	—	0.14	—
150 mm x 30 mm	—	0.21	—
60 mm x 60 mm	—	0.16	—
60 mm x 30 mm	—	0.24	—
30 mm x 30 mm	—	0.28	—
30 mm x 15 mm	—	0.39	—
15 mm x 15 mm	—	0.45	—
7.5 mm x 7.5 mm	—	0.68	—
3 mm x 3 mm	—	1.05	—
Glass			
Single glazing	5.0	5.6	6.7
Double glazing with 20 mm air space	2.8	2.9	3.2
Double glazing with 12 mm air space	2.8	3.0	3.3
Double glazing with 6 mm air space	3.2	3.4	3.8
Double glazing with 3 mm air space	3.6	4.0	4.4
Triple glazing with 20 mm air space	1.9	2.0	2.1
Triple glazing with 12 mm air space	2.0	2.1	2.2
Triple glazing with 6 mm air space	2.3	2.5	2.6
Triple glazing with 3 mm air space	2.8	3.0	3.3

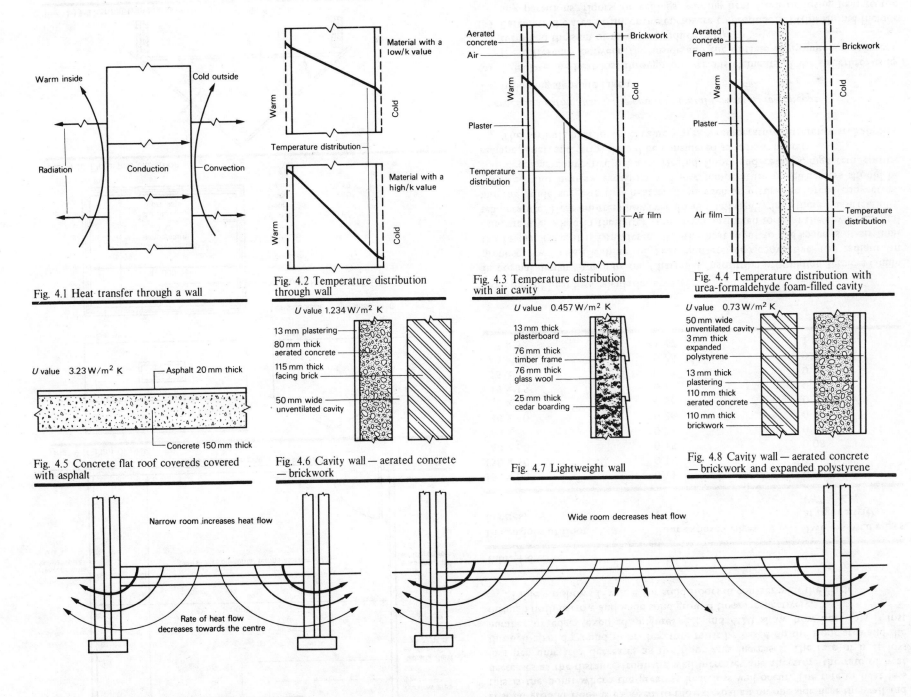

Fig. 4.1 Heat transfer through a wall

Fig. 4.2 Temperature distribution through wall

Material with a low/k value

Material with a high/k value

Temperature distribution

Fig. 4.3 Temperature distribution with air cavity

Aerated concrete

Air

Plaster

Temperature distribution

Brickwork

Air film

Fig. 4.4 Temperature distribution with urea-formaldehyde foam-filled cavity

Aerated concrete

Foam

Plaster

Air film

Brickwork

Temperature distribution

U value   3.23 W/m^2 K

Asphalt 20 mm thick

Concrete 150 mm thick

Fig. 4.5 Concrete flat roof covereds covered with asphalt

U value 1.234 W/m^2 K

13 mm plastering
80 mm thick aerated concrete
115 mm thick facing brick
50 mm wide unventilated cavity

Fig. 4.6 Cavity wall — aerated concrete — brickwork

U value   0.457 W/m^2 K

13 mm thick plasterboard
76 mm thick timber frame
76 mm thick glass wool
25 mm thick cedar boarding

Fig. 4.7 Lightweight wall

U value   0.73 W/m^2 K

50 mm wide unventilated cavity
3 mm thick expanded polystyrene
13 mm thick plastering
110 mm thick aerated concrete
110 mm thick brickwork

Fig. 4.8 Cavity wall — aerated concrete — brickwork and expanded polystyrene

Narrow room increases heat flow

Wide room decreases heat flow

Rate of heat flow decreases towards the centre

Fig. 4.9

Fig. 4.10 Heat flow through solid ground floor

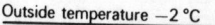

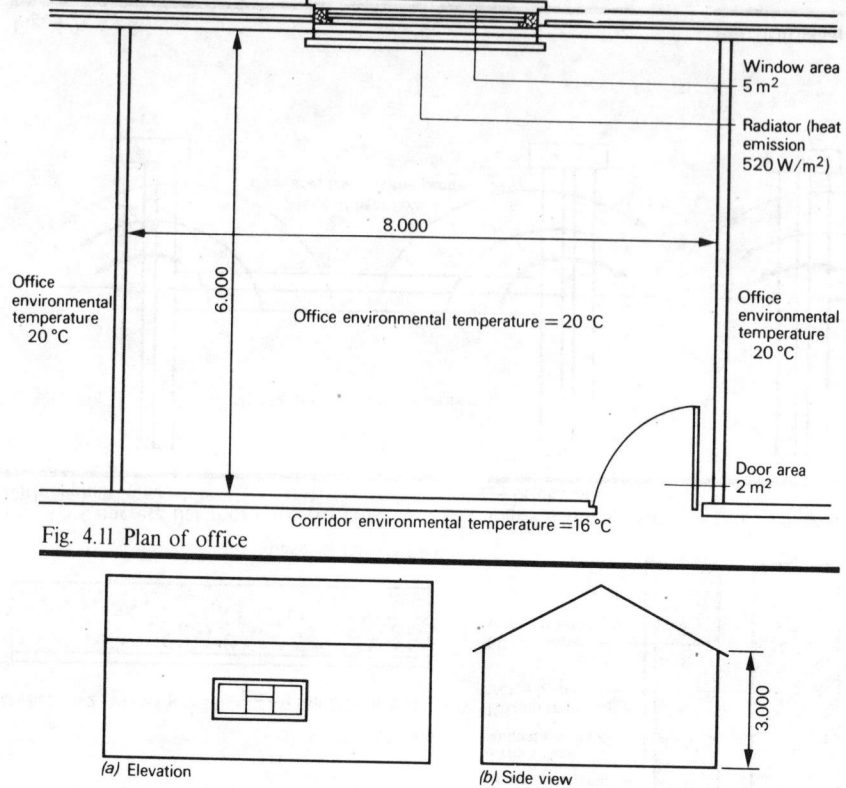

Fig. 4.11 Plan of office

Fig. 4.12 Detached classroom

Fig. 4.13 Thermal transmission (U value) pitched roof

Heat loss through solid ground floor

A solid ground floor is exposed to the external air on one side near the wall and this is the point where the greatest heat loss will occur. The rate of heat loss decreases as the distance from the wall increases and therefore the rate of heat loss per unit area decreases as the floor area increases. The rate of heat loss through a solid ground floor therefore must be based on the floor area and the number of edges exposed. Figures 4.9 and 4.10 show how heat loss is lost through both narrow and wide solid ground floors respectively.

Table 4.6 shows U values for solid floors in contact with the earth.

Table 4.6

Dimensions of floor (metres)	Four exposed edges (W/m² K)	Two exposed edges at right angles (W/m² K)
150 × 60	0.11	0.06
150 × 30	0.18	0.10
60 × 60	0.15	0.08
60 × 30	0.21	0.12
30 × 30	0.26	0.15
30 × 15	0.36	0.21
15 × 15	0.45	0.26
15 × 7.5	0.62	0.36
7.5 × 7.5	0.76	0.45
3 × 3	1.47	1.07

Examples of heat loss calculations

In normal design situations the Chartered Institution of Building Services Guide proposes two classifications of heat emitters: (a) convective, (b) radiant. In the former type, it is considered that the heat input to the space is at air tem-temperature, while in the latter type it is considered to be at the environmental temperature. Heat emitters however do not fall rigidly into these classifications, but the guide suggests that heat emitters such as natural or forced convectors, overhead unit heaters, radiators and low temperature panel heaters should be considered as convective heaters. Heated floors and ceilings, high temperature radiant panels and strips should be considered as radiant heaters.

The formula heat loss = U value x area x temperature difference still applies.

Convective heating, using the environmental temperature concept

The following steps are taken:

(a) Calculate the heat loss through the various elements of the structure, using the difference between the inside and the outside environmental temperatures and the sum of these to produce ΣQ_f.

(b) Calculate the area of the entire enclosure to produce ΣA (this should include any partitions, floors, or ceilings, gaining heat from or losing heat to the adjacent rooms).

(c) Calculate $\Sigma\, Q_f/\Sigma\, A$ and from the following equation find the difference between the inside air temperature and inside environmental temperature.

$$t_{ai} - t_{ei} = \Sigma\, Q_f/4.8\, \Sigma\, A$$

where

t_{ai} = inside air temperature °C

t_{ei} = inside environmental temperature °C

$\Sigma\, Q_f$ = rate of heat transfer through the building fabric W

$\Sigma\, A$ = total area of room surfaces in m²

(d) Calculate the heat loss by ventilation. The heat loss due to infiltration may be found from the formula:

$$Q_v = pVC(t_{ai} - t_{ao})$$

where

p = density of air which may be taken as 1.2 kg/m³

V = infiltration rate m³/s

C = specific heat capacity of air, which may be taken as 1000 J/kg K

t_{ai} = inside air temperature °C

t_{ao} = outside air temperature °C

Introducing the air infiltration rate N (air changes per hour) and room volume V (m³) the formula may be written,

$$Q_v = \frac{pNvC}{3600}(t_{ai} - t_{ao})$$

and substituting the above values for p and C,

$$Q_v = \frac{1.2 \times N \times V \times 1000}{3600}(t_{ai} - t_{ao})$$

$$Q_v = 0.33Nv(t_{ai} - t_{ao})$$

where

N = rate of air change per hour

v = volume of room in m³

t_{ai} = inside air temperature °C

t_{ao} = outside air temperature °C

(e) Find $\Sigma\, Q_f + \Sigma\, Q_v$ to give the total heat loss.

Worked examples on convective heating

Example 4.5. *Figure 4.11 shows the plan of an office on the second floor of a four-storey building. The other floors have the same construction and heating design conditions. From the data given below calculate the total rate of heat loss and the surface area of the radiator required.*

Office internal environmental temperature = 20 °C
Corridor environmental temperature = 16 °C
Design outside air temperature = −2 °C
Area of door = 2 m²
Area of window = 5 m²
Height of office = 3 m
Air change = 1 per hour
Ventilation allowance = 0.33 W/m³ K

U *values*

External wall = 1.5 W/m² K
Internal partition = 1.6 W/m² K
Door = 2.3 W/m² K
Window = 5.6 W/m² K
Heat emission from radiator = 520 W/m²

Heat loss through the structure

Type of structure	Area (m²)	U value (W/m² K)	Temperature difference (K)	Total heat loss (watts)
External wall (less window)	19	1.5	22	627
Window	5	5.6	22	616
Internal partition (less door)	22	1.6	4	140.8
Door	2	2.3	4	18.4
Floor	48	–	–	–
Ceiling	48	–	–	–
Other partition	36	–	–	–
ΣA	180		ΣQ_f	1402.2

Internal air temperature

$$t_{ai} - t_{ei} = \frac{\Sigma\, Q_f}{4.8\, \Sigma\, A}$$

$$t_{ai} - t_{ei} = \frac{1402.2}{4.8 \times 180}$$

$$t_{ai} - t_{ei} = 1.6$$

$$t_{ai} = 20 + 1.6$$

$$t_{ai} = 21.6\ °C$$

Heat loss by ventilation

$$Q_v = 0.33Nv(t_{ai} - t_{ao})$$

$Q_v = 0.33 \times 144[21.6 - (-1)]$

$Q_v = 0.33 \times 144 \times 22.6$

$Q_v = 1073.9W$

Total heat loss Q_t

$Q_t = \Sigma Q \Sigma Q_v$

$Q_t = 1402.2 + 1073.9$

$Q_t = 2476.1$

Area of radiator

$$\text{Area} = \frac{\text{Total heat loss } W}{\text{Heat emission W/m}^2}$$

$$\text{Area} = \frac{2426}{520}$$

$\text{Area} = 4.66 \text{ m}^2$

Example 4.6 (Convective heating). *Figure 4.12A, B, C shows the elevations and plan of a small detached classroom, to be heated by means of two convector heaters. Calculate the total heat loss for the classroom.*

Classroom internal environmental temperature 20 °C
Design outside air temperature   1 °C

		U values	
Area of door	2 m^2	External wall	1.40 W/m^2 K
Area of windows	12 m^2	Window	5.60 W/m^2 K
Height of classroom	3 m	Door	2.30 W/m^2 K
Air change	3 per hour	Floor	0.45 W/m^2 K
Ventilation allowance	0.33 W/m^3 K	Ceiling	0.49 W/m^2 K

Heat loss through structure

Type of structure	Area (m^2)	U value (W/m^2 K)	Temperature difference (K)	Total heat loss (watts)
External wall (less windows and doors)	(36 × 3) − (12 + 2) 94	1.4	21	2763.60
Window	12	5.6	21	1411.20
Door	2	2.3	21	96.60
Floor	80	0.45	21	756.00
Ceiling	80	0.49	21	823.20
ΣA	268		ΣQ_f	5850.60

Internal air temperature

$$t_{ai} - t_{ei} = \frac{\Sigma Q_f}{4.8 \, \Sigma A}$$

$$t_{ai} - t_{ei} = \frac{5850.6}{4.8 \times 268}$$

$t_{ai} - t_{ei} = 4.548$

$t_{ai} = 20 + 4.5$

$t_{ai} = 24.5 \,°\text{C}$

Heat loss by ventilation

$Q_v = 0.33Nv(t_{ai} - t_{ao})$

$Q_v = 0.33 \times 3 \times 240[24.5 - (-1)]$

$Q_v = 0.33 \times 3 \times 240 \times 25.5$

$Q_v = 6058.8 \text{ W}$

Total heat loss Q_t

$Q_t = \Sigma Q_f + \Sigma Q_v$

$Q_t = 5850.6 + 6058.8$

$Q_t = 11\,909.4 \text{ W}$

$Q_t = 12 \text{ kW}$

Each convector would require a heat output of 6 kW.

Radiant heating

In order to find the total heat requirements, the following steps are necessary.

(a) Calculate the heat loss through the structure as for convective heating to produce ΣQ_f.

(b) Calculate the surface area of the entire enclosure, as for convective heating to produce ΣA.

(c) Find the ventilation conductance C_v from:

$$\frac{1}{C_v} = \frac{1}{0.33Nv} + \frac{1}{4.8 \, \Sigma A}$$

where N = number of air changes per hour

where v = volume of enclosure m^2.

(d) Calculate ventilation loss Q_v from:

$$Q_v = C_v(t_{ei} - t_{eo})$$

(e) Add the results of steps (a) and (d) to give the total heat loss.

Example 4.7 (Radiant heating). *Calculate the total heat loss for the classroom given in Example 4.6 but assume that in this case the heating is to be in the form of a heated ceiling.*

Note: Since a heated ceiling is to be employed, there is no need to consider heat loss from the classroom through the ceiling.

$\Sigma \, Q_f = 5850.6 - 823.2$

$\Sigma \, Q_f = 5019.4 \text{ W}$

Also

$\Sigma \, A = 268 \text{ m}^2, v = 240, N = 3$

Ventilation conductance

$$\frac{1}{C_v} = \frac{1}{0.33 \times Nv} + \frac{1}{4.8 \, \Sigma \, A}$$

$$\frac{1}{C_v} = \frac{1}{0.33 \times 3 \times 240} + \frac{1}{4.8 \times 268}$$

$$\frac{1}{C_v} = 0.0042 + 0.000\,77$$

therefore

$C_v = 201.2 \text{ W K}$

Heat loss by ventilation

$Q_v = C_v(t_{ei} - t_{eo})$

$Q_v = 201.2\,[20 - (-1)]$

$Q_v = 201.2 \times 21$

$Q_v = 4225.2 \text{ W}$

Total heat loss Q_t

$Q_t = \Sigma \, Q_f + \Sigma \, Q_v$

$Q_t = 5850 + 4225.2$

$Q_t = 10\,075.8$

$Q_t = 10 \text{ kW}.$

Pitched roofs

The U values for common types of pitched roofs can be found in the Chartered Institution of Building Services Guide Book. There may be occasions, however, when some special type of construction is to be used and a U value for the roof will have to be calculated. The following equation gives the method of calculat-ing the U value for a pitched roof with a horizontal ceiling below. The values in the equation are shown in Figure 4.13.

$$U = \frac{U_r \times U_c}{U_r + U_c \, \text{Cos} \, \infty}$$

Example 4.8. *A pitched roof inclined at 40° has a horizontal ceiling below. If U_c and U_r are found to be 1.60 W/m² K and 3.20 W/m² K respectively, calculate the value of U.*

$$U = \frac{3.20 \times 1.60}{3.20 + (1.60 \times 0.766)}$$

$U = 1.157 \text{ W/m}^2 \text{ K}$

Example 4.9. *A pitched roof inclined at 45° has a horizontal ceiling below. If U_c and U_r are found to be 4.60 W/m² K and 7.38 W/m² K respectively, calculate the value of U.*

$$U = \frac{7.38 \times 4.60}{7.38 + (4.60 \times 707)}$$

$U = 3.19 \text{ W/m}^2 \text{ K}$

Alternatively, the U value for the total construction may be calculated by the use of the following formula:

$$U = \frac{1}{R_c + R_r \times \text{Cos} \, \infty}$$

where

R_c = thermal resistance of ceiling

R_r = thermal resistance of roof

The formula only applies where the roof space is unventilated: it can be used when considering the improvement made by adding an insulating lining to the roof by expressing the U value of the insulating material as a thermal resist-ance; the reciprocal of the total resistance gives the U value of the total construction.

Thermal insulation

Thermal insulation reduces the flow of heat through the structure of a building and its advantages may be summarised as follows:

1. The reduction in the size of the heating installation, resulting in the reduction in capital costs, fuel consumption and therefore running costs.

2. Saving in space for plant and fuel.
3. Reduction, or the complete elimination, of condensation problems.
4. Reduction of unsightly pattern staining and redecoration costs.
5. Improved comfort levels for the occupants of the building.
6. Reduced pre-heating time.
7. Reduced rate of heat gain from solar radiation and therefore reduction in size of cooling plant.

The inclusion of thermal insulation within the structure of a heated building can be regarded as an investment, from which an annual return on capital can be derived. The cost of insulation can also be largely offset by the saving in the cost of the heating installation.

The heat losses from a building having a compact layout are less than those from a straggling layout of the same floor area and volume. A room of square proportions on plan also has less heat losses than a room of rectangular proportions on plan. For example, a room measuring 6 m x 6 m has a perimeter of 24 m, while a room measuring 9 m x 4 m will have a perimeter of 26 m. The square room therefore will have less heat losses through the walls and floor perimeter than the rectangular room of the same floor area and volume.

Building Regulations 1985 Conservation of Fuel and Power Exposed elements means:

1. In relation to a dwelling:

 (*a*) exposed to the outside air: or
 (*b*) separating a dwelling from part of a building which is ventilated by means of an opening or duct to the outside air, the aggregate area of which exceeds 5 per cent of the area of the walls enclosing the spare, and which allows a passage of air at all times.

2. In relation to a building not consisting of a dwelling:

 (*c*) exposed to outside air; or
 (*d*) separating a part of the building which is heated from a part which is not and which is wholly or partly exposed to the outside air.

 The area of perimeter wall used to calculate the areas of windows should include all openings in the wall, and

 (*a*) an external door with 2 m² or more of glazed area should be counted as part of the window area; and
 (*b*) any part of a roof which has a pitch of 70° or more should be counted as walling: and
 (*c*) any opening in a wall other than a window opening and a meter cupboard recess may be counted as part of the wall area; and
 (*d*) lintels, jambs and sills associated with windows and roof lights may be counted as part of the window and roof light area, and
 (*e*) areas of walls, floors and roofs should be between finished infernal faces of the building and in the case of a roof, in the plane of the insulation.

 '*U* value' means thermal transmittance coefficient, that is to say, the rate of heat transfer in watts through 1 m² of a structure when the combined radiant and air temperatures at each side of the structure differ by 1 °C and is expressed in W/m² K

 'ventilated space' means a space which —

Note: A good deal of heat loss through window openings may be saved if the windows having the larger areas have a southern aspect.

In winter, south-facing windows during daytime may provide a useful degree of solar heating and buildings are sometimes constructed having most of the windows facing south.

This type of solar heating is known as 'passive' and another type, which circulates water through a solar collector, is known as 'active'. Both types of solar heating are dealt with in detail in Vol. 3 *Services and Equipment*.

Table 4.7 Maximum U value of walls, fllors and roofs

Type of building	Exposed walls	Roofs	Exposed floors
Dwellings	0.6	0.35	0.6
Other residential buildings including hotels and institutional	0.6	0.6	0.6
Places of assembly offices and shops	0.6	0.6	0.6
Industrial and storage	0.7	0.7	0.7

Note: Display windows in shops are not included.

Requirements for dwellings

1. The calculated rate of heat loss (WK) through any windows and roof lights shall be no greater than it would be if:

 (*a*) the aggregate of the areas of windows and roof lights were 12 per cent of the area of the walls bounding the dwelling, and
 (*b*) the windows and roof lights had a *U* value of 5.7

2. The calculated rate of heat loss through the solid parts of the exposed elements shall be not greater than it would be if:

 (*a*) the exposed walls and exposed floors had a *U* value of 0.6 and
 (*b*) the roof had a *U* value of 0.35.

3. The extent that the calculated rate of heat loss through the solid parts of exposed elements is less than the maximum permitted under sub-paragraph (2) the calculated rate of heat loss through windows and roof lights may be greater than the maximum permitted under paragraph (1).

Example 4.10 *A detached house has an area of exposed walls of 200 m² and the area of windows is 50 m². The walls and ceiling have* U *values of 0.5 W/m²K and 0.4 W/m²K respectively.*

If the area of the ceiling is 84 m² and double glazing is to be used, calculate if the levels of insulation meet the requirements of the Regulations.

Area of exposed walls

Area of exposed walls including windows	= 200 m²
Area of exposed walls to be insulated	
= 96 − area of windows = 200 − 50	= 150 m²
Area of roof to be insulated	= 84 m²

From Table 4.7 the *U* value of exposed walls and roof are 0.6 W/m² K and 0.35 W/m² K respectively.

Allowable rate of heat loss through exposed walls	
(150 X 0.6)	= 90 W/K
Allowable rate of heat loss through roof (84 X 0.35)	= 29.4 W/K
Total allowable rate of heat loss through walls and	
roof	= 119.4 W/K

In proposed house

Rate of heat loss through exposed walls	
= (150 X 0.5)	= 75
Rate of heat loss through roof (84 X 0.4)	= 33.6
Total heat loss through walls and roof	109.6 W/K

(The proposed insulation levels are satisfactory)

It is sometimes necessary to check if the type of construction has a *U* value that does not exceed the *U* value specified in the Regulations.

Example 4.11 The exposed wall of a dwelling consists of: 113 mm brick outer leaf: 25 mm air space; 25 mm insulation; 100 mm aerated concrete inner leaf; and 16 mm internal plastering. Calculate the thermal transmission

Example 4.11 *The exposed wall of a dwelling consists of: 113 mm brick outer leaf: 25 mm air space; 25 mm insulation; 100 mm aerated concrete inner leaf; and 16 mm internal plastering. Calculate the thermal transmission* U *value for the wall,*

Thermal conductivities

Brick	0.8 W/m K
Aerated concrete	0.14 W/m K
Insulation	0.04 W/m K
Plaster	0.20 W/m K

Surface resistances

R_{si}	External surface layer	0.12 m² KW
R_{so}	Internal surface layer	0.08 m² KW
R_a	Air space	0.18 m² KW

The thermal is obtained by dividing the thickness of the material in (m) by the thermal conductivity and the *U* value is obtained by adding the thermal resistances and taking the reciprocal of the result.

$$U = \frac{1}{R_{si} + R_{so} + R_a + \dfrac{L_1}{k_1} + \dfrac{L_2}{k_2} + \dfrac{L_3}{k_3} + \dfrac{L_4}{k_4}}$$

$$U = \frac{1}{0.12 + 0.08 + 0.18 + \dfrac{0.113}{0.8} + \dfrac{0.025}{0.04} + \dfrac{0.1}{0.14} + \dfrac{0.016}{0.2}}$$

$$U = \frac{1}{0.12 + 0.08 + 0.18 + 0.14 + 0.625 + 0.714 + 0.08}$$

$$U = \frac{1}{}$$

$$U = \frac{1}{1.939}$$

$$U = 0.516 \text{ (approx)}$$

The construction is below 0.6 (see Table 4.7) and is therefore satisfactory.

Insulating materials

Insulating materials may be divided broadly into two groups:

1. Non-loadbearing.
2. Loadbearing.

Non-loadbearing materials generally have a low density and make use of still air cells and usually possess a greater thermal resistance than loadbearing materials. The rate of heat loss through modern non-loadbearing lightweight structures is less than the heat losses through heavier loadbearing structures and this has led to a wider use of lightweight prefabricated sections in buildings. Curtain wall structures use lightweight non-loadbearing panels between heavy frames of reinforced concrete, or structural steel. Some materials are organic in origin, or contain organic materials and the subject of fire spread must be considered in conjunction with that of insulation when proposing to use these materials. The hazards of smoke and toxic fumes produced in fires where foamed plastics are used for insulating materials must also be considered. The Building Regulations (England and Wales) 1985 covers the spread of fire in buildings and the local Fire Prevention Officer should also be consulted where any doubt exists regarding the choice of materials. When selecting a suitable insulating material, the following points must be considered:

1. The risk of spread of fire, or the production of toxic fumes when fixed.
2. Whether the material required, is non-loadbearing or loadbearing.
3. The cost and thickness in relation to the saving in fuel and capital cost of the heating installation.
4. The availability of the material and the ease of fixing.

Foamed plastics

Foamed plastics are the most effective of all materials used for thermal insulation and their thermal conductivities are as follows:

Foamed polyurethane	0.020−0.025 W/m K
Expanded polystyrene	0.029−0.04 W/m K
Foamed urea formaldehyde	0.030−0.036 W/m K

Cavity fill

The insulation of an external wall can be improved by about 50 per cent by blowing non-toxic insulating material into the cavity (see Fig. 4.14)

Mineral wool may be used which does not shrink or crack and also prevents

the penetration of water. The material is fire resisting and can be used with absolute safety.

Resin-coated expanded polystyrene beads can be blown into the cavity to improve the insulation. The beads bond together by the resin and they are resistant to water.

On new work, expanded polystyrene board 25 mm thick may be fixed to the inner leaf of the cavity, thus leaving 25 mm of an air cavity (see Fig. 4.15). Alternatively, on new work, the cavity may be filled with mineral wool quilt, or semi-rigid glass wool slab treated with a water-repellent binder.

Insulating boards or slab are used for insulating all various types of structures and these include the following types:

1. Wood wool slabs.
2. Compressed strawboard.
3. Insulating plasterboard.
4. Aluminium foil-backed plasterboard.
5. Mineral wool slabs.
6. Expanded polystyrene slabs.
7. Corkboard.
8. Insulating wood fibre board.
9. Cellular glass rigid slabs.
10. Semi-rigid slabs of glass fibre, treated with a water repellent.

Before the boards or slabs are fixed they must be conditioned for at least 24 hours, by exposing them on all sides to the same air temperature and humidity as would exist when they are fixed on site or otherwise distortion of the boards would occur after fixing.

Internal linings

Insulation board fixed to battens on the inside walls enables a room to warm up quickly and helps to prevent condensation when intermittent heating is used. The insulation will reduce the radiant heat loss from the body and therefore provides better conditions of thermal comfort. The insulation, however, should be considered with the risk of spread of fire since the temperature of the combustible materials inside the room will reach ignition point more quickly if a fire occurs.

Figures 4.16, 4.17 and 4.18 show the methods of fixing internal linings to brick and corrugated asbestos cement sheet walls.

Insulating plasters

Insulating plasters containing lightweight perlite and vermiculite as aggregates are produced ready for use and require only the addition of clean water before use. They have three times the insulating value and are only one-third the density of ordinary plasters.

Vapour barriers

In conditions where there is a risk of condensation, moisture vapour will rise through the structure and, being unable to disperse, may condense within the structure and saturate the insulating material. If this occurs the insulating material will rapidly lose its insulating properties and if the material is organic in origin it will decompose. In such situations a suitable vapour barrier should be included in the structure and should be inserted on the warmest side of the structure.

Roof vents

In concrete roofs moisture may be present in the screed below the mastic asphalt or felt covering. During hot weather this moisture may vaporise and cause an upward pressure, resulting in the lifting of the roof covering. To prevent this occurrence roof ventilators are fixed to relieve this pressure. It is usual to fix one ventilator to every 20 m^2 of roof area (see Fig. 4.19).

Solid ground floors

Since the greatest heat loss is around the perimeter of the floor, it is essential to provide insulation at this location. For wide floors, a horizontal strip of insulation, may be extended 1 m from the wall and a vertical strip of insulation material should be extended through the thickness of the floor slab, around all exposed edges (see Fig. 4.20). For narrow floors, the horizontal strip insulation may be carried under the entire floor slab, with vertical strip insulation material extended through the thickness of the floor slab, as before. If the cavity is to be filled with semi-rigid slabs of glass fibre, specially treated with a water-repellent binder, horizontal strip insulation may be omitted (see Fig. 4.21). The slabs are available in three thicknesses, 50, 65 and 75 mm, and they provide a moisture barrier between the outer and inner leaf of the cavity wall. If required, a horizontal strip 1 m from the wall may also be included, to provide additional insulation for the floor.

Suspended timber ground floor

A suspended ground floor above an enclosed air space is exposed to air on both sides, but the air temperature below the floor is higher than the external air because the ventilation rate below the floor is very low. Additional thermal insulation of a suspended timber ground floor may be provided either in the form of a continuous layer of semi-rigid or flexible material laid over the joists (see Figs 4.22 and 4.23). Alternatively, a semi-rigid material may be laid between the joists (see Fig. 4.24). A vapour barrier should never be included in suspended timber ground floors as this would prevent the escape of water above and evaporation, which could lead to the dangerous conditions conductive to fungal attack on the timber floor.

Flat roofs

These are usually constructed of reinforced concrete or timber and may be covered with mastic asphalt, roofing felt, lead, copper, aluminium or zinc. To prevent radiant heat gain, mastic asphalt and felt roofs have a layer of reflective chippings spread over the external surface. Figure 4.25 shows a section through a concrete flat roof, insulated with mineral wool slabs and Fig. 4.26 shows an alternative method of insulating a concrete flat roof with an aerated concrete screed. Figure 4.27 shows a section through a timber roof, insulated with aluminium foil-backed plasterboard. The aluminium foil acts as both a thermal insulator and a vapour barrier.

Aerated concrete

Aerated concrete may be used for inside work as a screed to a concrete roof, or as concrete blocks for walls. The concrete uses lightweight aggregates such as foamed slag, sintered pulverised fuel ash, exfoliated vermiculite, expanded perlite or expanded clay. Alternatively, aerated concrete may be made by

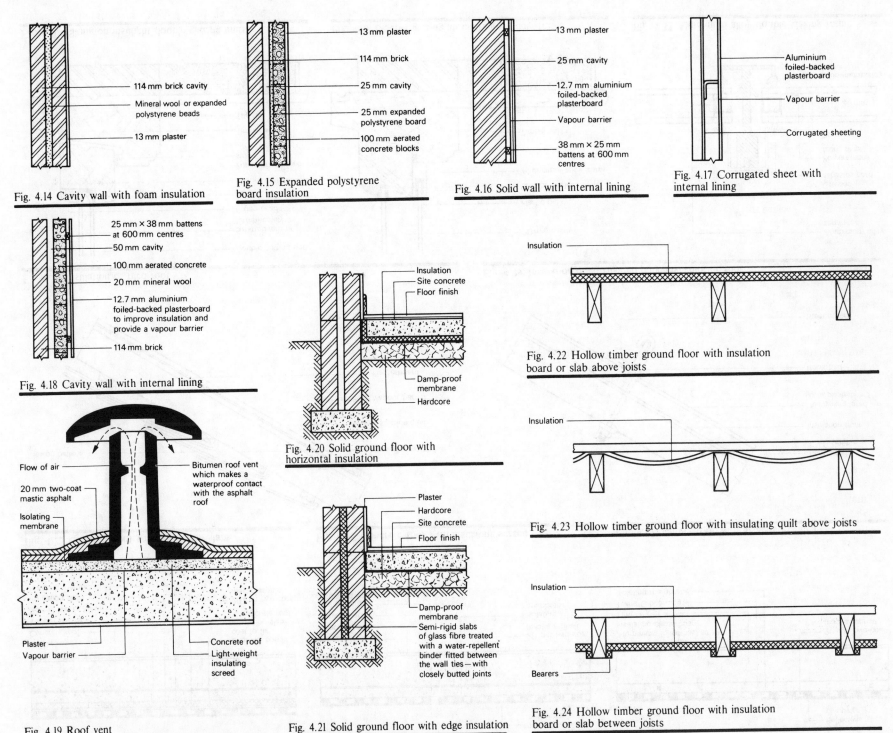

114 mm brick cavity

Mineral wool or expanded polystyrene beads

13 mm plaster

Fig. 4.14 Cavity wall with foam insulation

13 mm plaster

114 mm brick

25 mm cavity

25 mm expanded polystyrene board

100 mm aerated concrete blocks

Fig. 4.15 Expanded polystyrene board insulation

13 mm plaster

25 mm cavity

12.7 mm aluminium foiled-backed plasterboard

Vapour barrier

38 mm × 25 mm battens at 600 mm centres

Fig. 4.16 Solid wall with internal lining

Aluminium foiled-backed plasterboard

Vapour barrier

Corrugated sheeting

Fig. 4.17 Corrugated sheet with internal lining

25 mm × 38 mm battens at 600 mm centres

50 mm cavity

100 mm aerated concrete

20 mm mineral wool

12.7 mm aluminium foiled-backed plasterboard to improve insulation and provide a vapour barrier

114 mm brick

Fig. 4.18 Cavity wall with internal lining

Insulation
Site concrete
Floor finish

Damp-proof membrane

Hardcore

Fig. 4.20 Solid ground floor with horizontal insulation

Plaster
Hardcore
Site concrete
Floor finish

Damp-proof membrane

Semi-rigid slabs of glass fibre treated with a water-repellent binder fitted between the wall ties — with closely butted joints

Fig. 4.21 Solid ground floor with edge insulation

Flow of air

Bitumen roof vent which makes a waterproof contact with the asphalt roof

20 mm two-coat mastic asphalt

Isolating membrane

Plaster
Vapour barrier

Concrete roof

Light-weight insulating screed

Fig. 4.19 Roof vent

Insulation

Fig. 4.22 Hollow timber ground floor with insulation board or slab above joists

Insulation

Fig. 4.23 Hollow timber ground floor with insulating quilt above joists

Insulation

Bearers

Fig. 4.24 Hollow timber ground floor with insulation board or slab between joists

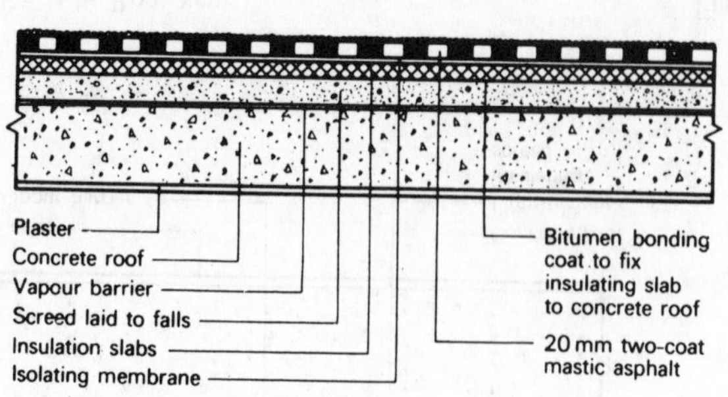

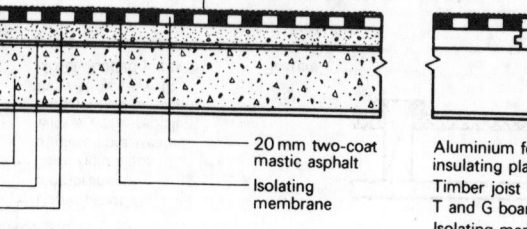

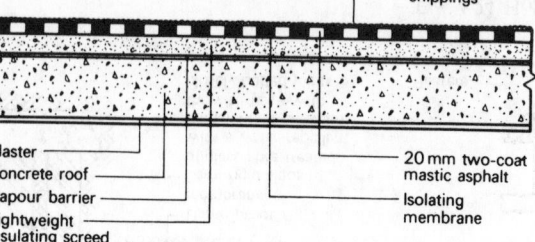

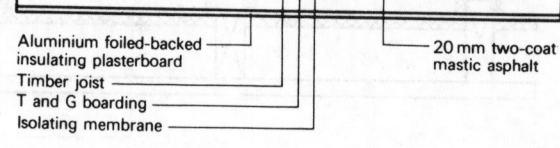

Plaster
Concrete roof
Vapour barrier
Screed laid to falls
Insulation slabs
Isolating membrane

Bitumen bonding coat to fix insulating slab to concrete roof
20 mm two-coat mastic asphalt

13 mm reflective chippings

Plaster
Concrete roof
Vapour barrier
Lightweight insulating screed

20 mm two-coat mastic asphalt
Isolating membrane

Reflective chippings

Aluminium foiled-backed insulating plasterboard
Timber joist
T and G boarding
Isolating membrane

20 mm two-coat mastic asphalt

Fig. 4.25 Concrete roof with insulation slabs

Fig. 4.26 Concrete roof with insulating screed

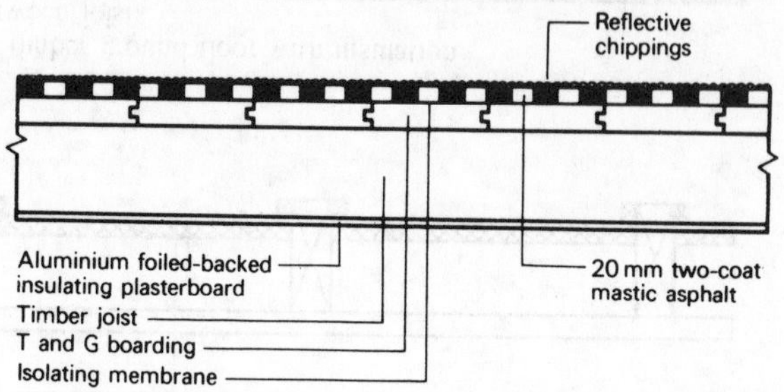

Fig. 4.27 Timber roof with insulated ceiling

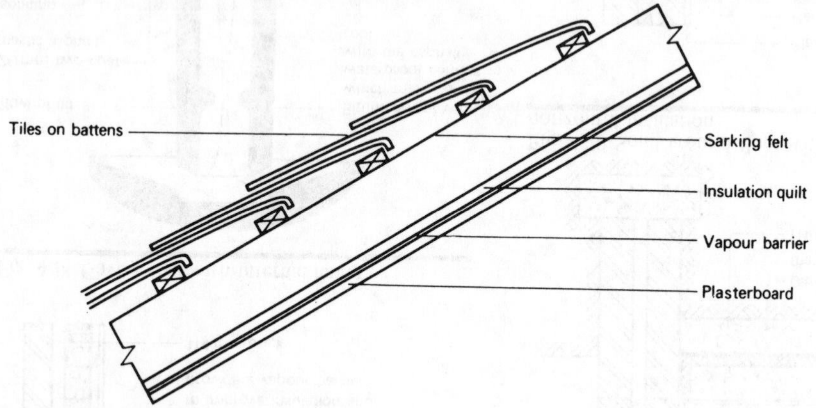

Tiles on battens
Sarking felt
Insulation quilt
Vapour barrier
Plasterboard

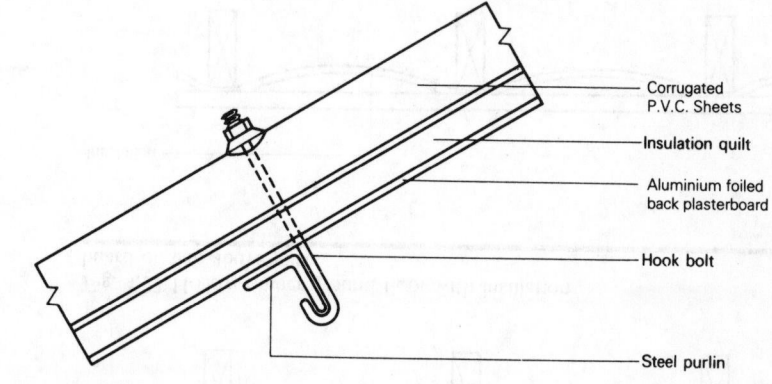

Corrugated P.V.C. Sheets
Insulation quilt
Aluminium foiled back plasterboard
Hook bolt
Steel purlin

Fig. 4.28 Insulation of tiled roof

Fig. 4.29 Insulation of corrugated polyvinyl chloride roof

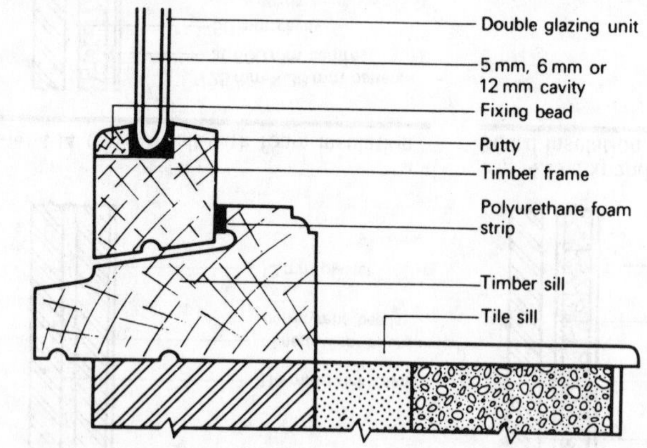

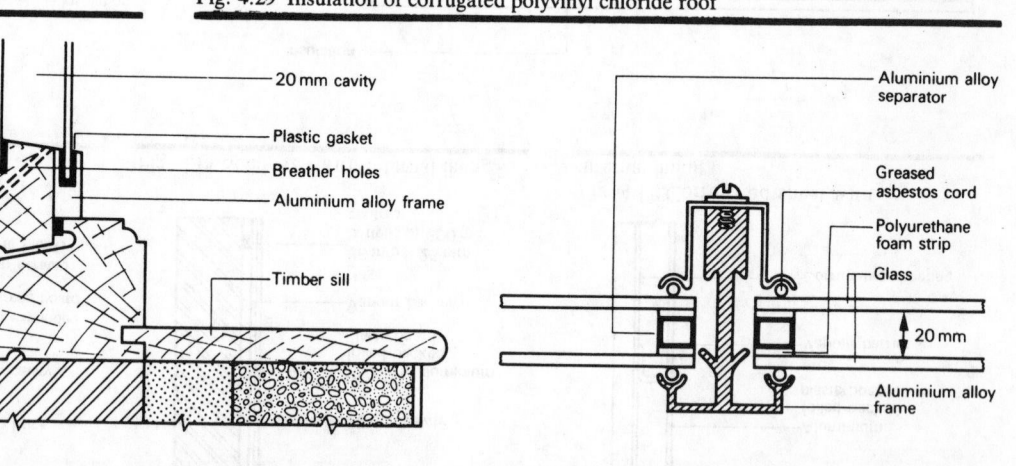

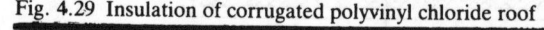

Double glazing unit
5 mm, 6 mm or 12 mm cavity
Fixing bead
Putty
Timber frame
Polyurethane foam strip
Timber sill
Tile sill

20 mm cavity
Plastic gasket
Breather holes
Aluminium alloy frame
Timber sill

Aluminium alloy separator
Greased asbestos cord
Polyurethane foam strip
Glass
20 mm
Aluminium alloy frame

Fig. 4.30 Pilkington insulight double glazing units

Fig. 4.31 Double glazing an existing window frame

Fig. 4.32 Aluminium alloy double glazing frame

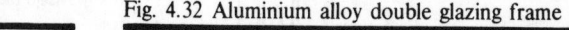

forcing air or gas into a concrete mix, and in both types the concrete is of a low density, containing a large number of still air pockets which provide good insulating properties.

Pitched roofs

Pitched roofs with a ceiling below may have a glass fibre or mineral wool quilt laid over the ceiling joists, or between them. If the roof space is to be used for habitation, the insulating quilt is laid between the roof rafters under the tiles, with a vapour barrier laid between the quilt and the plasterboard (see Fig. 4.28).

Figure 4.29 shows a method of insulating a corrugated asbestos cement roof. When the insulation is laid between the roof rafters, the cold water storage cistern may be left without insulation. Exfoliated vermiculite, or granulated cork, may also be used for insulating a ceiling, by pouring this loose fill material between the ceiling joists directly from a paper bag and levelling off with a shaped template. The position of electric ceiling roses and junction boxes should be marked by painting arrows on the roof timbers prior to insulating the ceiling. The thickness of ceiling insulation should be at least 50 mm, but preferably 76 mm.

Reflective insulation

Reflective insulation consists of reinforced aluminium foil, which may be used for ceiling and walls to reflect radiant heat. The foil is also bonded to one surface of plasterboard, which increases thermal insulation of the board.

Double glazing

Double glazing for thermal insulation consists of separate panes of glass 20 mm apart, with each pane in a separate frame, or factory-made hermetically sealed units. For sound insulation, sealed glazing units having double panes 200 mm apart should be used, but this distance is not so effective for thermal insulation as the 20 mm air space. Figure 4.30 shows a detail of a factory-made all glass unit. The internal air is dried to such a degree that condensation within the cavity wall will not occur in any condition likely to be encountered when the glass is in service. Figure 4.31 shows a detail of a double glazing unit, which may be either screwed, or hinged, to an existing window frame. Figure 4.32 shows a detail of an aluminium alloy double glazing bar. The bars are fitted at the top and bottom ends, with slipper sections for fixing to a steel or timber frame. For older houses, complete replacement window frames which have double glazing are very popular.

Chapter 5

Ventilation

Air filters

The purpose of an air filter is to free the air of as much of the airborne contaminants as is practicable. Filters will justify their cost by a reduction in the cost of cleaning and decorating the building and the protection of the heating or air conditioning equipment. Filters are also required for various dust-free industrial processes. The main types of filters are:

1. *Dry:* in which the contaminants are collected in the filter medium.
2. *Viscous or Impingement:* in which the contaminants adhere to a special type of oil.
3. *Electrostatic:* in which the contaminants are positively charged with electricity and collected on negative earthed plates.

Dry filters

These use materials such as cotton wool, glass fibre, cotton fabric, treated paper, foamed polyurethane as the cleaning medium. The efficiency of the filter depends largely upon the area of medium offered to the air stream, and for this reason the filter can be arranged in a V formation which increases the area.

Figures 5.1, 5.2 and 5.3 show the arrangements of V type dry filter.

After a period of use (depending upon the atmospheric pollution) the contaminants retained by the filter will increase and this will increase the resistance to the flow of air through the filter.

In the case of an automatic roller type filter, when the filter is dirty, a

46

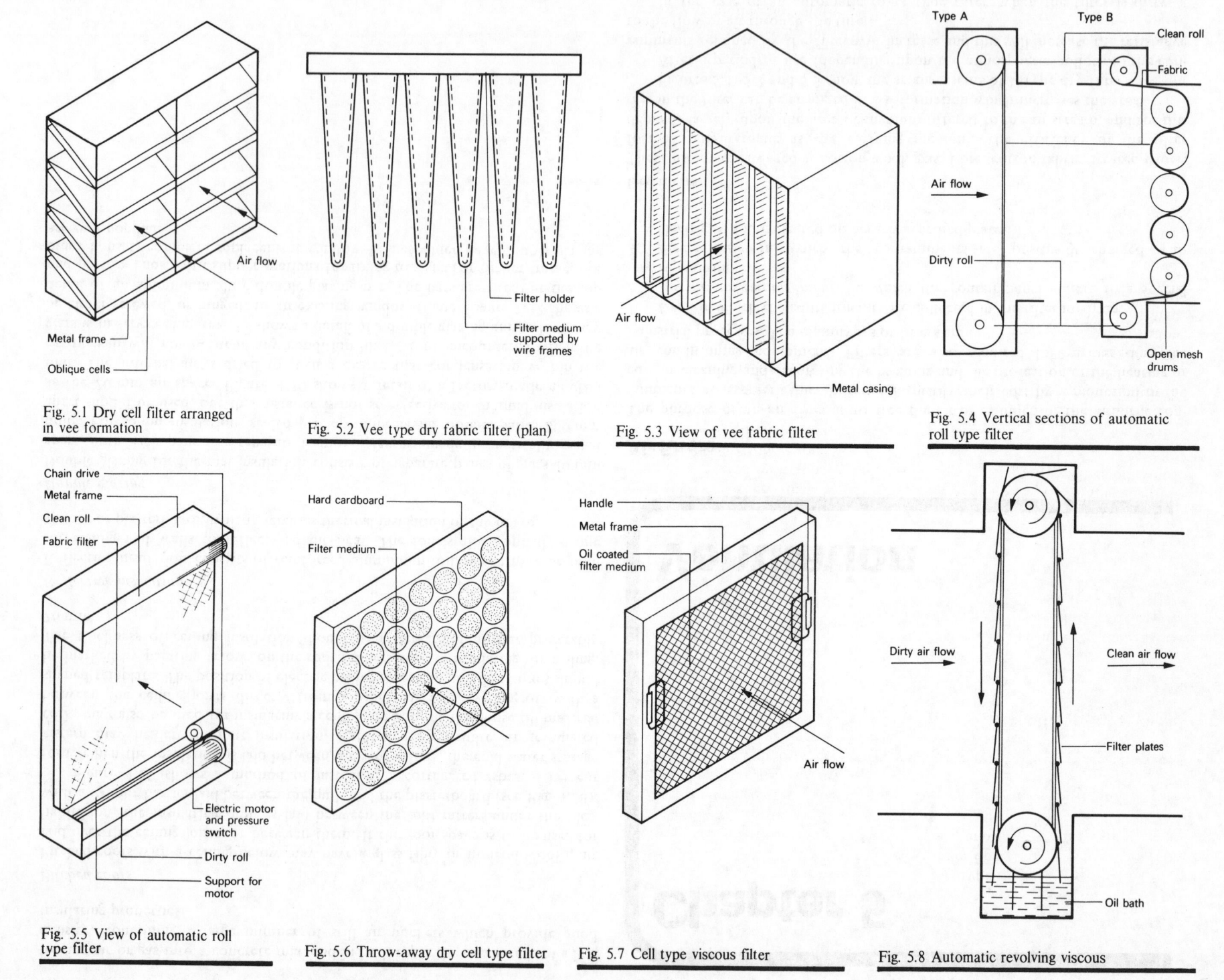

Fig. 5.1 Dry cell filter arranged in vee formation

Metal frame

Oblique cells

Air flow

Fig. 5.2 Vee type dry fabric filter (plan)

Filter holder

Filter medium supported by wire frames

Fig. 5.3 View of vee fabric filter

Air flow

Metal casing

Fig. 5.4 Vertical sections of automatic roll type filter

Type A

Type B

Clean roll

Fabric

Air flow

Dirty roll

Open mesh drums

Fig. 5.5 View of automatic roll type filter

Chain drive

Metal frame

Clean roll

Fabric filter

Electric motor and pressure switch

Dirty roll

Support for motor

Fig. 5.6 Throw-away dry cell type filter

Hard cardboard

Filter medium

Fig. 5.7 Cell type viscous filter

Handle

Metal frame

Oil coated filter medium

Air flow

Fig. 5.8 Automatic revolving viscous

Dirty air flow

Clean air flow

Filter plates

Oil bath

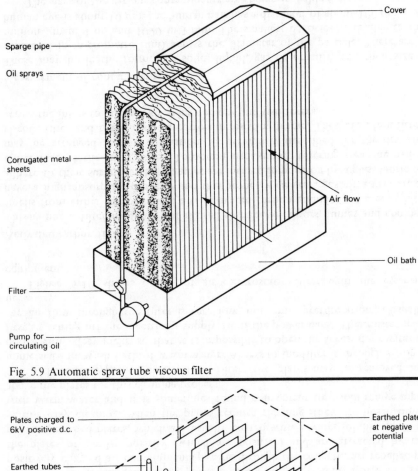

Fig. 5.9 Automatic spray tube viscous filter

Labels on figure:
- Cover
- Sparge pipe
- Oil sprays
- Corrugated metal sheets
- Air flow
- Filter
- Oil bath
- Pump for circulating oil

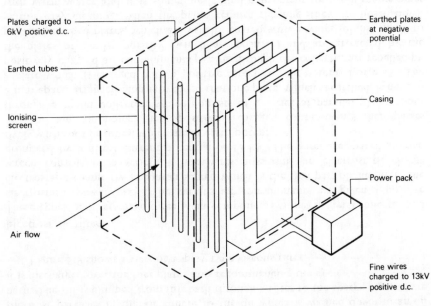

Fig. 5.10 Electrostatic filter

Labels on figure:
- Plates charged to 6kV positive d.c.
- Earthed plates at negative potential
- Earthed tubes
- Casing
- Ionising screen
- Power pack
- Air flow
- Fine wires charged to 13kV positive d.c.

pressure switch will switch on an electric motor which will turn the dirty spool and allow clean fabric to enter the filter chamber.

Figure 5.4 shows sections of two types of automatic roller type filter. Type A does not have as much filter area as type B.

Figure 5.5 shows a view of an automatic roll type filter.

With other types of dry filters an electrically operated pressure switch may be installed, which switches on a warning light to draw attention to a dirty filter. One type of dry filter is made up of a square cell of sizes from 254 by 254 mm to 600 by 600 mm. The filter medium, which is either 25 or 50 mm thick, is held in a metal or cardboard frame.

The cheaper type may be thrown away when dirty and the more expensive, vacuum-cleaned two or three times before being discarded.

Figure 5.6 shows a view of a disposable, or what is often called a 'throw-away dry cell type filter'.

Absolute filters

These are of dry fabric type and are very efficient in moving even the smallest particle from the air. This high performance is obtained by close packing of a very large number of small diameter fibres, but this unfortunately results in a high resistance to air passing through the filter.

Viscous filters

These have a large dust-holding capacity and are therefore often used in industrial areas where there is a high degree of atmospheric pollution. The filter medium is coated with a non-inflammable, non-toxic and odourless oil, which the contaminants adhere to as they pass through the filter.

There are two types of viscous filters:

1. *Cell type:* which consists of a metal frame into which wire mesh, industrial metal swarf, metal stampings or a combination of these materials are inserted and coated with the special oil. The cells are placed across the air stream in a V formation, similar to the dry cell type filter (shown in Fig. 1.1) which will allow the maximum area of the filter to be in contact with the air flow. When dirty, the cells are removed, washed in hot water, allowed to dry and re-coated with clean oil for further use.

Figure 5.7 shows a view of a cell type viscous filter.

2. *Automatic types:* one type uses a moving curtain, consisting of filter plates hung from a pair of chains. The chains are mounted on sprockets located at the top and bottom of the filter housing, so that the filter plates can be moved as a continuous curtain, up one side and down the other side of the sprockets. The arrangement is such that at the bottom the filter plates pass through a bath of special oil, which cleans the dirty oil from the filter plates and re-coates them with cleaner oil.

Figure 5.8 shows a vertical section of an automatic revolving type viscous filter. An electric motor is used to turn the sprockets and this may be arranged to move the curtain continuously, or periodically, depending upon the degree of atmospheric pollution. Another type has closely spaced corrugated metal plates, which are continuously coated by oil from a sparge pipe at the top of the filter.

The air passing through the plates has to take a tortuous route and, in so doing, the dust particles in the air adhere to the oil which is washed down to an oil bath. The oil is pumped from this bath through a filter to the sparge pipe, where it is discharged over the filer plates thus recommencing the cycle.

Figure 5.9 shows a view of a spray type viscous filter.

Electrostatic filters

These types of filters have three main components: ioniser, metal collector and electrostatic power pack. The various air contaminants are given a positive electrostatic charge by an ioniser screen which is the first part of the filter. The screen consists of a series of fine wires possessing an electrostatic charge produced by a direct current potential of 13 kV. The wires are spaced alternatively with rods or tubes, which are at earth potential.

The air containing these positively charged contaminants then passes through a metal collector, which consists of a series of parallel plates about 6 mm apart, arranged alternatively so that one plate, which is earthed, is next to a plate which is charged with a positive direct current potential of 6 kV. The positively charged air contaminants passing through the collector are repelled by the plates of similar polarity (which are positive) and are attracted by the negative earthed plates, which are usually coated with a water-soluble liquid to permit easy cleaning. When the plates require cleaning, they are hosed down with warm water and it is sometimes found convenient to install sparge pipes above the plates for this cleaning process.

With some types of electrostatic filters the plates may be removed and immersed in a wash bath of warm water. Access to the filter is through air-tight doors with safety locks, so that it is impossible to open the access door without first switching off the electricity supply from the power pack. The resistance to the air flow through the filter is very low, but some prefilteration is usually necessary.

Figure 5.10 shows details of an electrostatic filter with the essential equipment.

Activated carbon filters

This is not a filter medium, but is used to remove odours, fumes and cooking smells from either dry or humid air. Activated carbon granules are used, which have a high absorptive capacity and this property causes a gas, or a vapour, to adhere to their surfaces. The granules are located in position by a glass fabric on either face of the filter panel. To obtain a maximum filtering area, the panels may be arranged in a V formation, as previously described for the dry and viscous filter and fitted inside a corrosion-resisting frame. They are often fitted across the inlets to ductwork serving cookers and fish fryers.

Measurement of contaminants

These may be in the form of a solid, liquid, gas or organic, and their size is measured in micrometre (μm) across the diameter of the particle. There are 1 million μm in 1 m, and 1000 in 1 mm. The smallest particle discernible to the human eye is about 10 μm; the human hair has a diameter of about 100 μm.

The sizes of the various contaminants are shown in Table 5.1.

Table 5.1

Contaminants	Average diameter (μm)
Tobacco smoke	0.05—0.1
Industrial smoke	0.1—1.0
Mist and fog	2.5—40
Atmospheric dust	1.0—100
Bacteria	0.3—10
Pollen	20—50
Virus	≥1.0

Filter efficiencies

There are two main tests to determine the efficiency of a filter:

1. The determination of methylene blue efficiency, which uses the staining effect as a criterion. This test is applicable to any type of filter. A solution of methylene blue dye in water in injected into the air stream before the filter is tested. When the solution passes through the duct, the water is evaporated, leaving behind the blue dye in the form of small solid particles. A sample of the air containing the particles is drawn off before and after passing through the filter by means of a vacuum pump. The vacuum pump contains white filter paper, which is stained by the blue particles when the air is drawn through it. The extent of this staining is measured by means of a photosensitive cell. The values obtained are used to find the filter efficiency. The test is a stringent one and gives very low efficiency values for many filters.

2. The determination of the dust-holding capacity and gravimetric efficiency. This test does not apply to filters that require the accurate weighing of the filter, before and after the collection of the dust. The dust-holding capacity is rated in g/m^2 of filter face at a specified air velocity and the rating is only valid if the dust-holding capacity quoted is the same dust as used for the efficiency test.

Filter efficiencies

Dry: from 50 to 95 per cent in the 0.1 to 5 μm range.
Viscous: from 85 to 95 per cent in the 5 μm range.
Electrostatic: up to 99 per cent in the 0.1 μm range.

Ventilation

Principles

The purpose of ventilation of buildings is to remove high concentrations of body odours, carbon dioxide and water vapours and also to remove dust, fumes and smoke (which may be toxic) and excess heat. The air in the room containing these contaminants is replaced by fresh air and this creates air movement inside a building, so that the occupants obtain a feeling of freshness without draughts. At

one time the concentration of carbon dioxide was used as a criterion of good ventilation, but even in very badly ventilated rooms the carbon dioxide rarely rises to a harmful level. The absence of body odours, dust and fumes in the air is a better criterion of good ventilation; also if the air movement is too low the air in the room will feel 'stuffy'. An air velocity of between 0.15 and 0.5 m per second is acceptable to most people under normal circumstances and higher velocities may be used for heavy manual work; to prevent monotony, a variable air speed is preferable to a constant air speed. The design of a ventilation system must be considered with the design of the heating system and it may be necessary to comply with the various public health legislations; for example, Factories Act 1961, Offices, Shops and Railway Premises Act 1963. A minimum ventilation rate of 28 m^3 of fresh air per hour, per person, is also required by most licensing authorities for theatres, cinemas and dance halls.

Systems of ventilation

Ventilation can be achieved by either natural or mechanical means. Natural ventilation for its operation depends on one of the following; (a) wind pressure, (b) stack effect, (c) a combination of wind pressure and stack effect. Wind causes a positive pressure to act on the windward side of the building and a negative pressure to act on the leeward side. Figures 5.11, 5.12 and 5.13 show the wind pressure distribution diagrams for pitched and flat roofs.

The inlet openings in the room should be well distributed and should be located on the windward side near the bottom. The outlet openings should also be well distributed and located on the leeward side near the top; this will allow cross ventilation in the room. Stack effect is created by the difference in temperature between the air inside and the air outside a building; the warmer, less dense air inside is displaced by the cooler denser air from outside. Figures 5.14 and 5.15 show upward and cross ventilation created by stack effect and Fig. 5.16 shows the stack effect in a tall building. The higher the column or 'stack' of warm air the greater will be the air movement inside the building; therefore in order that the force created by stack effect may be operated to its maximum advantage, the vertical distance between inlet and outlet openings should be as great as possible. If a wind is present and the air inside a building is warmer than the air outside, ventilation will take place, by both wind pressure and stack effect. Figure 5.17 shows how ventilation takes place through a casement window.

Unfortunately, natural ventilation cannot ensure a specified air change, nor is it possible to filter the air before it enters the building, also if the air inside is at the same temperature as the air outside, and there is no wind, natural ventilation will be non-existent. For the removal of excessive heat from founderies, bakeries, welding and plastic moulding shops natural ventilation is usually very successful, providing that the temperature differential is about 10 °C.

Mechanical ventilation

These systems employ an electrically driven fan or fans to provide the necessary air movement; they have the advantage over natural ventilation in providing positive ventilation at all times, irrespective of outside conditions. They also ensure a specified air change and the air under fan pressure can be forced through filters. There are three types of mechanical ventilation systems, namely:

1. *Natural inlet and mechanical extract* (exhaust system). This is the most common type of system and is used for kitchens, workshops, laboratories, internal sanitary apartments, garages and assembly halls. The fan creates a negative pressure on its inlet side, and this causes the air inside the room to move towards the fan, and the room air is displaced by the fresh air from outside the room.

Figure 5.18 shows a ducted exhaust system for an assembly hall. The air inlets can be placed behind radiators, so that the air is warmed before it enters the hall.

Internal bathrooms and W.C.s. Ventilation for these rooms should provide a minimum extract rate of 20 m^3/h from a W.C. cubicle, or a bathroom without a W.C. and a minimum extract rate of 40 m^3/h from a bathroom with a W.C. The ventilation system must be separate from any ventilation plant installed for any other purpose. In the common duct system the inlets from the bathrooms or W.C. compartments should preferably be connected to the main vertical duct by a shunt duct at least 1 m long. This shunt duct will offer better sound attenuation between the dwellings and also tends to prevent the spread of smoke and fumes in the event of a fire. The fans must be capable of extracting the total flow of air, plus an allowance on the fan static pressure, to counteract wind pressures. In order to keep the system operating in the event of failure of a single fan, it is recommended that two fans and motors are installed, with an automatic change-over damper. To replace air extracted from the rooms, air should be drawn from the entrance lobby through a wall grill, or a 19 mm gap left under the door of the room.

Figures 5.19 and 5.20 show the methods of ventilating internal bathrooms and W.C. cubicles in multi-storey blocks of flats. Figure 5.21 shows the method of duplicating the extract fans, including an automatic change-over damper.

2. *Mechanical inlet and natural extract.* It is essential with this system that the air is heated before it is forced into the building. The system may be used for boiler rooms, offices and certain types of factories. The air may be heated in a central plant and ducted to the various rooms, or a unit fan convector may be used. Figure 5.22 shows a ducted system for a multi-storey office block and Fig. 5.23 shows a system using a sill-level fan-heater unit.

3. *Mechanical inlet and extract.* This provides the best possible system of ventilation, but it is also the most expensive and is used for many types of buildings including cinemas, theatres, offices, lecture theatres, dance halls, restaurants, departmental stores and sports centres. The system is essential for operating theatres and sterilising rooms.

The air is normally filtered and provision is made for recirculation of the heated air which reduces fuel costs and, in order to save further fuel costs, the air may be extracted through the electric light fittings, which also increases the lighting efficiency about 14 per cent. Slight pressurisation of the air inside the building is achieved by using an extract fan smaller than the inlet fan; this requires the windows to be sealed and swing or revolving doors to be used. The slight internal air pressure and sealed windows prevent the entry of dust, draughts and noise.

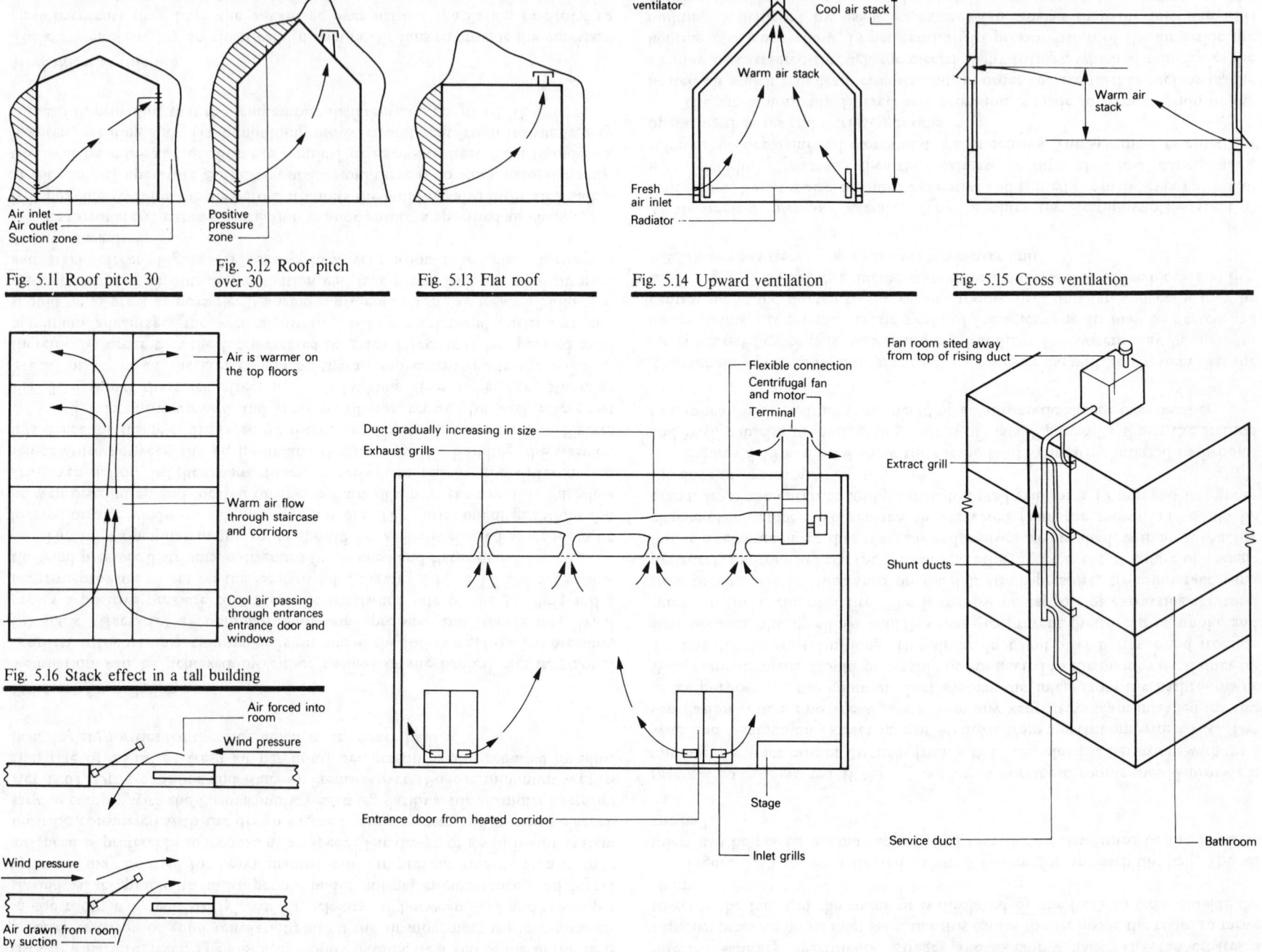

Direction of wind

Air inlet
Air outlet
Suction zone

Fig. 5.11 Roof pitch 30

Positive
pressure
zone

Fig. 5.12 Roof pitch
over 30

Fig. 5.13 Flat roof

Roof extract
ventilator

Cool air stack

Warm air stack

Fresh
air inlet

Radiator

Fig. 5.14 Upward ventilation

Warm air
stack

Fig. 5.15 Cross ventilation

Air is warmer on
the top floors

Warm air flow
through staircase
and corridors

Cool air passing
through entrances
entrance door and
windows

Fig. 5.16 Stack effect in a tall building

Flexible connection
Centrifugal fan
and motor
Terminal

Duct gradually increasing in size

Exhaust grills

Entrance door from heated corridor

Stage

Inlet grills

Fig. 5.18 Ducted exhaust system for an assembly hall

Air forced into
room

Wind pressure

Wind pressure

Air drawn from room
by suction

Fig. 5.17 Plan of casement window

Fan room sited away
from top of rising duct

Extract grill

Shunt ducts

Service duct

Bathroom

Fig. 5.19 Common duct system for internal bathrooms
and WC cubicles

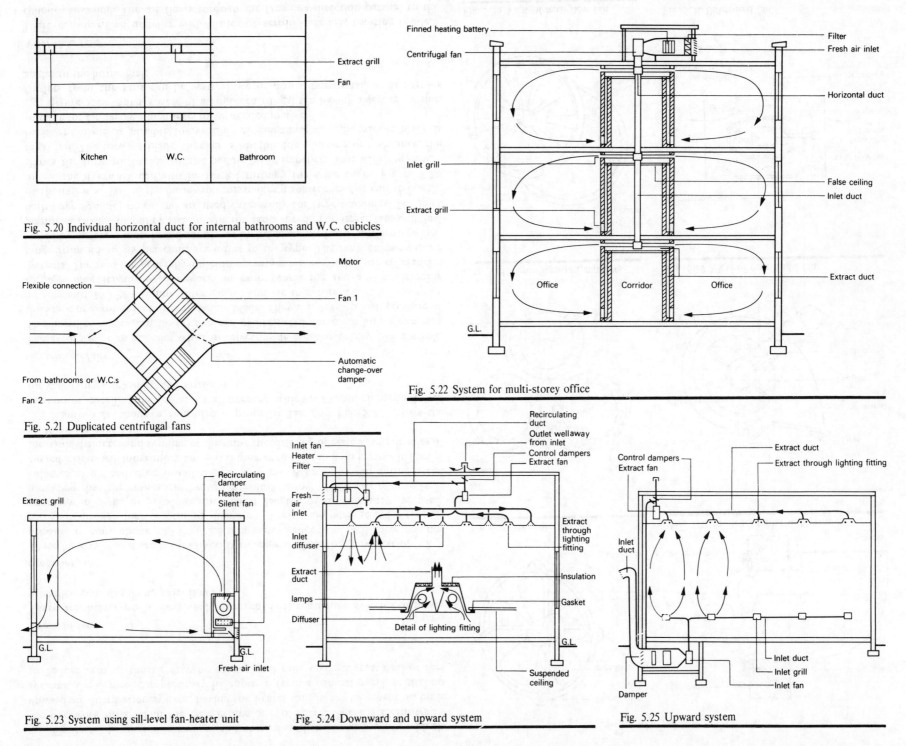

Fig. 5.20 Individual horizontal duct for internal bathrooms and W.C. cubicles

Fig. 5.21 Duplicated centrifugal fans

Fig. 5.22 System for multi-storey office

Fig. 5.23 System using sill-level fan-heater unit

Fig. 5.24 Downward and upward system

Fig. 5.25 Upward system

Figure 5.24 shows a downward and upward air distribution system, with all the ductwork installed in the false ceiling or roof space. Figure 5.25 shows an upward air distribution system, having the heater unit in the basement. In both systems, the control dampers may be adjusted from a control panel, so that up to 75 per cent of the air may be recirculated thus saving a great deal of fuel costs.

Types of fans

There are three types of fans used for mechanical ventilation systems, namely: propeller, centrifugal and axial flow.

Propeller fans

These have two or more blades fixed at an angle to the hub. They develop a low pressure of only about 125 Pa (12.5 mm water gauge) and will therefore not force air through long lengths of ductwork. Their main purpose is for free air openings in walls or windows, but short lengths of ducting may be used, providing that the resistance through the duct is low. They can remove large volumes of air and their installation cost is low. A propeller fan having broad curved blades will move more air, and is quieter than a fan with narrow blades of the same diameter and running at the same speed. They operate at efficiencies of up to 50 per cent.

Figure 5.26 shows a four-bladed propeller fan and Fig. 5.27 shows the method of installing the fan at a wall opening, which is a common method used for ventilating workshops and kitchens.

Centrifugal fans

These consist of an impeller which revolves inside a casing shaped like a scroll. The impeller blades can be (a) forward or backward curved of either constant thickness or aerofoil section, (b) paddle blade. They can develop high pressure of up to about 760 Pa (76 mm water gauge) and are therefore used for forcing air through long lengths of ductwork, in both ventilating and air conditioning systems. The fans are generally quiet in operation but are bulky, the efficiencies range from 45 to 85 per cent, according to the type. The fan's output can be varied by different motors and speeds, or by being coupled to a bolt drive, which permits a change of pulley size to suit the fan's speed. The larger fans will deal with large volumes of air and are used extensively for large buildings. The inlet of the fan is at 90° to the outlet and this makes it sometimes difficult to install, unless the ductwork can also be turned through the same angle. Figure 5.28 shows the forward, backward and paddle blade impellers used with the centrifugal fan. The forward blade impeller is suitable for constant air resistance, the backward blade is suitable for variable air resistance and the paddle blade is suitable for air having a high level of suspended matter.

Figure 5.29 shows a view of a large centrifugal fan having a vertical outlet. Outlets from the fans can be vertical, up or down, horizontal, or at various angles to the horizontal.

Axial flow fans

These consist of an impeller with blades of aerofoil section, rotating inside a cylindrical casing. The air flows through the fans in a direction parallel to the

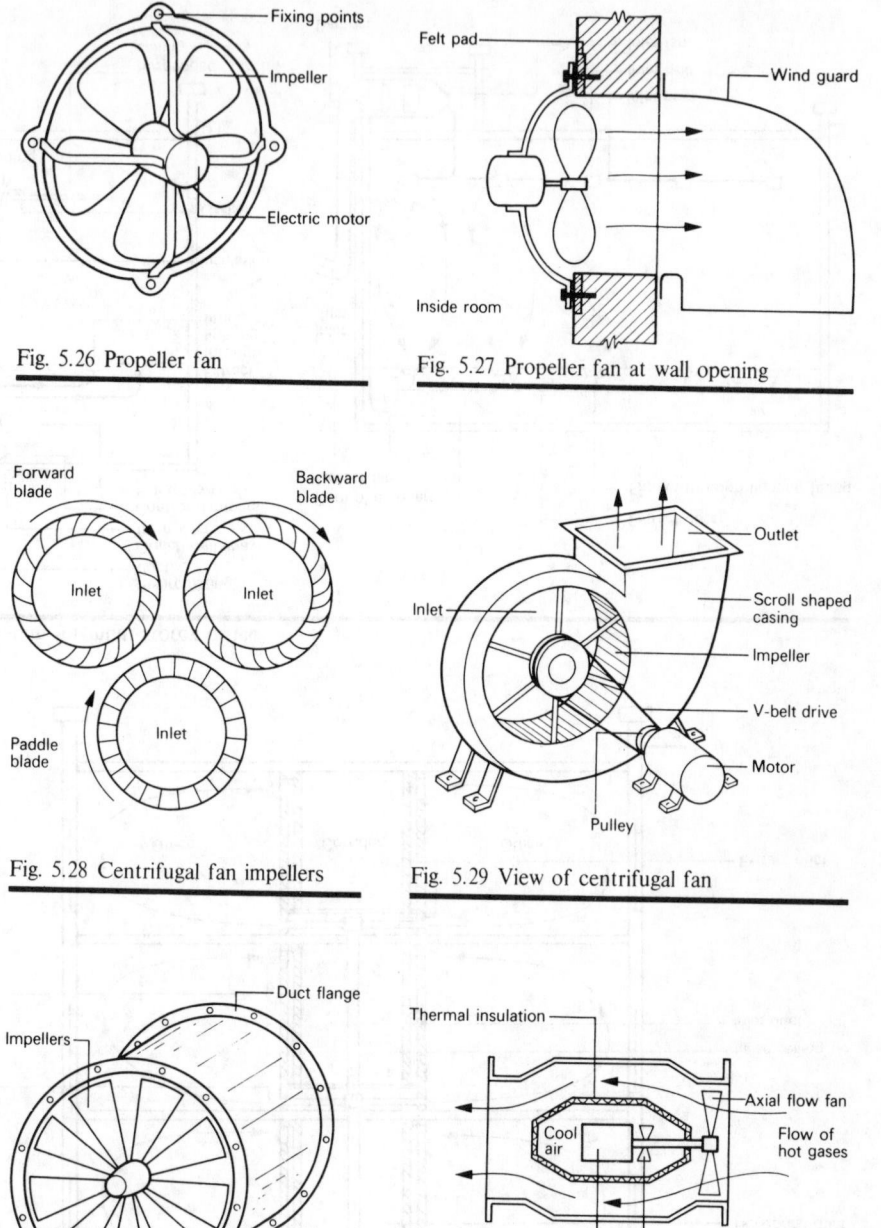

Fig. 5.26 Propeller fan

Fig. 5.27 Propeller fan at wall opening

Fig. 5.28 Centrifugal fan impellers

Fig. 5.29 View of centrifugal fan

Fig. 5.30 View of axial flow fan

Fig. 5.31 Bifurcated fan

propeller shaft and can be installed without a fan base. The efficiencies of the fans range between 60 and 75 per cent and they can develop pressures of up to about 1500 Pa (150 mm water gauge). The fan can be driven directly by a motor mounted in the air stream, or by a belt drive from a motor mounted outside the duct. Like the centrifugal fan, it can be used to deal with large volumes of air, and is sometimes used in preference to the centrifugal fan, due to the ease of installation.

Figure 5.30 shows a view of an axial flow fan and Fig. 5.31 shows a bifurcated axial flow fan, which is used to extract hot air from founderies, or flue gases in the chimney.

Fan laws with constant air temperature

1. The volume of air delivered varies as the fan speed.
2. The pressure developed varies as the square of the fan speed.
3. The power absorbed varies as the cube of the fan speed.

Table 5.2 gives the minimum ventilation rates for different types of buildings: Code of Practice Mechanical Ventilation and Air Conditioning of Buildings.

Table 5.2

Type of building	Air changes per hour	Type of building	Air changes per hour
Schools		Operating theatres	10
Classrooms	6	X-ray rooms	6
Assembly halls	3	Recovery rooms	3
Changing rooms	3	Entrances	3
Cloakrooms	3	Staircases	2
Dining rooms	3	Lavatories and bathrooms	2
Dormitories	3	Internal lavatories and	
Gymnasia	3	bathrooms	3
Common rooms	2	Duplicated fans required,	
Lavatories	3	and system distinct from	
Staff rooms	2	any other ventilating	
Laboratories	4	system	
Hospitals		Kitchens	20–40
Wards	3	Varying according to	
Dormitories	3	volume of air required	
Day rooms	3	through canopy	
Dining rooms	3	Laundries	10–20
Staff bedrooms	2	Boiler houses	10–15
Corridors	2	Smoking rooms	10–15

Note: For places of public entertainment such as cinemas, theatres, concert halls, assembly halls and dance halls the ventilation rate depends upon the Local Authorities Regulations, usually 28 m^3 per hour, per person.

Table 5.3 gives the Chartered Institution of Building Services recommended outlet velocities from grills, for acceptable noise levels.

Table 5.3

Application	Maximum outlet velocity (m/s)
Libraries, sound studios, operating theatres	1.75–2.5
Churches, residences, hotel bedrooms, hospital rooms and wards, private offices	2.5–4.0
Banks, theatres, restaurants, classrooms, small shops, general offices, public buildings, ballrooms	4.0–5.0
Arenas, stores, industrial buildings, workshops	5.0–7.0

Chapter 6

Air conditioning

Principles

The term 'air conditioning' has often been misused; for example, denoting a system of heating combined with mechanical ventilation. It should, however, be defined as a system giving automatic control, within predetermined limits of the environmental conditions, by heating, cooling, humidification, dehumidification, cleaning and movement of air in buildings. The control of these conditions may be desirable to maintain the health and comfort of the occupants, or to meet the requirements of industrial processes irrespective of the external climatic conditions.

Advantages

1. In factories and offices the working efficiency of personnel is improved and work output is known to increase. There is also a reduction in illness and absenteeism.
2. Shops and departmental stores have increased sales due to customers and staff being able to enjoy greater comfort. The cost of the plant will therefore be offset by extra income from customers.
3. There is a reduction in cleaning and decorating; also fabrics and furnishings last longer.
4. Hotels, restaurants, theatres and cinemas received better patronage.
5. Many industrial premises need air conditioning to keep plant and processes working at maximum efficiency; these include computer rooms, food production rooms, electronic laboratories, textile factories, printing rooms and laboratories.
6. Sealed windows reduce the entry of noise from aircraft and traffic; also the entry of fumes, smoke, dust and draughts.
7. There is less risk of fire due to static electricity caused by dry air.

Relative humidity

The relative humidity of the air is usually expressed as a per cent and is a ratio between the actual amount of moisture in a given volume of air, and the amount of moisture that would be necessary to saturate that volume. It may also be expressed as a ratio between the actual vapour pressure and the saturated vapour pressure. Most people feel comfortable when the relative humidity is between 30 and 70 per cent and an air conditioning plant usually operates to maintain a relative humidity of between 40 and 50 per cent. When the air is too dry, moisture evaporates more readily from the skin and this produces a feeling of chilliness, even if the air temperature is satisfactory. Dry air also removes moisture from the nose, throat and eyes, causing these to be irritated. When the air is too damp, moisture cannot readily evaporate from the skin and this causes the body to become overheated, resulting in a feeling of drowsiness.

$$\text{Relative humidity} = \frac{\text{mass of water vapour in a given volume of air}}{\text{mass of water vapour required to saturate the same volume of air at the same temperature}}$$

$$\text{Relative humidity} = \frac{\text{actual vapour pressure}}{\text{saturation vapour pressure at the original air temperature}}$$

The psychrometric chart

This shows, by graphical representation, the relationship that exists between wet and dry bulb temperatures at different relative humidities. The usefulness of the chart is extended by the addition of moisture content values and specific enthalpy, or total heat of air. The chart is based on a barometric pressure of 101.325 kPa. The following illustrates the use of the chart (see Fig. 6.1).

1. *Dry bulb temperature*: read vertically downwards to 25 °C.
2. *Wet bulb temperature*: trace parallel to the sloping wet bulb lines to intersect the wet bulb temperature on the 100 per cent relative humidity curve at 18 °C.
3. *Relative humidity*: read directly from the point of intersection of the dry and wet bulb temperatures, the relative humidity curve at 50 per cent.
4. *Dew point*: travel horizontally to the left of the intersection, to intersect the 100 per cent curve at 14 °C.
5. *Moisture content*: read horizontally to the right at 0.010 kg/kg (dry air).
6. *Specific enthalpy or total heat*: this is found by drawing a line to the total heat lines at the chart extremities, which reads 50.1 kJ/kg.

Note: The total heat lines are not parallel to the wet bulb lines.

The following examples will show how the chart is used to solve various problems.

Example 1. *In winter, air at a dry bulb temperature of 5 °C and 60 per cent relative humidity enters a building through a heating battery and is heated to a dry bulb temperature of 20 °C without adding moisture.*

From the chart find:

1. Wet bulb temperature of the incoming air.
2. The relative humidity of the heated air.

Figure 6.2 shows how these values are found from the chart and it will be seen that the wet bulb temperature of the incoming air will be 2.2 °C and the relative humidity of the heated air 25 per cent.

Note: This is sensible heating of the air and it shows that without humidification the relative humidity of the air is too low.

Example 2. *In summer, air at a dry bulb temperature of 25 °C and a wet bulb temperature of 21 °C enters a building through a cooling coil and is cooled to a dry bulb temperature of 20 °C.*

From the chart find:

1. The relative humidity of the incoming air.
2. The relative humidity of the supply air after cooling.

Figure 6.3 shows how these values are found from the chart and it will be seen that the relative humidity of the incoming air is 70 per cent and the relative humidity of the supply air after cooling is 95 per cent.

Note: This is sensible cooling of the air in summer and it shows that without dehumidification the relative humidity of the supply air is too high.

Need for humidifying and dehumidifying

It will be clear from Figs 6.2 and 6.3 that if air enters a building at a low temperature in winter and is passed through a heating battery, its relative humidity will be reduced and may be below that required for human comfort. Also if air enters a building at a high temperature in summer and is passed through a cooling battery and cooled above its dew point, its relative humidity may be increased above that required for human comfort.

Example 3. *The air in a room has a dry bulb temperature of 22 °C and a wet bulb temperature of 16 °C.*

From the chart find:

1. The relative humidity of the air.
2. The temperature of the walls when condensation would occur.

Figure 6.4 shows how these values are found from the chart and it will be seen that the relative humidity is 52 per cent and condensation will occur when the temperature of the walls reach dew point at 12.5 °C.

Dehumidification by cooling and reheating

In summer, the air passing through an air conditioning plant is cooled in the spray to below its dew point and then reheated.

Example 4 (see Fig. 6.5). *Air enters the plant at a dry bulb temperature of 25 °C and 70 per cent relative humidity and is required to be cooled to a dry bulb temperature of 20 °C and 50 per cent relative humidity.*

From the chart find (assuming washer efficiency of 100 per cent):

1. The temperature of air in the washer.
2. The reduction in the moisture content of the supply air.

Note: The air is first cooled to below a wet bulb temperature of 10 °C in the spray and then reheated to 20 °C when since no more moisture is added, the relative humidity will have been reduced to 50 per cent.

Example 5 (see Fig. 6.6). *If air enters the plant in winter at a dry bulb temperature of 10 °C and 60 per cent relative humidity, the amount of moisture in the air would be very low and if the air was heated (without adding moisture) to a supply temperature of 25 °C dry bulb, its relative humidity would be reduced to about 22 per cent. If the air is preheated to a dry bulb temperature of 20 °C before entering the washer, the resultant condition would be a wet bulb temperature of 11.2 °C, and if the air is reheated to a dry bulb temperature of 25 °C its relative humidity would be about 41 per cent, which has increased the relative humidity from 22 to 41 per cent.*

Adiabatic saturation

Evaporation of moisture in the air takes place without a change of temperature and if unsaturated air is passed over a thin film of water the air will evaporate moisture for as long as it remains in an unsaturated state. If there is no other source of heat, the latent heat of evaporation will be supplied from the air and its temperature will be lowered.

In Example 5 air at a dry bulb temperature of 10 °C and 60 per cent relative humidity is passed through a washer with 100 per cent efficiency and not preheated; its resultant condition would be a wet bulb temperature of 6.5 °C. This process is known as 'adiabatic saturation', being a term used to describe a process where no heat is added or taken from the air.

Air conditioning systems

The type of system depends upon the type of building, and if it is necessary to vary the air temperatures and relative humidities also upon the space within the building for the plant ductwork and pipework.

The systems may be divided into three categories:

1. All air systems where the conditioned air is treated in a central plant and ducted to the various rooms. It requires large duct spaces and plant rooms, but very little is taken up inside the rooms.

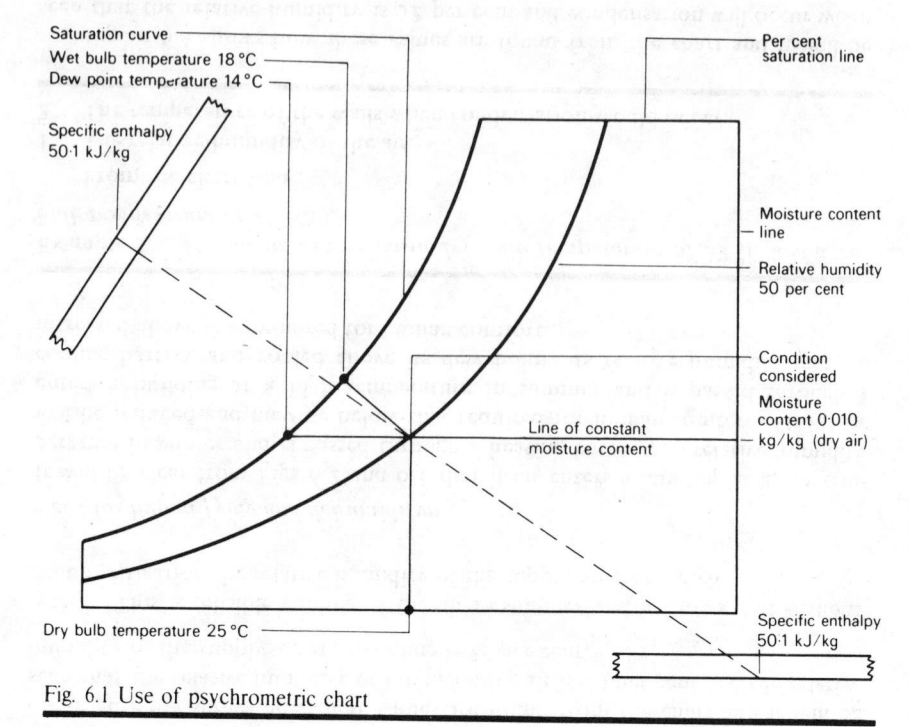

Saturation curve
Wet bulb temperature 18 °C
Dew point temperature 14 °C

Specific enthalpy 50·1 kJ/kg

Per cent saturation line

Moisture content line
Relative humidity 50 per cent

Condition considered
Moisture content 0·010 kg/kg (dry air)

Line of constant moisture content

Dry bulb temperature 25 °C

Specific enthalpy 50·1 kJ/kg

Fig. 6.1 Use of psychrometric chart

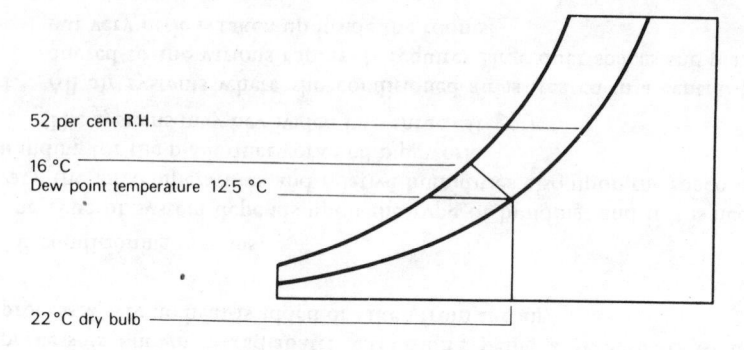

52 per cent R.H.

16 °C
Dew point temperature 12·5 °C

22 °C dry bulb

Fig. 6.4 Condensation on wall surface

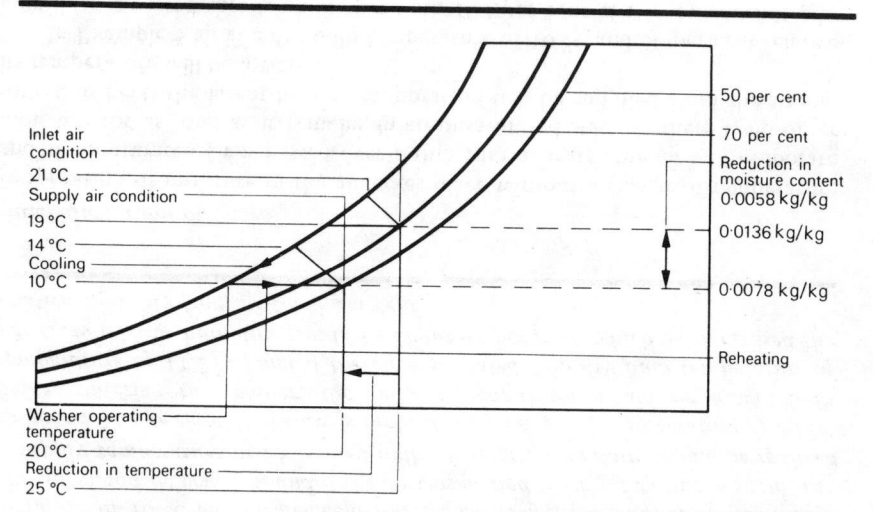
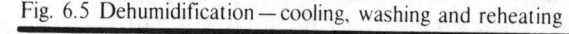

50 per cent
70 per cent

Inlet air condition
21 °C
Supply air condition
19 °C
14 °C
Cooling
10 °C

Reduction in moisture content
0·0058 kg/kg
0·0136 kg/kg
0·0078 kg/kg

Reheating

Washer operating temperature
20 °C
Reduction in temperature
25 °C

Fig. 6.5 Dehumidification — cooling, washing and reheating

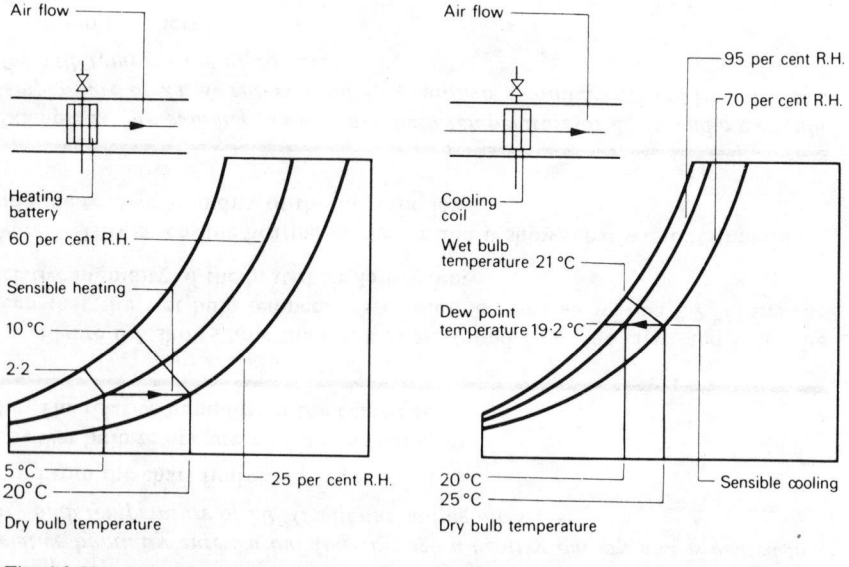

Air flow

Heating battery

60 per cent R.H.

Sensible heating

10 °C

2·2

5 °C
20 °C

25 per cent R.H.

Dry bulb temperature

Fig. 6.2 Heating without adding moisture

Air flow

Cooling coil

Wet bulb temperature 21 °C

Dew point temperature 19·2 °C

95 per cent R.H.
70 per cent R.H.

20 °C
25 °C

Sensible cooling

Dry bulb temperature

Fig. 6.3 Cooling without dehumidifying

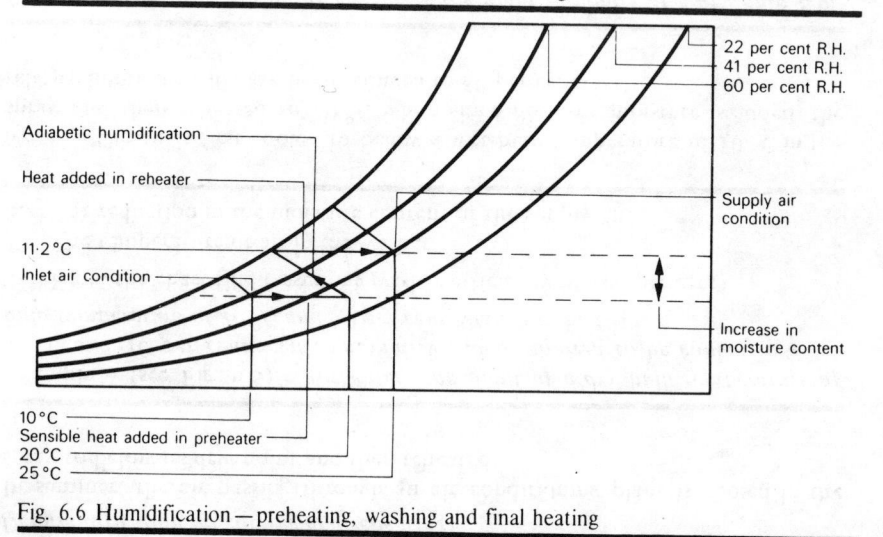

Adiabatic humidification

Heat added in reheater

11·2 °C

Inlet air condition

10 °C
Sensible heat added in preheater
20 °C
25 °C

22 per cent R.H.
41 per cent R.H.
60 per cent R.H.

Supply air condition

Increase in moisture content

Fig. 6.6 Humidification — preheating, washing and final heating

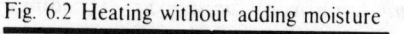

2. Air and water systems in which the air is treated in a central plant, but the bulk of heating and cooling is done in the rooms by passing the air over hot or cold water coils supplied from a central boiler and refrigeration plant. The central plant is smaller and less ductwork is required, but space will be taken up by the room air heating and cooling units.

3. Self-contained units in which the complete air conditioning process is carried out and sited inside the various rooms against the outside wall, so that fresh air can be drawn into the units through a short duct in the wall. There is less duct space and the central plant is small or non-existent, but some space is taken up inside the room for the units.

Central plant system

This is suitable for large spaces where the air temperature and relative humidity is constant. The system is suitable for large factory spaces, open plan offices, theatres, cinemas, supermarkets and assembly halls. The main air conditioning unit may be sited at ground floor level, in a basement, or in a plant room on the roof.

Figure 6.7 shows a longitudinal diagram of a central plant system, which operates as follows:

1. Fresh air is drawn in from the side of the building where the air is likely to be the cleanest and at this stage is mixed with a proportion of recirculated air, which will reduce the load on the heating or cooling plants.

2. The air passes through a dry or viscous filter to remove suspended matter in the form of dust. If required, the air can be further cleaned by passing it through an electrostatic filter.

3. In winter, the air is heated by a preheating coil formed of finned copper pipe, heated by steam, electricity or hot water. Preheating of the air will allow the air to absorb more moisture in the washer and also prevent freezing of the water in the sprays, should the temperature of the incoming air fall below 0 °C. The spray water itself may also be heated.

4. In summer, the air is cooled by a cooling coil, or by passing the air through cold water in the spray. The cooling coil or spray water may be cooled by a refrigeration plant or by water pumped from a deep borehole. The air at this stage is cooled to below dew point and will therefore be saturated. The washer also cleans the air. The warming or cooling of the water in the washer forms an important factor in the air conditioning process. Provision is made for changing the spray water and cleaning out the cistern when necessary.

5. After leaving the washer, the air passes through two sets of glass or galvanised steel plates, made up in a zig-zag formation. The first set is known as scrubber plates which are washed by a continuous stream of water, so that any particles of dust still held in suspension are washed down to the cistern. The second set are known as eliminator plates and are designed to intercept any droplets of water held in the air, so that only absorbed moisture is carried forward with the air to the final or reheater.

6. In both winter and summer the air may require reheating in the final heater, which brings the air up to the required temperature with a corresponding reduction in the relative humidity. This heater is similar in construction to the preheater, but has a greater surface area.

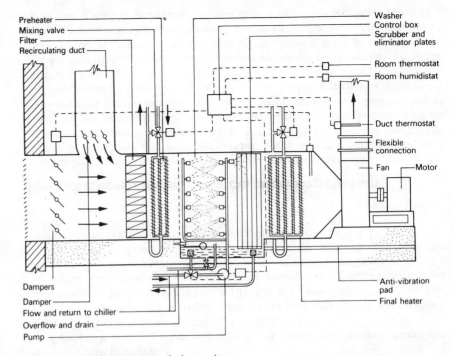

Fig. 6.7 Air conditioning central plant unit

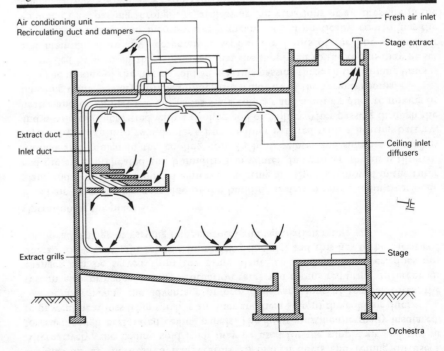
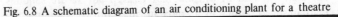

Fig. 6.8 A schematic diagram of an air conditioning plant for a theatre

7. The inlet fan forces the conditioned air into the building via the inlet ductwork and diffusers.

8. Vitiated air is extracted from the building by the extract fan via the extract ductwork and grills. The extract fan is smaller than the inlet fan and this causes slight pressurisation of the air inside the building, which helps to prevent the entry of draughts and dust.

If required, up to about 75 per cent of the extract air may be recirculated via the recirculating duct, also the air may be drawn through the lighting fitting, which would extract heat from the lamps and also improve lighting efficiency by about 14 per cent.

Automatic control

Motorised dampers automatically control the flow of air through the fresh air inlet and recirculating duct. The temperatures of the preheater, washer and reheater are thermostatically controlled, to give the required supply air temperature and relative humidity.

Figure 6.8 shows a central plant system installed in a theatre, using downward air distribution.

Variable air volume system

In this system, the air is supplied from a central plant unit at a temperature and relative humidity, which vary with the weather conditions. An insulated single duct carries the conditioned air at high velocity to variable volume units, usually positioned in the ceiling. The units are provided with thermostatically controlled actuators, which vary the quantity of air supplied to each space. The amount of air passing into the room from the units, vary with the room temperature and the system is particularly suitable for a building having fairly evenly distributed cooling loads.

The dual duct system of air conditioning

This system is similar to the central plant system, in that it uses air as a heating or cooling medium and there is no pipework in the rooms being air conditioned. The system has two ducts, for hot and for cold air respectively. The use of these air ducts provides a means of controlling varying temperatures in different rooms and it is possible to meet the most exacting room climatic requirements.

All the main air conditioning units can be housed in the plant rooms and the hot and cold air ducts may be installed in the ceiling void to mixing or blending units. The air in the ducts flows at a higher pressure and velocity than the central plant system and this allows the ducts to be smaller. They must be fully airtight, or otherwise there is a risk of whistling noises resulting from small leaks. Circular ducts are easier to seal than rectangular ones, but because of the greater size they are more difficult to house in the ceiling void.

Because of the higher pressure and velocity of the air, the fan must operate at a higher speed and this requires a larger fan motor, which uses more electrical power. The system requires more duct space than the central plant system, but this can usually be arranged in a false ceiling, which will take up to 300 mm from the head room of each storey. If required, the units may be installed under the windows, to intercept the cold air in winter, or inside the ceiling void, which supplies air at the desired temperature to branch ducts and ceiling diffusers. Alternatively, the ceiling void itself may be used for the supply air and the air forced through perforated ceiling panels. The ducts must be thermally insulated, to prevent heat loss from the hot air duct and heat gain to the cold air duct.

Besides having the advantage of providing variations of temperatures, the system also provides varying ventilation rates and rooms requiring different air changes can be served from the same plant. The system, however, does not operate on a constant relative humidity principle, and this has to be sacrificed for the advantage of varying temperatures and ventilation rates.

Operation of the plant

Fresh air from the cleanest side of the building is drawn into a main treatment plant and is mixed as required with recirculating air. Dust is removed in the filter and the air is preheated and humidified in winter. In summer, the air is filtered, washed and cooled in the cooling coil. In both winter and summer, the air is forced by a fan into two ducts; a hot air duct is fitted with a heating battery, and a cold-air duct fitted with a cooling coil or chiller. After passing through the heater and or chiller, the air passes via a cold air and a hot air duct to mixing or blending units, usually installed under the windows of the various rooms.

The mixing of the hot or cold air streams passing through the mixing units is controlled by a room thermostat, so that the supply air to the rooms may be set and maintained at any temperature within the limits of the hot or cold air streams. The unit also incorporates a means of automatically controlling the volume of air passing through, regardless of the variations of air pressure in the ducts. This is known as a 'constant volume controller' and this is an essential part of the dual duct system.

Figure 6.9 shows a schematic diagram of a dual duct system, where the vitiated air is extracted through the light fittings.

Figure 6.10 shows a sectional elevation of one type of constant volume room mixing or blending unit, which is fitted below a window.

Figure 6.11 shows an elevation of the mixing unit, with the method of making the flexible connections from the main hot and cold air ducts.

Figure 6.12 shows how mixing units may be installed, to force the conditioned air through a perforated ceiling into a room.

The induction convector air conditioning system

Most rooms in a building have different requirements so that the supply air will have to be conditioned to meet the comfort needs of each room. The induction system can be used as an alternative to the dual duct system for varying the supply air temperatures in multi-roomed buildings. Instead of hot and cold air ducts required for the dual duct system, the induction system utilises hot and or cold water pipes, which takes up less space, but a single duct is also required for the primary air from a main conditioning unit.

An induction system, as the name implies, causes the secondary air in the room to be recirculated by the primary air passing through nozzles in an induction unit, and this causes good air movement in the room. This induced air flows over a cooling or heating coil and cooled or heated secondary room air is

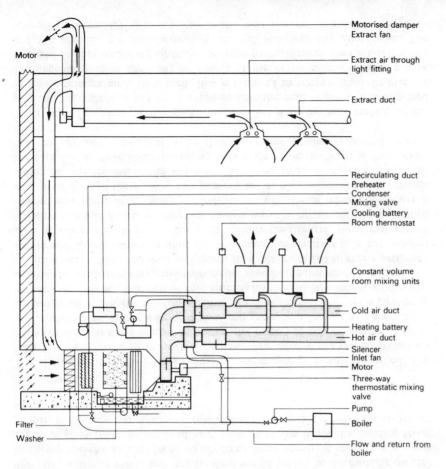

Fig. 6.9 Dual duct system

Labels for Fig. 6.9:
- Motorised damper
- Extract fan
- Motor
- Extract air through light fitting
- Extract duct
- Recirculating duct
- Preheater
- Condenser
- Mixing valve
- Cooling battery
- Room thermostat
- Constant volume room mixing units
- Cold air duct
- Heating battery
- Hot air duct
- Silencer
- Inlet fan
- Motor
- Three-way thermostatic mixing valve
- Pump
- Boiler
- Flow and return from boiler
- Filter
- Washer

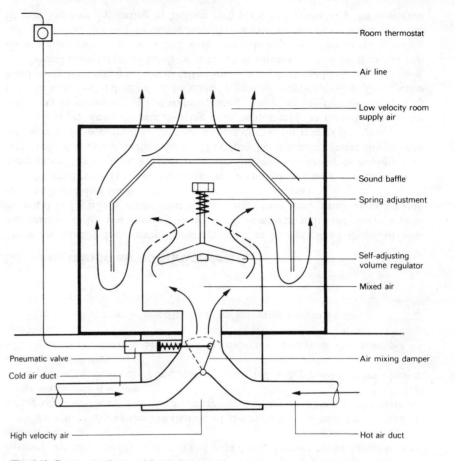

Fig. 6.10 Constant volume mixing unit

Labels for Fig. 6.10:
- Room thermostat
- Air line
- Low velocity room supply air
- Sound baffle
- Spring adjustment
- Self-adjusting volume regulator
- Mixed air
- Air mixing damper
- Hot air duct
- Pneumatic valve
- Cold air duct
- High velocity air

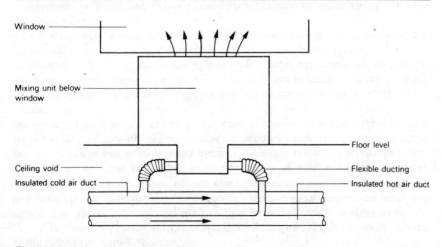

Fig. 6.11 Elevation of the mixing unit

Labels for Fig. 6.11:
- Window
- Mixing unit below window
- Ceiling void
- Insulated cold air duct
- Floor level
- Flexible ducting
- Insulated hot air duct

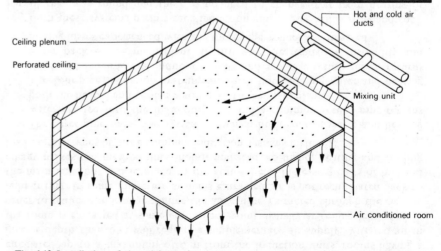

Fig. 6.12 Installation of the mixing unit

Labels for Fig. 6.12:
- Hot and cold air ducts
- Ceiling plenum
- Perforated ceiling
- Mixing unit
- Air conditioned room

mixed with the primary conditioned air before being discharged through a grill at the top of the induction unit.

A two-pipe induction system is the most common type of air/water system in use. This system operates on the two-pipe principle, and is referred to as 'the two pipe change over induction convector system'. Flow and return pipes are used leading to induction unit convectors, one inlet and one return pipe. During summer, chilled water is pumped from a refrigerating unit, whilst during winter hot water is pumped from a boiler plant. The system can easily be changed over from heating to cooling and vice versa, depending upon the external climatic conditions. The change over can be effected manually, or automatically from a control station.

With the aid of the system the rooms can be heated or cooled quite apart from the limited amount of heating or cooling by the primary air from the main air conditioning unit. With rooms of differing aspects, the water piping may be so arranged that if required chilled water may be pumped through to rooms having a southern aspect, and heated water pumped through to rooms having a northern aspect.

During the spring and autumn transitional periods, an important factor with the system is the date when a change over is made, from winter or summer operation and vice versa, and this decision would be left to the building owner. In theory the change over from winter to summer operation can be made when the heat gains from the sun, and heat generated from inside the building from people, machinery and lighting is equal to the heat losses through the structure to the outside atmosphere.

Operation of the plant

The air is treated in the main air conditioning unit as previously described. The inlet fan then forces air through the primary air duct to the various induction units fitted below the windows. In winter, the curtain of cold air descending past the windows can in this way be drawn into the induction unit and a warm air curtain formed over the window opening, so that the window area is pleasantly warmed. In summer, a stream of cold air blown vertically upwards cools the area around the window opening and reduces the solar heat gains inside the room.

In order to keep the duct size down to a minimum and to create sufficient air velocity in the induction unit, the air velocity in the duct is in the order of 20 m/s, whereas with the ordinary air conditioning central plant system this velocity is about 5 m/s. In summer, chilled water is circulated through finned copper pipe coils, fitted inside the induction units; during winter, hot water is circulated through these coils. The primary air, at high velocity, passes through induction nozzles inside the induction unit and in so doing reduces the air pressure below these nozzles. The secondary air inside the room is thus drawn into the unit and is mixed with the primary air before being discharged into the room.

If the occupants of the room are satisfied with the temperature of the primary air, a damper can be arranged so that the cold or hot coil is bypassed and only primary air is then used. If it is required to warm or cool the primary air, depending upon the time of year and use of room, the damper can be arranged so that secondary air from the room passes through the cold or hot coil and the cooled or heated secondary air will mix with the primary air. The mixture of secondary to primary air in the induction unit is usually three to six

volumes of secondary air, to one volume of primary air. Air in the room can be circulated up to a maximum of 6 m from the induction units; rooms having a greater depth than 6 m will require a supplementary air supply. Vitiated air in the room is extracted at some convenient point, usually at ceiling level, and in order to reduce the load in winter the air can be extracted through the electric light fitting. In some induction systems extracted air is not recirculated back to the main unit, but is discharged to the atmosphere. Increasing costs of fuel and energy makes this method uneconomic and leads to higher heating and cooling loads. Other types of more complex induction systems include:

1. The three-pipe induction system, in which both cooling water and heating water are fed into the induction unit and a common return pipe returns both the heated and cooled water back to the main plant.
2. The four-pipe induction system, in which both cooling water and heating water are fed into the induction units and separate return pipes, returns water back to the main plant. In this way the regulation instability and mixing losses experienced with the three-pipe system are obviated.

Both these systems permit each induction unit to give full heating and cooling all the year round and there is no need to change over from heating or cooling or vice versa, as with the two-pipe change over system. These systems, however, are more costly to install and would only be used for buildings such as first-class hotels.

Figure 6.13 shows the installation of the system for a two-storey building having separate zones. For clarity, only the primary air duct and the main air conditioning unit is shown.

Figure 6.14 shows the same system connected with pipework from a heater and a cooling battery.

Figure 6.15 shows a section of an induction unit and the principle of operation.

Figure 6.16 shows the installation of an induction unit in a room.

Self-contained air conditioners

There are usually for smaller buildings such as houses, small shops, offices, restaurants, clubs and hotels. They are a complete unit and only require electrical and perhaps water connections. Sizes vary from small units for one room to larger sizes for several rooms served from ductwork.

The smaller units contain a heater and a refrigerator, or a heat pump with compressor, fan or fans, evaporator or cooling coil, condenser or heating coil, filter and, if required, a humidifier. They may be mounted against the outside wall so as to draw in fresh air through a short duct at the rear of the unit.

The larger units contain heating coils, which may be heated directly by electricity or indirectly by hot water, cooling coils from a refrigerator or cooling tower, filter, fan and again, if required, means of humidification. These units usually require ducting for the distribution of the conditioned air.

Several manufacturers produce modules of various parts of the unit and it is therefore possible to install a unit with only a heating coil and filter; a cooling coil and humidifier may be added later. These modules are of standard sizes and fit into the unit by sliding or bolting into place and connecting the necessary

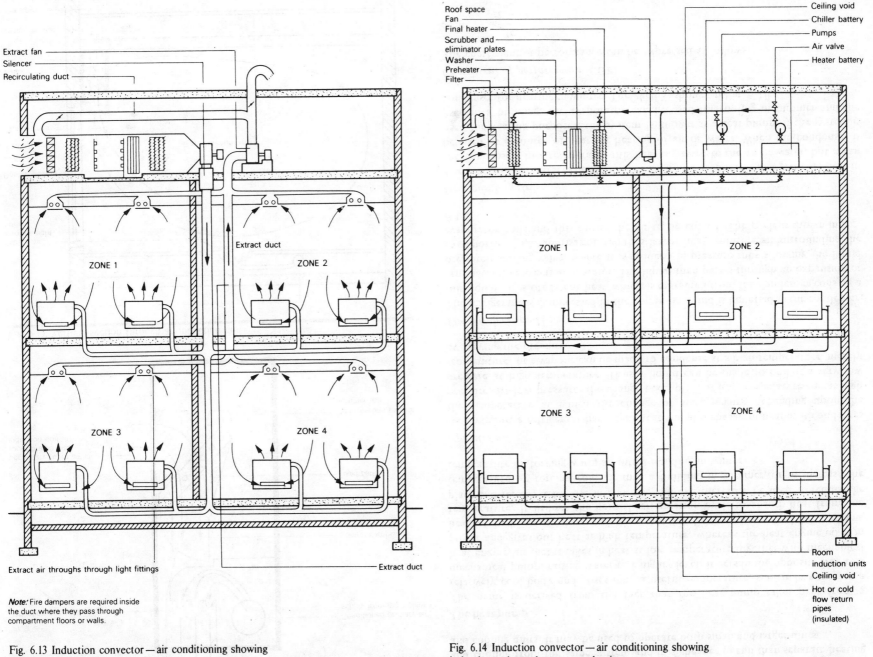

Extract fan

Silencer

Recirculating duct

Extract duct

ZONE 1

ZONE 2

ZONE 3

ZONE 4

Extract air throughs through light fittings

Extract duct

Note: Fire dampers are required inside the duct where they pass through compartment floors or walls.

Fig. 6.13 Induction convector — air conditioning showing induction units and primary air supply

Roof space

Fan

Final heater

Scrubber and eliminator plates

Washer

Preheater

Filter

Ceiling void

Chiller battery

Pumps

Air valve

Heater battery

ZONE 1

ZONE 2

ZONE 3

ZONE 4

Room induction units

Ceiling void

Hot or cold flow return pipes (insulated)

Fig. 6.14 Induction convector — air conditioning showing induction units and water supply pipes

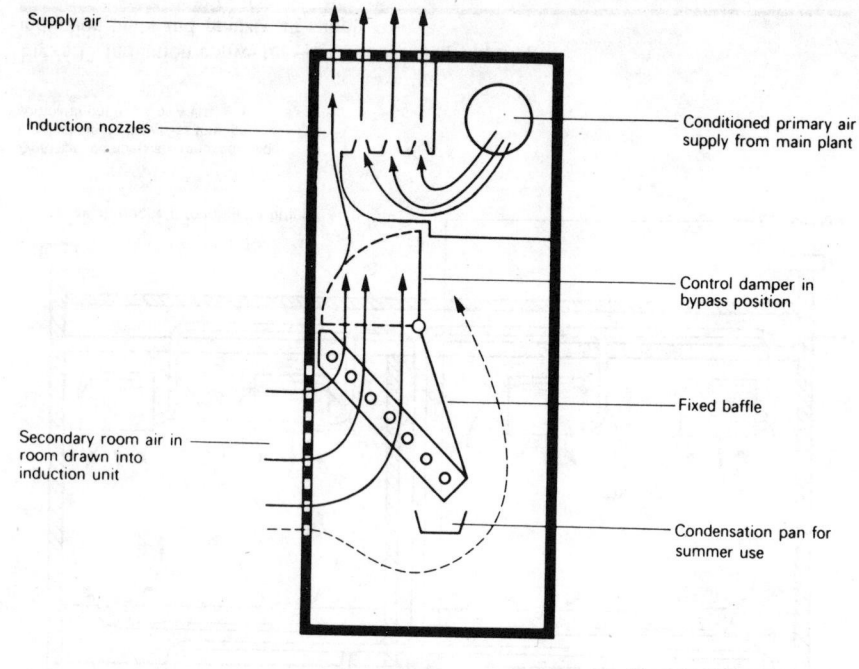

Supply air

Induction nozzles

Conditioned primary air
supply from main plant

Control damper in
bypass position

Fixed baffle

Secondary room air in
room drawn into
induction unit

Condensation pan for
summer use

Fig. 6.15 Vertical section through an induction unit

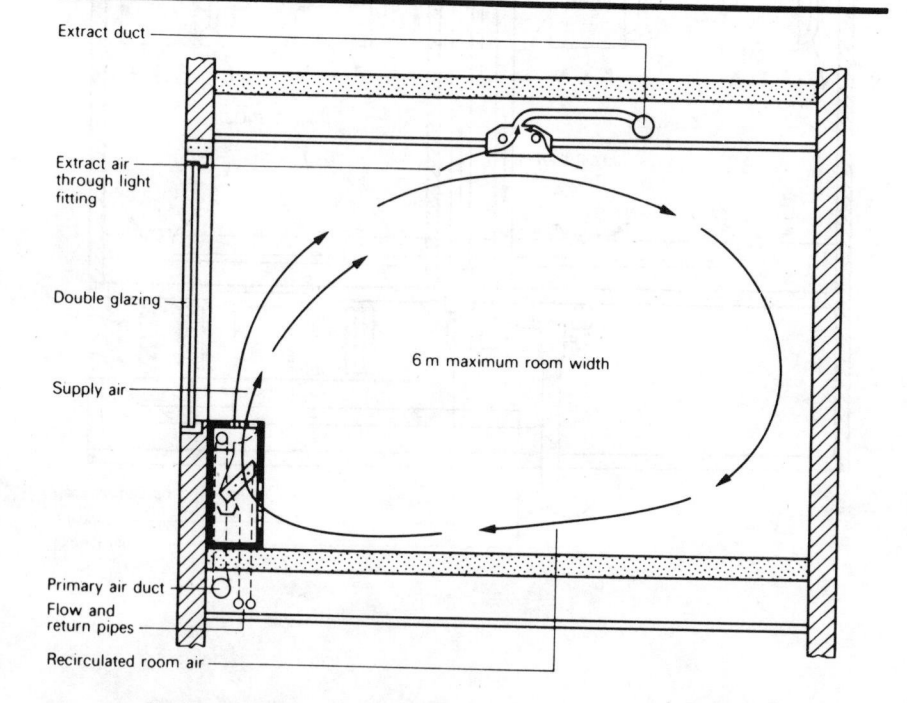

Extract duct

Extract air
through light
fitting

Double glazing

6 m maximum room width

Supply air

Primary air duct

Flow and
return pipes

Recirculated room air

Fig. 6.16 An induction unit fitted in a room

pipework or cables. The heat pump has the advantage of providing both heating and cooling from the same system and it is cheaper to run than separate heating and cooling units. It may be used to operate both small and larger units.

The heat pump

The name is derived from the fact that the heat pump takes heat from a relatively cool body and raises it to a useful temperature, similar in a way to a mechanical pump, raising water to a higher level. It acts in the opposite way to a heat engine, in that it takes in heat at low temperature together with mechanical power and gives out heat at high temperature, whereas the heat engine takes in heat at high temperature and gives out mechanical power and heat at low temperature. It utilises the normal refrigeration cycle to absorb heat from one place and release it for use in another, and can therefore be used to reduce the temperature by forcing cold air into a building in summer, or to increase the temperature by forcing warm air into a building in winter.

Operation

As a vapour, a refrigerant has latent heat and absorbs latent heat as it condenses the temperature at which the change of state occurs, depending upon the pressure. At low pressure, the change takes place at low temperature and at high pressure at high temperature. Thus a liquid can be made to boil at a very low temperature, or a vapour can be made to condense at a high temperature, merely by varying the pressure.

The refrigeration cycle

The refrigerant is compressed in the compressor and is liquefied in the condenser; this liquid gives off latent heat which is extracted from the condenser coils by a fan, natural convection or water. The liquid then passes through an expansion or pressure-reducing valve, where it is reduced in pressure thus allowing the liquid to vaporise. This vaporisation requires latent heat from the air surrounding the evaporator coils and this causes the air to be cooled. The cycle is shown in Fig. 6.17.

Heat pump cycle

The refrigerator extracts heat from the air in the evaporator and gives off heat in the condenser. In a refrigerator this heat is wasted to the air or water, but it can be used as a source of heat for heating of air or water. When the condenser is used for heating purposes, the system is known as a heat pump. If the system is used in a building, the evaporator can be used for cooling in summer and the condenser for heating in winter. The cycle is shown in Fig. 6.18.

Coefficient of performance (COP)

The coefficient of performance can be expressed as follows:

$$COP = \frac{tc}{tc - te}$$

where te = Evaporator temperature in degrees Kelvin

tc = Condenser temperature in degrees Kelvin

It will be clear from the above equation that, in common with all vapour

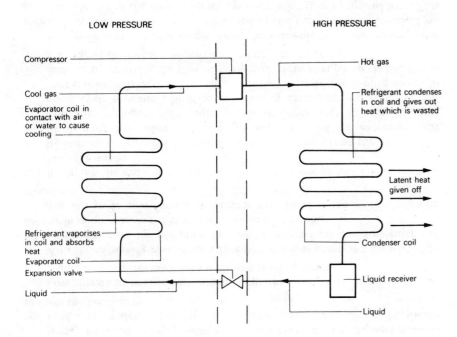

Fig. 6.17 Vapour refrigerator cycle

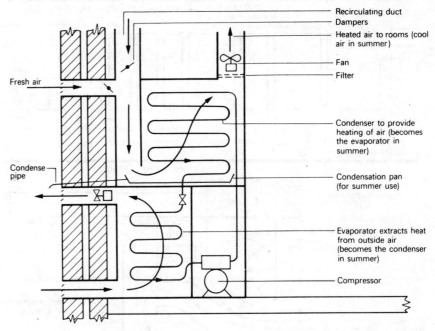

Fig. 6.19 Detail of heat pump for winter use (refrigerant flow reversed for summer use)

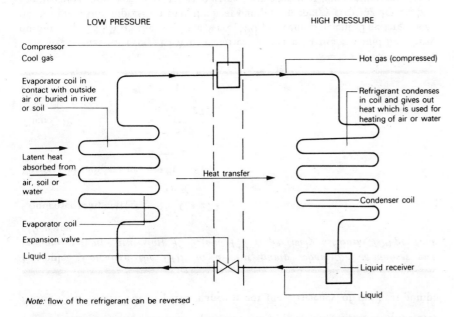

Note: flow of the refrigerant can be reversed

Fig. 6.18 Heat pump cycle

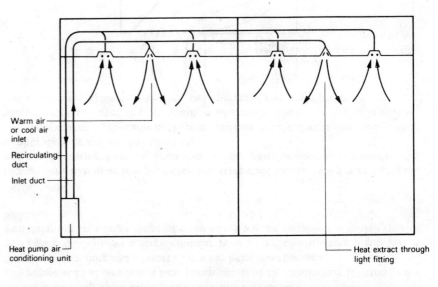

Fig. 6.20 Use of the heat pump (detail of inlet and recirculating ductwork)

refrigeration systems, the heat pump operates to a greater advantage when working with a low condenser temperature and a higher evaporator temperature.

Note: The coefficient of performance is not the efficiency of the heat pump.

Example. *Calculate the COP of a heat pump when the condenser and evaporator temperatures are 45 °C and 4 °C respectively, assuming 100 per cent efficiency.*

Condenser temperature = 45 °C + 273

$$t_c = 318 \text{ K}$$

Evaporator temperature = 4 °C + 273

$$t_e = 277 \text{ K}$$

Then COP $= \dfrac{318}{318 - 277}$

COP $= 7.75$

This means that for every 1 kW of electrical power used there would be a heat output of 7.75 kW if the machine were 100 per cent efficient. The large heat output gives the impression that the heat pump is an energy-creating device, but the machine, however, only takes existing heat energy at low temperature from any available source and raises it to a higher temperature. Allowing for loss of efficiency, the machine would show a coefficient of performance of between two and three.

It may be wondered why, with such a high performance, the heat pump is not used more widely. There are two possible reasons for the apathy shown towards the heat pump:

1. Low outside air temperature in winter months.
2. Higher cost of plant compared with traditional systems.

With the increasing cost of fuels, however, the heat pump as a big advantage for air conditioning by providing heating in winter from the condenser and cooling in summer from the evaporator at competitive costs.

The low-grade heat for the evaporator may be obtained by the following methods:

1. By laying the coils at the bottom of a river or stream and extracting heat from the water.
2. By burying the coils below ground and extracting heat from the soil.
3. By drawing atmospheric air over the evaporator coil and extracting heat from the air. Figure 6.19 shows a unit air conditioner, utilising the heat pump cycle to warm a building in winter by extracting heat from the low temperature outside air. The same unit may be used for cooling rooms in summer by reversing the flow of refrigerant in the heat pump circuit. Figure 6.20 shows the method of installing the unit in a building.

To supplement the heat output from the condenser coils during very cold spells, a boost heater may be incorporated in the unit, which is switched on automatically when the outside air temperature falls below a required level. This

supplementary heating will only be in use for a very small part of the heating season, depending upon the degree of thermal insulation.

Figure 6.21 shows how a heat pump-operated air conditioner is fitted below the window opening and a boost heater has been incorporated.

Figure 6.22 shows a large modular type of air conditioning plant that is manufactured in a large range of sizes which will be suitable for most types of buildings.

Refrigerants

Liquids which boil at low pressures and temperatures are used for refrigerants, and the following have all been used in the past: ammonia, carbon dioxide, sulphur dioxide and methyl chloride.

Halogenated hydrocarbons (Freons) are now nearly always used due to their many advantages over the previously used liquids. They are: non-inflammable, non-toxic, non-explosive, odourless and operate at moderate pressures.

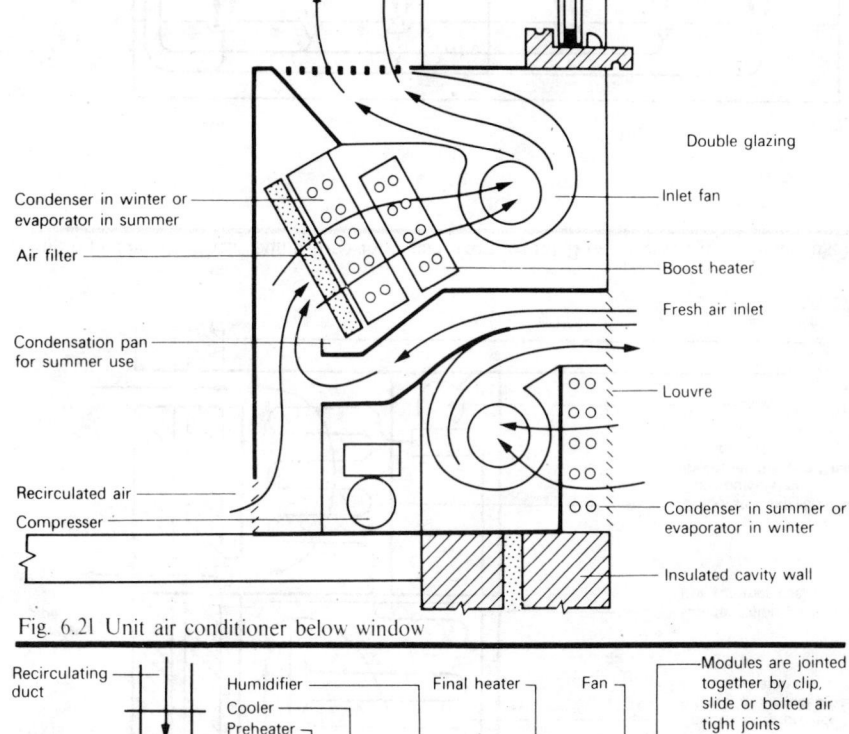

Fig. 6.21 Unit air conditioner below window

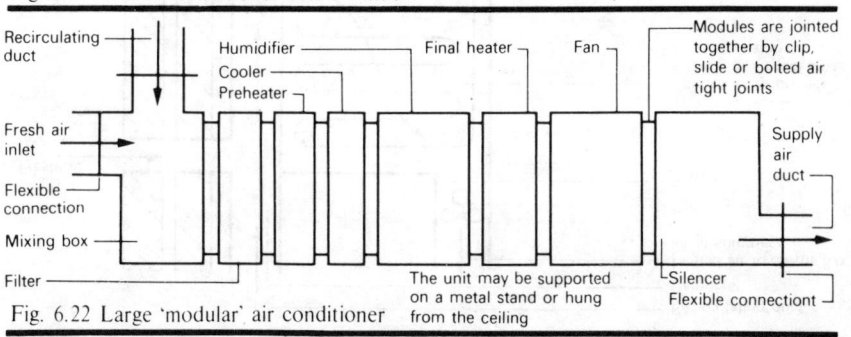

Fig. 6.22 Large 'modular' air conditioner

Chapter 7

Medium and high-pressure hot-water heating

Systems

In these systems the water is in a closed circuit and is subjected to pressure, by either steam or gas, so that its temperature may be raised above 100 °C. As the temperature of the water rises, the pressure in the system must also be raised to ensure that pressure is always above the evaporation pressure, otherwise the water would flash into steam and the systems would not function as high-temperature hot-water heating systems. The pressure on the water can be produced by steam maintained in the boiler, or a steam space inside an external drum. When a gas is used for pressurising, it is maintained in a pressure vessel and the boilers are completely filled with water. One of the simplest methods of pressurising a system is by means of a high-level head tank. This method however is limited, since few buildings are of suitable design. It would require a minimum height of 30 m above the highest main pipe in order to produce a pressure of 300 kPa.

Figure 7.1 shows a schematic diagram of a medium- or high-pressure system, using nitrogen inside a pressure cylinder. Nitrogen is preferable to air for this purpose because, unlike air, it is an inert gas, which prevents risk of corrosion. Furthermore it is less soluble in water than air.

Operation of the system

The pressure cylinder is maintained partly filled with water and partly with nitrogen. The boilers are fired and the expanded water enters the pressure cylinder until the nitrogen cushion is compressed to the working pressure of the system. At this pressure the pressure switch A is off, but if the pressure is exceeded the switch operates and opens the spill valve. Water is thus allowed to flow into the spill cistern and release the pressure in the cylinder to the correct level, when switch A closes the spill valve. If the pressure drops, due to leaks on packing glands, etc., the pressure switch B cuts in the feed pump, and water is forced into the cylinder until the correct water level and pressure is reached when switch B cuts out the pump.

Figure 7.2 shows a schematic diagram of a medium- or high-pressure boiler plant, utilising steam for pressurisation. The main flow and return pipes are taken below the water level in the boiler, so that steam cannot enter the pipework. In this system the steam and the water are at saturation temperature and any reduction of pressure on the water side will cause the water to flash into steam. To prevent this, a pipe is connected between the main flow and return pipes, so that cooled water from the return pipe can be allowed to mix with the hotter water in the flow pipe. A valve is fixed on this pipe to regulate the correct amount of mixing.

In another type of gas-pressurised system the cylinder is made large enough to accommodate all the expanded water without the need for a spill valve and cistern. The main reason behind the provision of a spill cistern however is the saving in cost of a large pressure cylinder. The circulation of the hot water is by means of a centrifugal pump, specially designed for use with high temperature water.

In both medium- and high-pressure systems it is essential that the feed water is treated, or otherwise scaling of boilers and equipment will occur. The workmanship and equipment must also be of a very high standard, with welded or flanged pipe joints, and bends must be of large radius and tees must be swept in the direction of flow. Allowances must be made for expansion and contraction of the pipework and air bottles or air valves fitted at the highest points of the circuits.

Table 7.1 shows the usual temperatures and pressures used for the systems.

Table 7.1   Temperatures and pressures used

Temperatures (°C)	Gauge pressure (kPa)	
150–200	600–1100	(high pressure)
120–135	300–400	(medium pressure)

Temperature control

The water temperature can be controlled at various points in the building by means of a three-way mixing valve. The valves allow cooler water in the return pipe to mix with the hotter water in the flow pipe at a controlled rate, before distribution to the various points in the building.

Types of heat emitters

If the water temperature exceeds 82 °C it is dangerous to install ordinary types of radiators that may be touched by occupants of the building. Above this temperature it is advisable to install overhead radiant strips or panels, unit heaters, convector type skirting heaters or fan convectors.

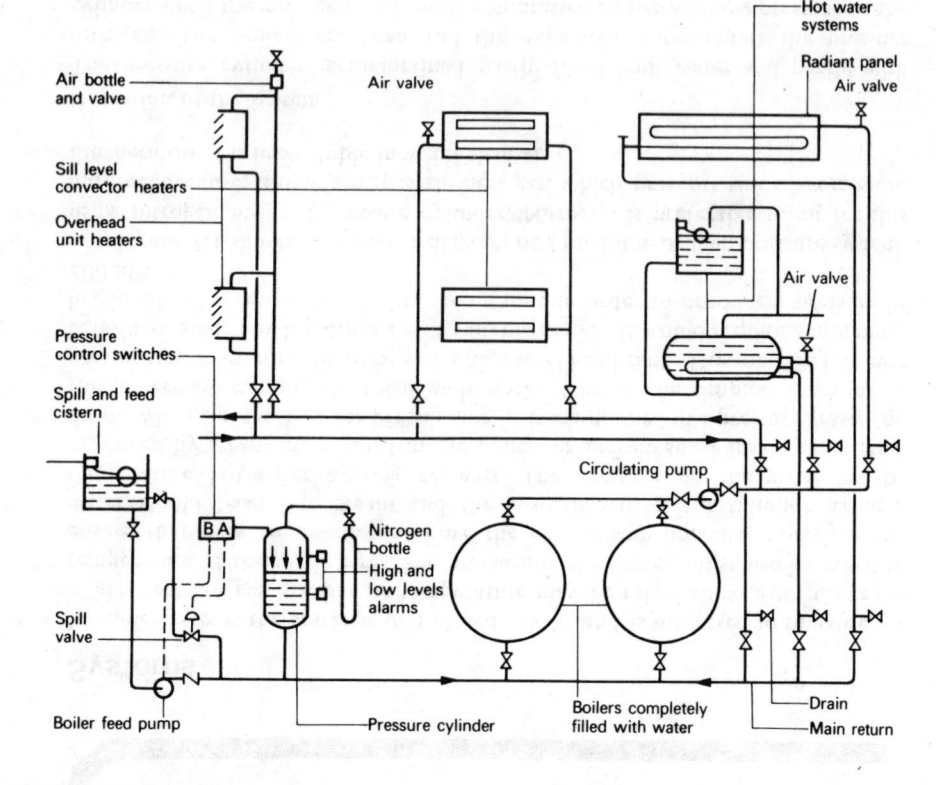

Air bottle and valve

Air valve

Sill level convector heaters

Overhead unit heaters

Pressure control switches

Spill and feed cistern

Spill valve

Boiler feed pump

B A

Nitrogen bottle

High and low levels alarms

Pressure cylinder

Boilers completely filled with water

Circulating pump

Hot water systems

Radiant panel

Air valve

Air valve

Drain

Main return

Fig. 7.1 Nitrogen pressurisation

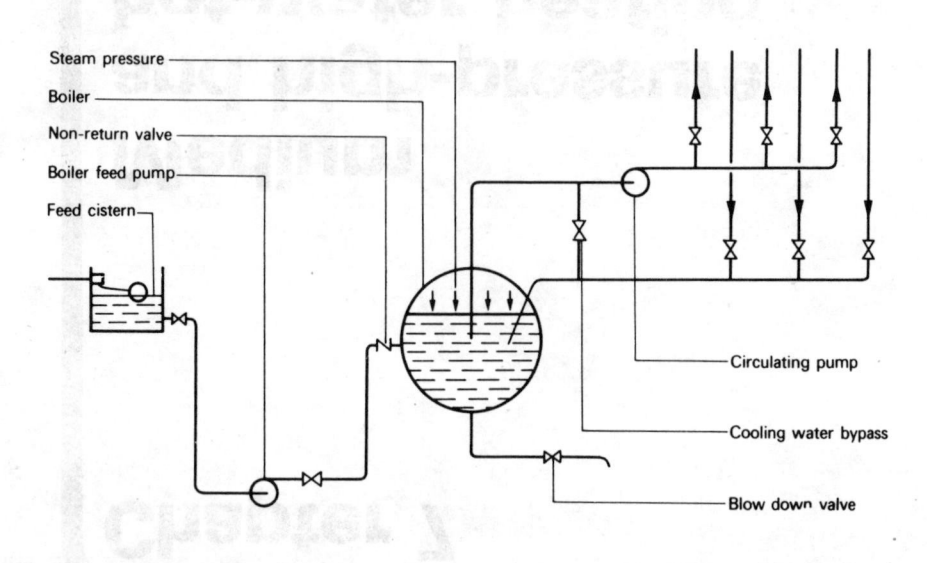

Steam pressure

Boiler

Non-return valve

Boiler feed pump

Feed cistern

Circulating pump

Cooling water bypass

Blow down valve

Fig. 7.2 Steam pressurisation

If it is required to have some part of a building heated by low-pressure hot water, which is easier to control by thermostats, non-storage type calorifiers may be used. These calorifiers have a battery of tubes inside which are heated by the medium- or high-pressure hot water. These tubes heat the water on the low-pressure side of the circuit, which is supplied with water from an open expansion and feed cistern.

Advantages over low-pressure systems

1. They contain a large thermal storage within the water, which will meet any sudden heat demands made upon the systems.
2. Because of the higher temperature of the water, smaller pipes and heat emitters may be used.
3. By use of smaller pipes, it is easier to distribute heat in very large buildings and less duct space is required.
4. The systems are very suitable for processing work in factories, district heating schemes and high-rise buildings.

Note: 1 Pascal = 1 N/m^2
           1 Bar   = 100 kPa

Steam heating

Steam as a heating medium is often used for the heating of industrial buildings, where the steam plant is also used for various processes requiring high temperatures. It is also used in hospitals, where it may provide a means of sterilising and of space heating and hot water supply indirectly by means of calorifiers. The system uses the latent heat of vaporation of water, which is equal to 2257 kJ/kg at atmospheric pressure, being greatly in excess of 419 kJ/kg, the sensible heat of water, to raise its temperature from 0° to 100 °C at the same pressure (see Table 7.2).

For this reason and because steam usually flows at a high velocity of between 24 to 36 m/s and at temperatures of between 100 °C and 198 °C, the pipework and heat emitters may be smaller than those used for hot water. Steam also flows under its own pressure and pumps are therefore not required. This is most useful for the heating of tall buildings, where the cost of pumping hot water is expensive. Steam systems, however, are more complicated than hot water and require more maintenance and supervision.

Terms relating to steam

Latent heat: heat which produces a change of state, without producing a change of temperature. In steam, it is the heat added to boiler water which converts some of the water into steam.

Sensible heat: heat which produces a rise in temperature of a substance without a change of state; in steam generation it is heat which has been added to water to raise its temperature to boiling point and over.

Saturated steam: steam which is generated in contact with the water.

Wet steam: steam which carries droplets of water in suspension.

Dry saturated steam: steam at the temperature of 100 °C which does not

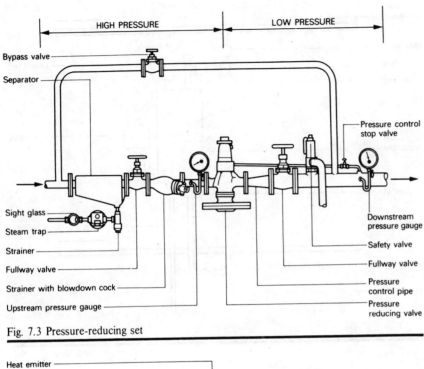

Fig. 7.3 Pressure-reducing set

Labels (Fig. 7.3):
HIGH PRESSURE LOW PRESSURE
Bypass valve
Separator
Pressure control stop valve
Sight glass
Steam trap
Strainer
Fullway valve
Strainer with blowdown cock
Upstream pressure gauge
Downstream pressure gauge
Safety valve
Fullway valve
Pressure control pipe
Pressure reducing valve

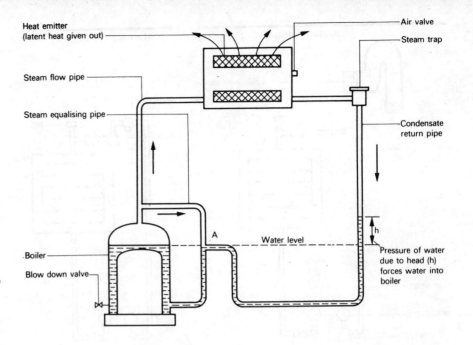

Fig. 7.4 Principle of steam heating

Labels (Fig. 7.4):
Heat emitter (latent heat given out)
Air valve
Steam trap
Steam flow pipe
Steam equalising pipe
Condensate return pipe
Boiler
Blow down valve
A
Water level
h
Pressure of water due to head (h) forces water into boiler

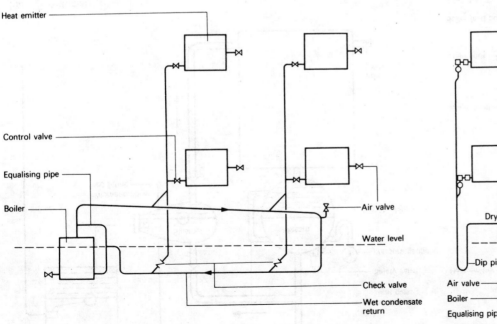

Fig. 7.5 One-pipe gravity system

Labels (Fig. 7.5):
Heat emitter
Control valve
Equalising pipe
Boiler
Air valve
Water level
Check valve
Wet condensate return

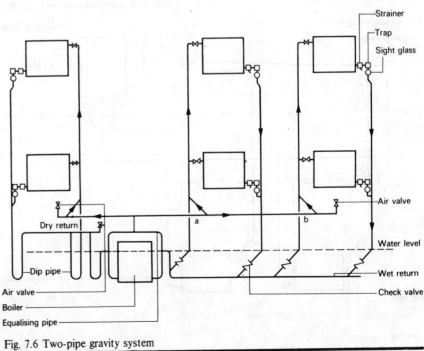

Fig. 7.6 Two-pipe gravity system

Labels (Fig. 7.6):
Strainer
Trap
Sight glass
Dry return
a
b
Dip pipe
Air valve
Boiler
Equalising pipe
Air valve
Water level
Wet return
Check valve

68

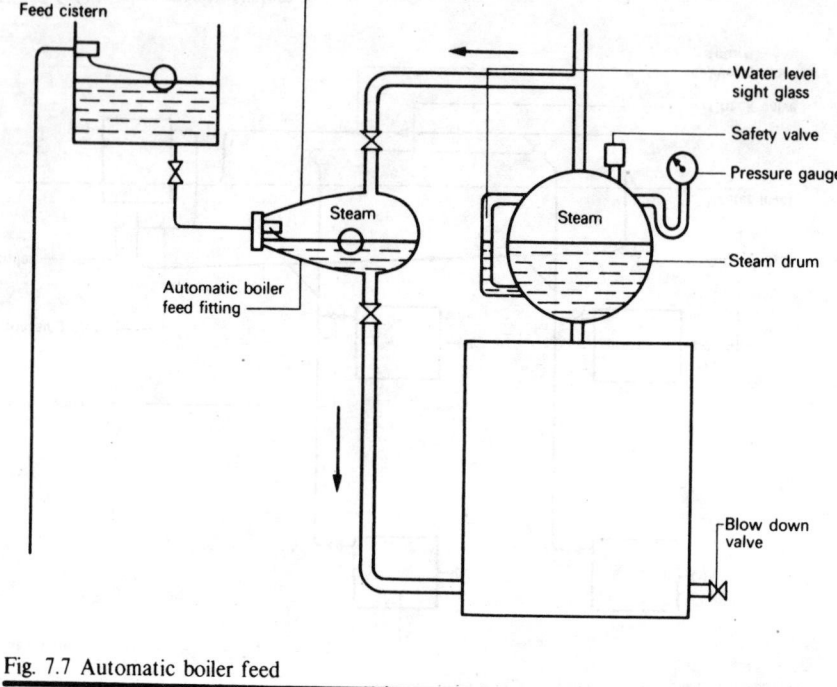

Fig. 7.7 Automatic boiler feed

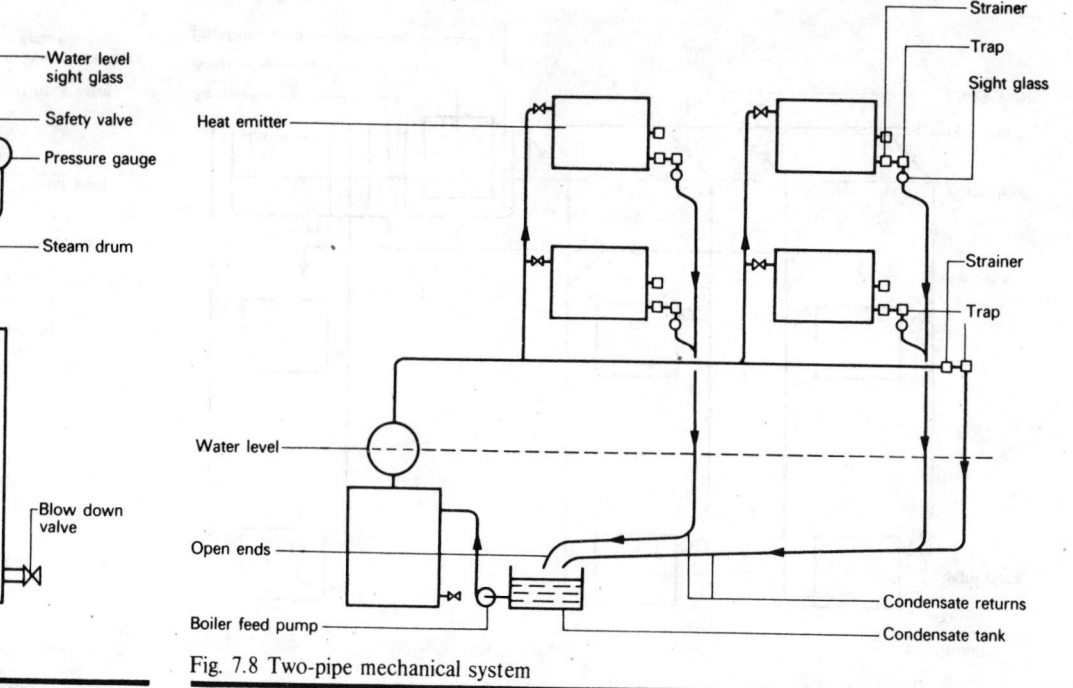

Fig. 7.8 Two-pipe mechanical system

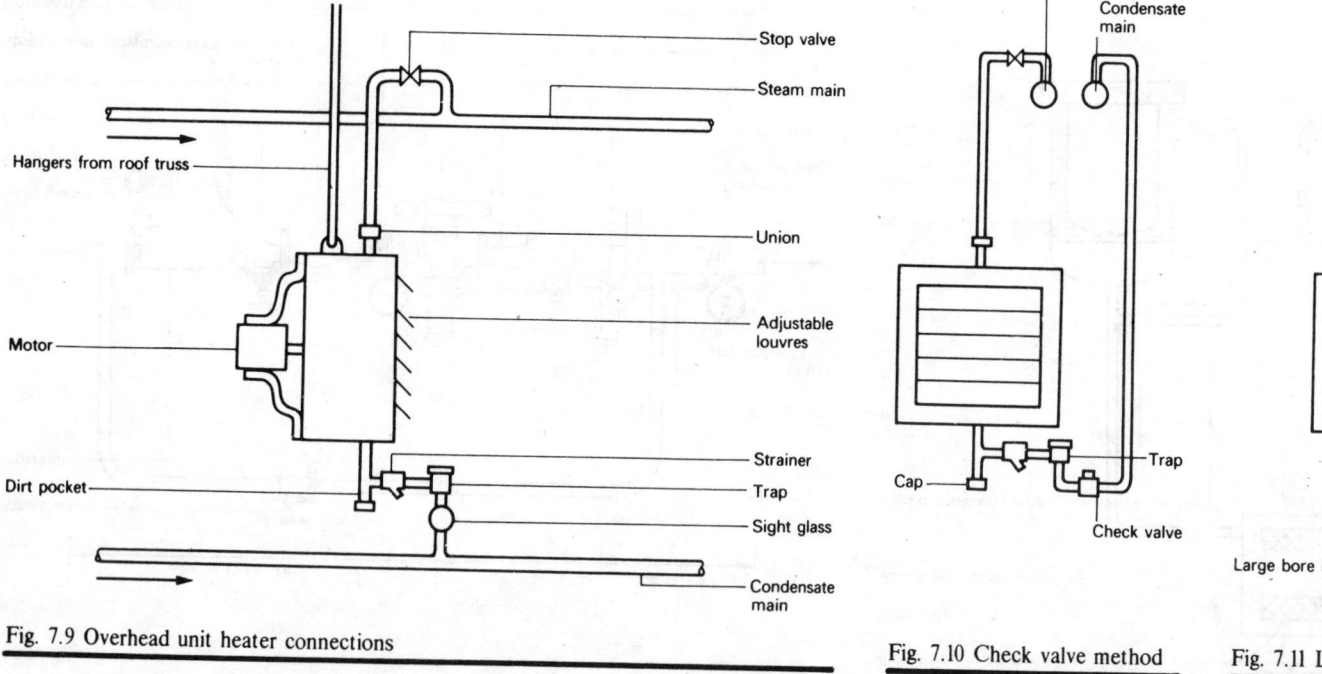

Fig. 7.9 Overhead unit heater connections

Fig. 7.10 Check valve method

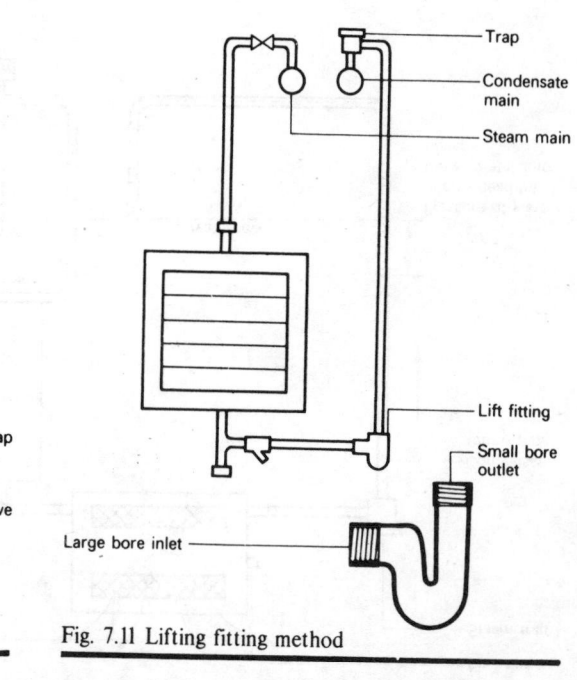

Fig. 7.11 Lifting fitting method

contain any free droplets of water. It is the ideal type of steam, but is very rarely, if ever, found in practice.

Enthalpy, the total heat of steam: sensible heat plus latent heat.

Super-heated steam: steam which has further heat added, after it has left the boiler in which it was generated.

Absolute pressure: gauge pressure plus atmospheric pressure.

Steam pressures

Steam may be used at the following gauge pressures:

Low pressure up to 35 kPa

Medium pressure 140 to 550 kPa

High pressure 550 to 1400 kPa

Steam has a higher latent heat content at low pressure than at high pressure. Because latent heat is given out in the heat emitters and unless high temperatures are required, it follows that low-pressure steam is to be preferred. Low pressure also causes less risk of noise and wear on valves and other equipment.

In large installations having long steam mains, it may be necessary to use either medium or high pressure for the mains to overcome the frictional resistances, and in order to obtain low pressure a pressure-reducing set may be fitted at branch connections to various rooms or appliances. Figure 7.3 shows the equipment required to reduce the steam pressure and is known as a 'pressure-reducing set'.

Figure 7.4 shows the principle of operation of a simple steam heating system. The boiler is partly filled with water and, when cold, the remaining space in the boiler pipes and heat emitter is filled with air. When the water is heated to 100 °C steam is produced and flows up to the heat emitter, pushing the air before it. The air is allowed to escape through the air valve until the emitter is filled with steam, which on condensing gives up its latent heat. Water enters the steam trap which opens and allows the water to flow back to the boiler for reheating. Air is heavier than steam at low pressure and therefore it will be noticed that the air valve is fitted near the bottom of the heat emitter.

In a gravity system an equalising pipe is required, which causes the steam pressure to act downwards against the water pressure in the boiler to equalise the pressure at point A. The gradual increase in the height of water in the vertical condensate pipe, due to the formation of condensation, will overcome the steam pressure and allow water to enter the boiler.

Systems used

There are two systems:

1. *Gravity:* in which condensate runs back to the boiler by gravity.
2. *Mechanical:* in which condensate is pumped back to the boiler.

Figure 7.5 shows a one-pipe gravity system, where the flow pipe to the emitter is also used to carry condensate back to the boiler. The system does not require the use of steam traps, but may be noisy, due to water hammer.

Figure 7.6 shows a two-pipe gravity system, with steam traps connected to the outlets of each heat emitter. Water is prevented from being forced back up to the condensate return pipe by the steam pressure by fitting a check valve or a

dip on the pipe. Both methods are shown in the diagram. In order to check if the steam traps are functioning correctly, a sight flow indicator or a sight glass may be fitted to the outlet side of the traps, so that the operator can see if water is flowing and not steam.

The systems illustrated require the boiler to be initially filled by means of a force pump connected to the drain and blow down valve.

Figure 7.7 shows an automatic means of filling the boiler with water by means of a ball valve inside a sealed chamber, supplied from a cold-water feed cistern. A steel drum shown may be used to contain the steam space instead of the upper part of the boiler.

Figure 7.8 shows a two-pipe mechanical steam heating system, which requires a condensate tank or hot well. The tank may be sited at any level providing it is below the lowest heat emitter.

Types of heat emitters

In older systems radiators were used, but it has become more popular to use convector heaters, overhead unit heaters, radiant panels or strips.

Figure 7.9 shows the method of installing an overhead unit heater, with the steam main and condensate main at high and low levels respectively. If required, both mains may be at high level, providing either a check valve or a lifting fitting is installed, as shown in Figs 7.10 and 7.11.

Steam traps

The purpose of a steam trap is to remove the water which condenses inside appliances or pipelines. They may be divided into four groups, namely:

1. Thermostatic.
2. Ballfloat.
3. Bucket.
4. Thermodynamic.

Figures 7.12, 7.13, 7.14, 7.15 and 7.16 show the various types of traps, which operate as follows:

Thermostatic type (see Fig. 7.12).   The closed bellows contain a volatile spirit which has a boiling point to suit the temperatures involved. When steam enters the trap the volatile spirit expands and opens the bellows, thus closing the valve. When water enters the trap, which is at a lower temperature than steam, the spirit contracts and closes the bellows, thus opening the valve and allowing the water to flow back to the boiler.

Ballfloat type (see Fig. 7.13).   When steam enters the trap the ballfloat is suspended and the weight of the float keeps the outlet valve closed. When water enters the trap the float becomes buoyant and opens the valve, thus allowing the water to flow back to the boiler.

Open bucket type (see Fig. 7.14).   The outlet valve is closed as long as the bucket floats, but when water enters the trap it eventually overflows into the bucket, which causes the bucket to sink, thus opening the valve. Steam forces the water out of the bucket through the tube, until the bucket is buoyant, thus closing the valve.

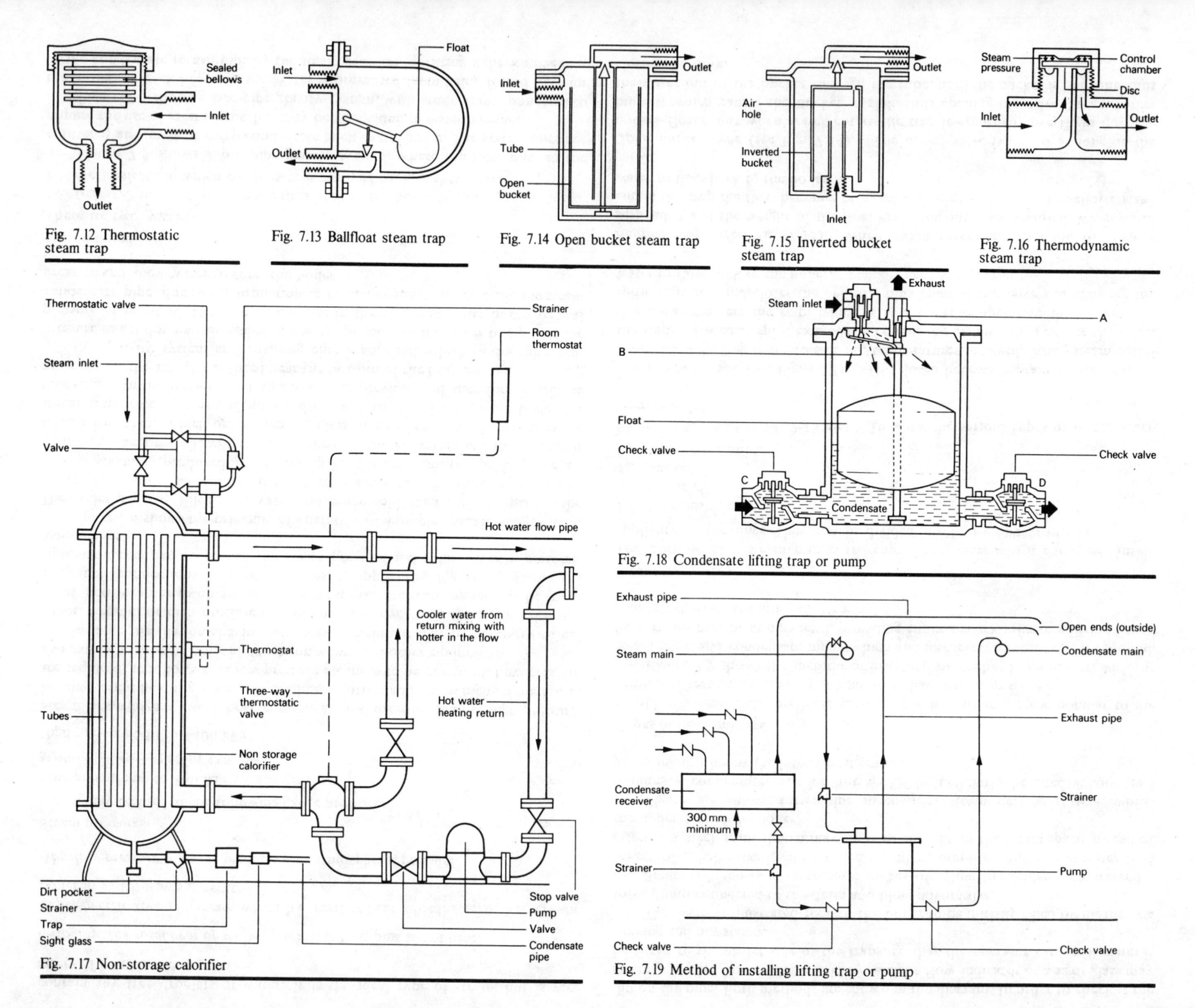

Fig. 7.12 Thermostatic steam trap

Fig. 7.13 Ballfloat steam trap

Fig. 7.14 Open bucket steam trap

Fig. 7.15 Inverted bucket steam trap

Fig. 7.16 Thermodynamic steam trap

Fig. 7.17 Non-storage calorifier

Fig. 7.18 Condensate lifting trap or pump

Fig. 7.19 Method of installing lifting trap or pump

Inverted bucket type (see Fig. 7.15).   If steam enters the trap the bucket is lifted and the valve is closed. When water enters the trap the bucket eventually falls under its own weight and the valve opens, allowing the water to be forced out by the steam pressure.

Thermodynamic type (see Fig. 7.16).   The trap operates on the Bernoulli principle which states that if no friction exists the total energy in a moving fluid is constant. The total energy is the sum of the kinetic pressure and potential energies of the moving fluid, so that an increase in one energy produces a decrease of another and vice versa.

When steam flows through the trap, an increase of kinetic energy is produced between the disc and the seating, which results in a reduction of pressure energy at this point and the disc moves nearer the seating until there is a reduction of kinetic energy. This reduction in kinetic energy produces an increase of pressure energy which tends to lift the disc from the seating, but is prevented from doing so by the steam pressure acting upon the top of the disc in the control chamber. Because the area at the top of the disc is greater than the area of the inlet underneath, the upper pressure forces the disc firmly on to its seat.

When water enters the trap the steam above the disc condenses, thus reducing the pressure, and the disc is forced up; this allows the water to flow through the trap. Water flows through the trap at a lower velocity than steam and does not cause sufficient reduction in pressure below the disc, so that the trap remains open until steam again enters. It will be clear that the trap operates on both kinetic energy and heat, hence the term, thermodynamic.

Non-storage calorifiers

These are used for providing hot water for space heating. The water in these calorifiers is heated indirectly by steam passing through a battery of pipes. To prevent overheating of the water, thermostatic control is required.

Figure 7.17 shows a method of installing a non-storage type calorifier, including details of thermostatic control of the calorifier and heating circuit.

Lifting trap

It may be necessary to lift condensate from low-level condensate branch pipes to a high-level condense main. This may be achieved by means of an automatic pump, as shown in Fig. 7.18. Figure 7.19 shows the method of installing the pump. The pump operates as follows:

1. Condensate enters the pump and forces air out of the chamber through the exhaust pipe.
2. The float rises to the top of the chamber and at this point closes exhaust valve A and opens the steam valve B.
3. Steam enters and forces the water out of the chamber to the main condensate pipe. On filling the chamber, check valve C is opened and D is closed. On emptying, check valve C is closed and D is opened.

Steam tables

These **are** used to find the properties at various pressures. Table 7.2 shows part of the steam tables, showing the values that may be obtained from them. A complete table may be obtained from the manufacturers of steam equipment. It will

Table 7.2   Steam tables

Gauge pressure (kPa)	Absolute pressure (kPa)	Temperature (°C)	Specific enthalpy			Specific volume
			Water	Evaporation	Steam	
			Sensible heat (kJ/kg)	Latent heat (kJ/kg)	Total heat (kJ/kg)	Steam (m³/kg)
–	5.0	32.88	137.82	2423.7	2561.5	28.192
–	10.0	45.81	191.83	2392.8	2584.7	14.674
–	15.0	53.97	225.94	2373.1	2599.1	10.022
–	20.0	60.06	251.40	2358.3	2609.7	7.649
–	25.0	64.97	271.93	2346.3	2618.2	6.204
–	30.0	69.10	289.23	2336.1	2625.3	5.229
–	35.0	72.70	304.30	2327.2	2631.5	4.530
–	40.0	75.87	317.58	2319.2	2636.8	3.993
–	45.0	78.70	329.67	2312.0	2641.7	3.580
–	50.0	81.33	340.49	2305.4	2645.9	3.240
–	55.0	83.72	350.54	2299.3	2649.8	2.964
–	60.0	85.94	359.86	2293.6	2653.5	2.732
–	65.0	88.01	368.54	2288.3	2656.9	2.535
–	70.0	89.95	376.70	2283.3	2660.0	2.365
–	75.0	91.78	384.39	2278.6	2663.0	2.217
–	80.0	93.50	391.66	2274.1	2665.8	2.087
–	85.0	95.14	398.57	2269.8	2668.4	1.972
–	90.0	96.71	405.15	2265.7	2670.9	1.869
–	95.0	98.20	411.43	2261.8	2673.2	1.777
–	100.0	99.63	417.46	2258.0	2675.5	1.694
0	101.3	100.00	419.04	2257.0	2676.0	1.673
5.0	106.3	101.40	424.9	2253.3	2678.2	1.601
10.0	111.3	102.66	430.2	2250.2	2680.4	1.533
15.0	116.3	103.87	435.6	2246.7	2682.3	1.471
20.0	121.3	105.10	440.8	2243.4	2684.2	1.414
25.0	126.3	106.26	445.7	2240.3	2686.0	1.361
30.0	131.3	107.39	450.4	2237.2	2687.6	1.312
35.0	136.3	108.50	455.2	2234.1	2689.3	1.268
40.0	141.3	109.55	459.7	2231.3	2691.0	1.225
45.0	146.3	110.58	464.1	2228.4	2692.5	1.186
50.0	151.3	111.61	468.3	2225.6	2693.9	1.149
55.0	156.3	112.60	472.4	2223.1	2695.5	1.115
60.0	161.3	113.56	476.4	2220.4	2696.8	1.083
65.0	166.3	114.51	480.2	2217.9	2698.1	1.051
70.0	171.3	115.40	484.1	2215.4	2699.5	1.024
75.0	176.3	116.28	487.9	2213.0	2700.9	0.997
80.0	181.3	117.14	491.6	2210.5	2702.1	0.971
85.0	186.3	117.96	495.1	2208.3	2703.4	0.946
90.0	191.3	118.80	498.9	2205.6	2704.5	0.923
95.0	196.3	119.63	502.2	2203.5	2705.7	0.901
100.0	201.3	120.42	505.6	2201.1	2706.7	0.881
105.0	206.3	121.21	508.9	2199.1	2708.0	0.860
110.0	211.3	121.96	512.2	2197.0	2709.2	0.841
115.0	216.3	122.73	515.4	2195.0	2710.4	0.823
120.0	221.3	123.46	518.7	2192.8	2711.5	0.806
125.0	226.3	124.18	521.6	2190.7	2712.3	0.788
130.0	231.3	124.90	524.6	2188.7	2713.3	0.773
135.0	236.3	125.59	527.6	2186.7	2714.3	0.757
140.0	241.3	126.28	530.5	2184.8	2715.3	0.743
145.0	246.3	126.96	533.3	2182.9	2716.2	0.728
150.0	251.3	127.62	536.1	2181.0	2717.1	0.714
155.0	256.3	128.26	538.9	2179.1	2718.0	0.701

Table 7.2 – *continued*

Gauge pressure (kPa)	Absolute pressure (kPa)	Temperature (°C)	Specific enthalpy Water Sensible heat (kJ/kg)	Specific enthalpy Evaporation Latent heat (kJ/kg)	Specific enthalpy Steam Total heat (kJ/kg)	Specific volume Steam (m³/kg)
160.0	261.3	128.89	541.6	2177.3	2718.9	0.689
165.0	266.3	129.51	544.4	2175.5	2719.9	0.677
170.0	271.3	130.13	547.1	2173.7	2720.8	0.665
175.0	276.3	130.75	549.7	2171.9	2721.6	0.654
180.0	281.3	131.37	552.3	2170.1	2722.4	0.643
185.0	286.3	131.96	554.8	2168.3	2723.1	0.632
190.0	291.3	132.54	557.3	2166.7	2724.0	0.622
195.0	296.3	133.13	559.8	2165.0	2724.8	0.612
200.0	301.3	133.69	562.2	2163.3	2725.5	0.603
205.0	306.3	134.25	564.6	2161.7	2726.3	0.594
210.0	311.3	134.82	567.0	2160.1	2727.1	0.585
215.0	316.3	135.36	569.4	2158.5	2727.9	0.576
220.0	321.3	135.88	571.7	2156.9	2728.6	0.568
225.0	326.3	136.43	574.0	2155.3	2729.3	0.560
230.0	331.3	136.98	576.3	2153.7	2730.0	0.552
235.0	336.3	137.50	578.5	2152.2	2730.7	0.544
240.0	341.3	138.01	580.7	2150.7	2731.4	0.536
245.0	346.3	138.53	582.8	2149.2	2732.0	0.529
250.0	351.3	139.02	585.0	2147.6	2732.6	0.522
255.0	356.3	139.52	586.9	2146.3	2733.2	0.515

Produced by permission of Spirax–Sarco Ltd, Charlton House, Cheltenham.

be clear from studying the table that the pressure of steam increases as the temperature increases and also specific latent heat increases as the pressure decreases. To find mass flow rate of steam passing through a pipe, the heat load is divided by the specific latent heat of steam.

Example 7.1. *Calculate the mass flow rate through a steam main, when the heat load is 2000 kW and the steam pressure is 100 kPa gauge.*

From the tables:

The specific latent heat of steam at 100 kPa gauge is 2201 kJ/kg

and the mass flow rate will be:

2000 ÷ 2201 = 0.9086 kg/s

In terms of volume, this mass flow rate will be:

0.9086 x 0.881 m³/kg = 0.8 m³/s

Feed water

The water used will require treatment to prevent corrosion and scaling of the boiler, pipework and equipment. A base exchange softening plant may be required, but the water will require partial re-hardening to prevent acidity which would otherwise cause corrosion.

Pipework

All steam and condensate pipes should be suitably insulated to prevent loss of heat. The pipework will require expansion joints, or loops, to relieve stresses due to expansion and contraction. All pipework must be provided with a fall of about 1 in 300, and at low points where condensate may be retained a drainage point must be fitted with a suitable steam trap.

District heating

District heating is an extension of the provision of space heating and hot water supply from a central boiler plant for one building, to the provision of these services from a central boiler plant for large development schemes, which may incorporate dwellings, offices, schools, factories and public buildings.

The term 'district' is a general term and covers the following types of heat distribution systems:

Block: a heat distribution system to a 'block' of similar buildings, such as a housing estate, an industrial estate, a shopping centre and offices.

Group: a heat distribution system to a group of the type of buildings described for the block system.

District: a heat distribution system for an entire city or town.

Figure 7.20 shows the distribution system for a group heating scheme.

The central station should, as far as possible, be sited near to the centre of the buildings being supplied with heat. This will reduce the frictional losses on the mains and also provide a better balanced distribution system. Although district heating can improve the environmental conditions in which people live and work, and also save fuel, the idea has not received the same consideration in this country as it has in many other countries. This is probably due to our temperate climate, plentiful supplies of cheap fuel and the high capital cost of a district heating installation.

With the development of new towns in Britain, there is now much greater interest being shown in district heating and there are now many schemes in this country. Nottingham has a scheme which uses combustible refuse to reduce the use of coal. Pimlico and Bankside of London have thermo-electric schemes which use waste heat from the steam turbines at Battersea and Bankside electric power stations. There are other schemes in London, and also at Billingham, Manchester, Sunderland, Oldham, Leicester and Paisley, to name just a few.

Bretton, a new town development by Peterborough Corporation, has a district heating scheme serving about 5000 homes, schools, factories, offices, library, cafés and health and social centres. The scheme uses natural gas and the system operates with low temperature hot water, at 83 °C during summer and 95 °C during winter. The system is pressurised by applying nitrogen gas to the water in a pressure vessel.

The USA pioneered district heating and there are schemes in New York, Washington, Detroit, Philadelphia, Cleveland, Chicago and many other cities. In Europe, Manchester installed the first district heating scheme and there are now schemes in Germany, Norway, Denmark, Sweden, Holland, Belgium, France and Italy.

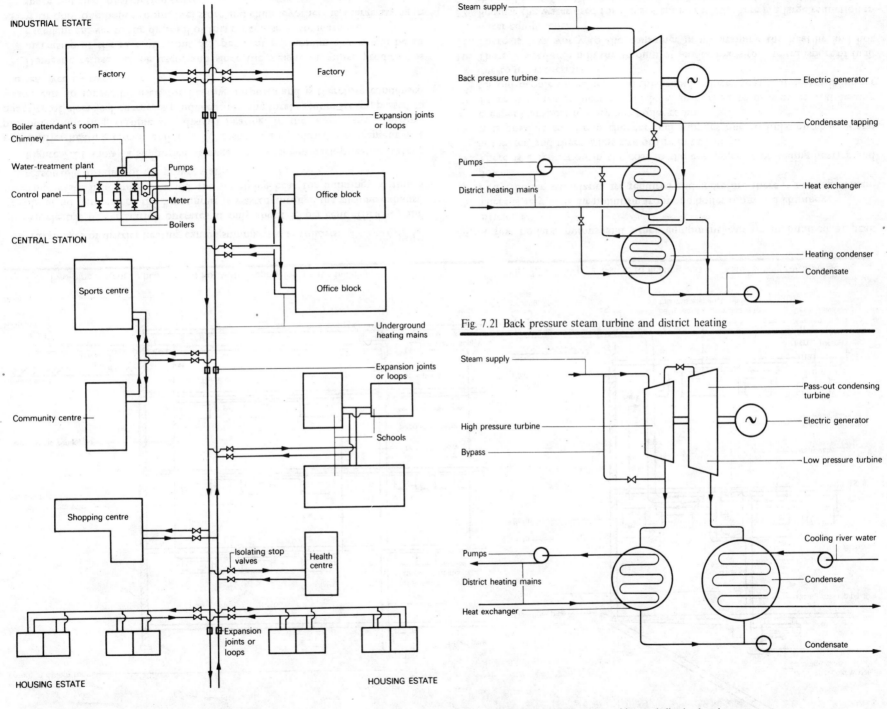

INDUSTRIAL ESTATE

Factory

Factory

Expansion joints
or loops

Boiler attendant's room
Chimney

Water-treatment plant

Pumps

Control panel

Meter

Boilers

CENTRAL STATION

Office block

Office block

Sports centre

Underground
heating mains

Community centre

Expansion joints
or loops

Schools

Shopping centre

Isolating stop
valves

Health
centre

Expansion
joints or
loops

HOUSING ESTATE

HOUSING ESTATE

Fig. 7.20 Group heating scheme

Steam supply

Back pressure turbine

Electric generator

Condensate tapping

Pumps

District heating mains

Heat exchanger

Heating condenser

Condensate

Fig. 7.21 Back pressure steam turbine and district heating

Steam supply

Pass-out condensing
turbine

High pressure turbine

Electric generator

Bypass

Low pressure turbine

Cooling river water

Pumps

District heating mains

Condenser

Heat exchanger

Condensate

Fig. 7.22 Pass-out condensing turbine and district heating

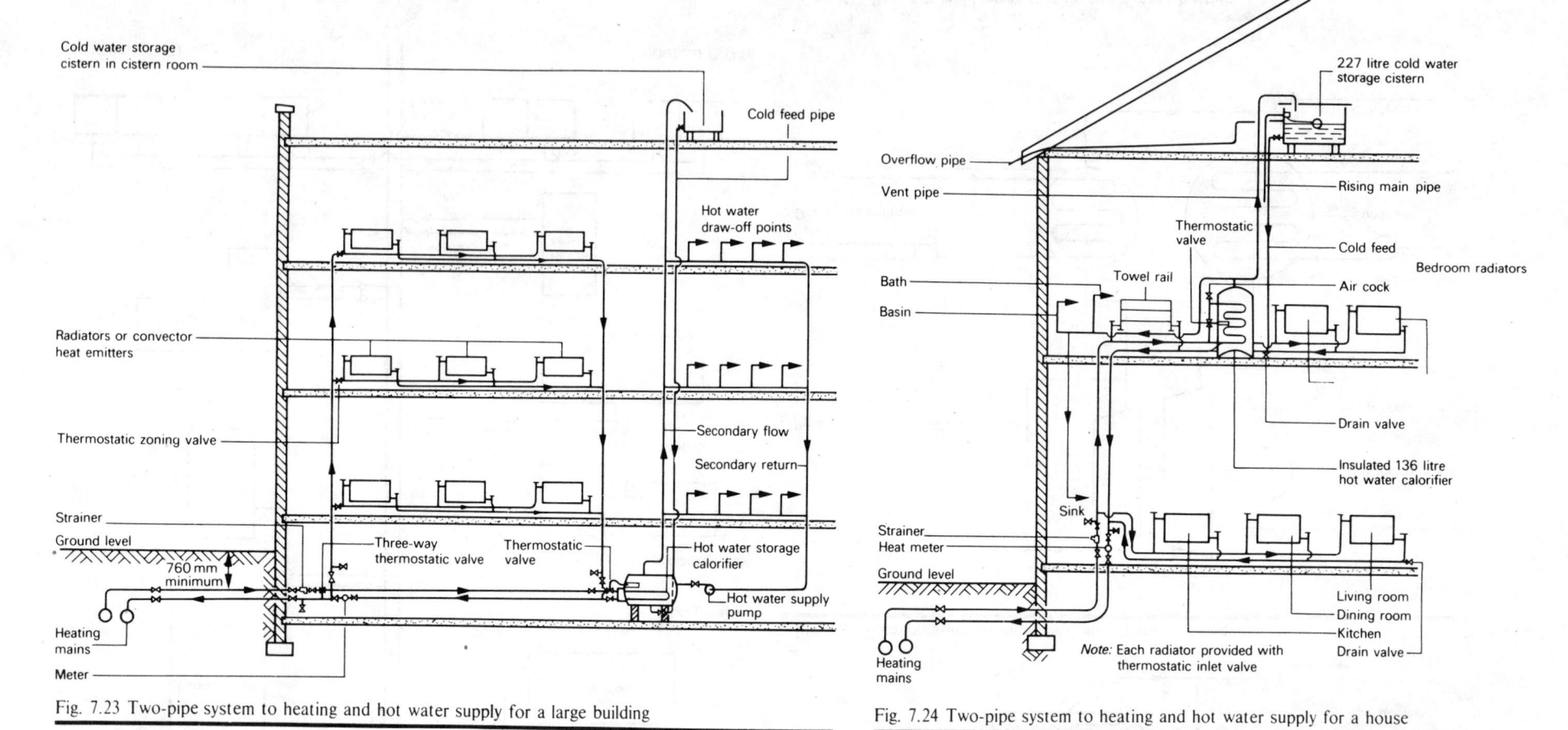

Fig. 7.23 Two-pipe system to heating and hot water supply for a large building

Fig. 7.24 Two-pipe system to heating and hot water supply for a house

Advantages of district heating can be summarised as follows:

1. An electric power station operates at only about 35 per cent efficiency and about 60 per cent of the heat input is wasted through the turbine exhaust. This waste heat may be used for supplying heat for buildings, within a reasonable distance of the power station.

Figure 7.21 shows a simplified arrangement of steam back-pressure turbine and district heating circuit and Fig. 7.22 shows a simplified arrangement of a pass-out condensing turbine and district heating. In the latter type, a greater quantity of electrical energy is produced, but the total vacuum steam, passed to waste, can no longer be used for heating purposes and is therefore condensed and removed by river water.

2. Domestic refuse can be used, and since the weight of refuse produced is increasing at the rate of about 3.5 per cent per annum this would be an excellent answer to the disposal of this otherwise waste material.

3. With only one boiler room, fuel store and chimney, there is a large saving in space and also construction costs.

4. A few boilers, pumps and other equipment can be maintained at peak efficiency.

5. There is very little air pollution from one boiler room and chimney.

6. Fewer boiler attendants are required and therefore there is a saving in manpower.

7. There is a reduction in the cost to the consumer of providing heating and hot water, and there is less risk of fire in buildings.

8. It is possible to obtain cheaper fuel supplies due to bulk purchasing and cheaper transport to only one boiler room.

9. There is a greater incentive for local authorities to provide central heating for old people's dwellings and this would reduce the risk of deaths in winter due to hyperthermia.

10. There is a reduction in the amount of service roads for the transport of fuel.

11. There is less worry to the consumer in maintaining the heating and hot water equipment.

12. Because the water used for the system is treated, there is a large reduction in the problem of scaling and rusting of boilers, heat emitters and pipework.

13. There are no house or factory chimneys to spoil the landscape. The only chimney is at the central station and this may be hidden from view by the discreet siting, or by installing the chimney, in a nearby office block.

Heat distribution

In this country the medium used for the supply of heat is hot water pumped through a distribution main (some countries, including the USA, use steam as the heat medium). The water is usually at high or medium pressure of 300 to 1000 kPa and a temperature of 120° to 150 °C. Low-pressure systems operating at about 150 kPa and 80° to 100 °C have also been used.

The hot water is circulated through a number of insulated mains, each of which serves an individual consumer, or a sector of the district. Industrial and other large consumers take their heating requirements direct from the mains and storage type calorifiers are installed, which are thermostatically controlled, to provide hot water at 60 °C. Domestic and smaller consumers may also use this arrangement, or alternatively, both the heating and hot water supply may be provided indirectly from thermostatically controlled calorifiers.

There are three types of heat distribution systems, namely:

Two-pipe: this is the most common type of system and the heating mains serve both space heating and hot water supply. The power used for the pumps may be reduced during the summer months by using variable speed drives, or by using smaller pumps than are used in the winter months.

Figure 7.23 shows a two-pipe heat distribution system for a large building, such as a hospital, office block, factory or a university.

Figure 7.24 shows a two-pipe system for a small house.

Three-pipe system: this has both a small diameter and a large diameter flow main and a common large diameter return main. During summer, when only hot water is required, the small diameter flow pipe and the large diameter return pipe are used, which will reduce the pumping and heat losses on the large diameter flow main, which is only used during winter.

Four-pipe system: this system has both heating and hot water flow and return mains, which save the cost of individual hot water calorifiers for each building, but increase the cost of the distribution mains. This system is often used for a block scheme.

Figures 7.25 and 7.26 show schematic diagrams of the three-pipe and four-pipe distribution systems respectively, and Fig. 7.27 shows a plan of a central plant station for a block scheme, using the four-pipe distribution system.

Figure 7.28 shows a section through an oil tank storage room and the method of installing the oil tank. The boilers may be fired by solid fuel or natural gas.

Figure 7.29 shows the method used to supply the boilers with solid fuel and Fig. 7.30 shows a section through an 'economic' boiler, which is often used for district heating.

Charges for heat supply

The consumers may be charged by the amount of heat passing through a heat meter installed inside the building, or by a flat rate charge, based usually upon the number of occupants, or the number and size of rooms in the building. It may be necessary to meter the heat supply to large consumers, such as offices, factories and schools, and charge a flat rate for domestic consumers.

The charge for heat made by the amount passing through a meter has the advantage of providing a means of using the heat more economically. Some domestic consumers may require less heat, due to being out at work during the day, or requiring low air temperatures. Both domestic and large consumers are encouraged to economise in the use of heat when a meter is installed. Meters, however, add to the capital and maintenance cost of the installation and also require reading, which increases the administrative costs. It is also sometimes difficult to gain entry to dwellings when people may be out at work, but this problem may be solved by installing the meter in a locked cupboard, in which the door can be opened by the meter reader and the readings taken from outside the building. Meters should always be installed in the central station, to record the heating output from the boilers.

Distribution mains

The success of a district heating system depends largely upon the correct installation of the heating mains. It is essential that adequate provision is made for expansion and contraction of the pipework, and the pipe joints should be thoroughly made and tested. The type of ductwork for the pipes and method of thermal insulation are important considerations. It is also essential that the pipe insulation is kept dry, or otherwise its efficiency would be seriously impaired and therefore, if possible, the pipe duct should be laid above the ground water table and drainage below the duct pipes should be laid below the duct. These drainage pipes should be carried to a soakaway or a pumping sump.

Various methods are used for insulating the pipes against heat losses and the type used will depend upon the size of the heating mains, cost and speed of laying, accessibility and type of ground.

Aerated concrete-filled duct

A pre-cast concrete duct is placed in the excavated trench and the heating pipes are then placed on to concrete blocks, which rest on the base of the concrete duct. After welding and testing, the pipes are wrapped with corrugated cardboard, which prevents bonding between the pipes and the insulation. The duct is then filled with aerated concrete slurry, which surrounds the heating pipes. The insulation is completed by placing a pre-cast concrete cover over the duct and backfilling the trench.

Figure 7.31 shows a cross section through the concrete duct and the method of insulating the pipes.

Arched duct

When the heating mains are to be laid below roads where there is a possibility of heavy traffic, an arched duct will provide greater strength than the rectangular duct. The duct is easily laid and instead of filling with cellular concrete the pipes are insulated with glass fibre or asbestos.

Figure 7.32 shows the arched conduit method, with the heating pipes supported on rollers and guides to allow for expansion and contraction.

Conduit casing

Since ground water is one of the main problems of laying the underground

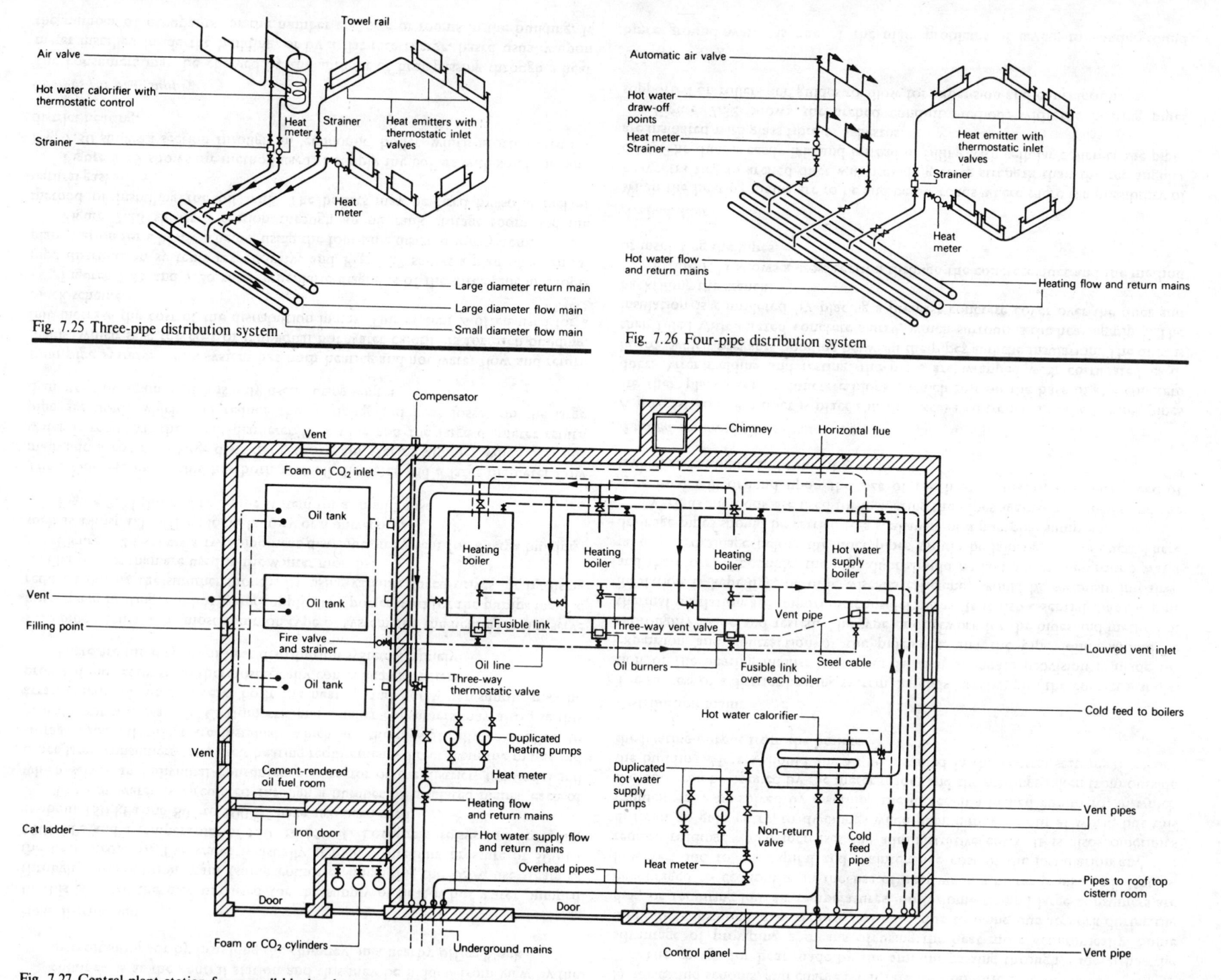

Fig. 7.25 Three-pipe distribution system

Fig. 7.26 Four-pipe distribution system

Fig. 7.27 Central plant station for a small block scheme

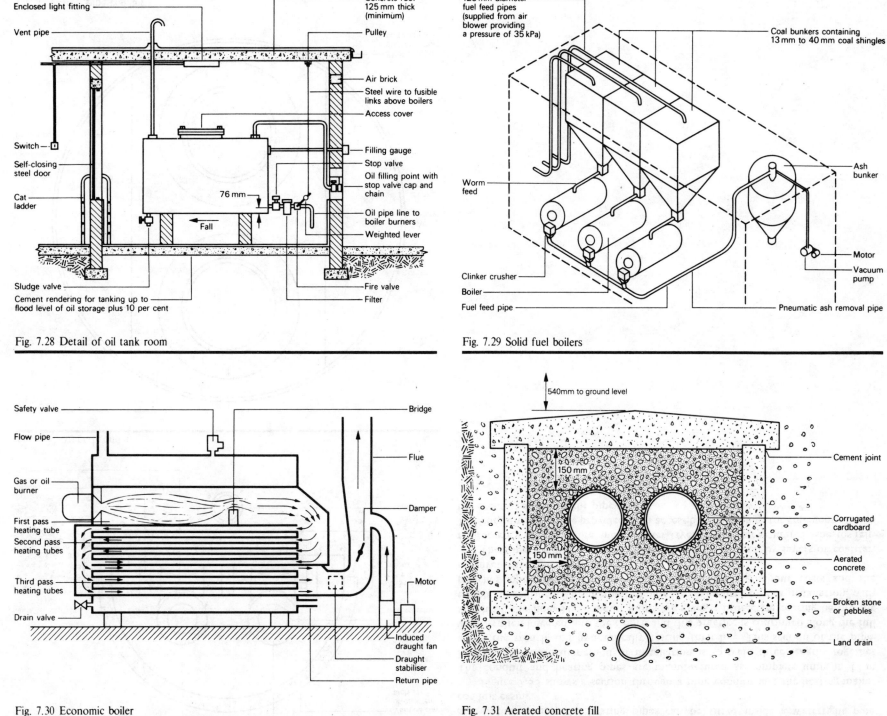

Fig. 7.28 Detail of oil tank room

Enclosed light fitting

Vent pipe

Switch

Self-closing steel door

Cat ladder

Sludge valve

Cement rendering for tanking up to flood level of oil storage plus 10 per cent

Concrete roof 125 mm thick (minimum)

Pulley

Air brick

Steel wire to fusible links above boilers

Access cover

Filling gauge

Stop valve

Oil filling point with stop valve cap and chain

Oil pipe line to boiler burners

Weighted lever

Fire valve

Filter

76 mm

Fall

Fig. 7.29 Solid fuel boilers

125 mm diameter fuel feed pipes (supplied from air blower providing a pressure of 35 kPa)

Coal bunkers containing 13 mm to 40 mm coal shingles

Worm feed

Clinker crusher

Boiler

Fuel feed pipe

Ash bunker

Motor

Vacuum pump

Pneumatic ash removal pipe

Fig. 7.30 Economic boiler

Safety valve

Flow pipe

Gas or oil burner

First pass heating tube

Second pass heating tubes

Third pass heating tubes

Drain valve

Bridge

Flue

Damper

Motor

Induced draught fan

Draught stabiliser

Return pipe

Fig. 7.31 Aerated concrete fill

540mm to ground level

150 mm

150 mm

Cement joint

Corrugated cardboard

Aerated concrete

Broken stone or pebbles

Land drain

77

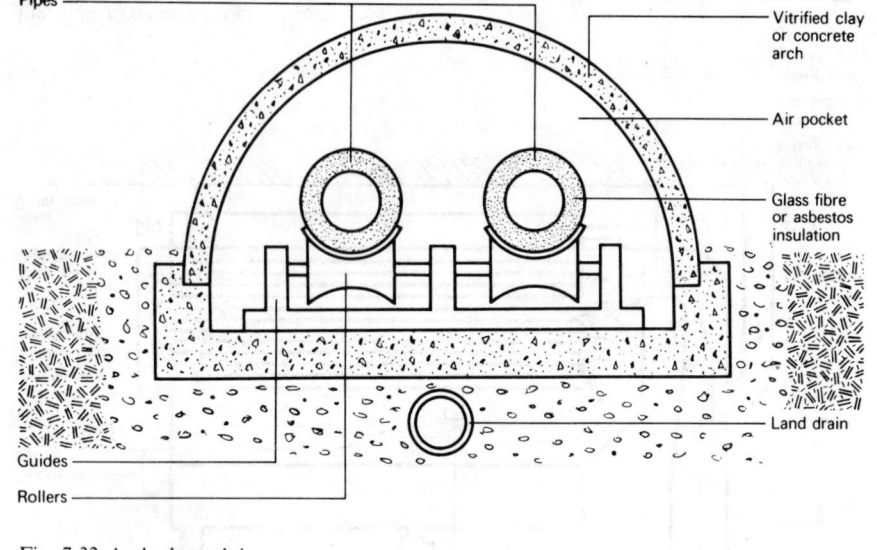

Pipes

Vitrified clay
or concrete
arch

Air pocket

Glass fibre
or asbestos
insulation

Land drain

Guides

Rollers

Fig. 7.32 Arched conduit

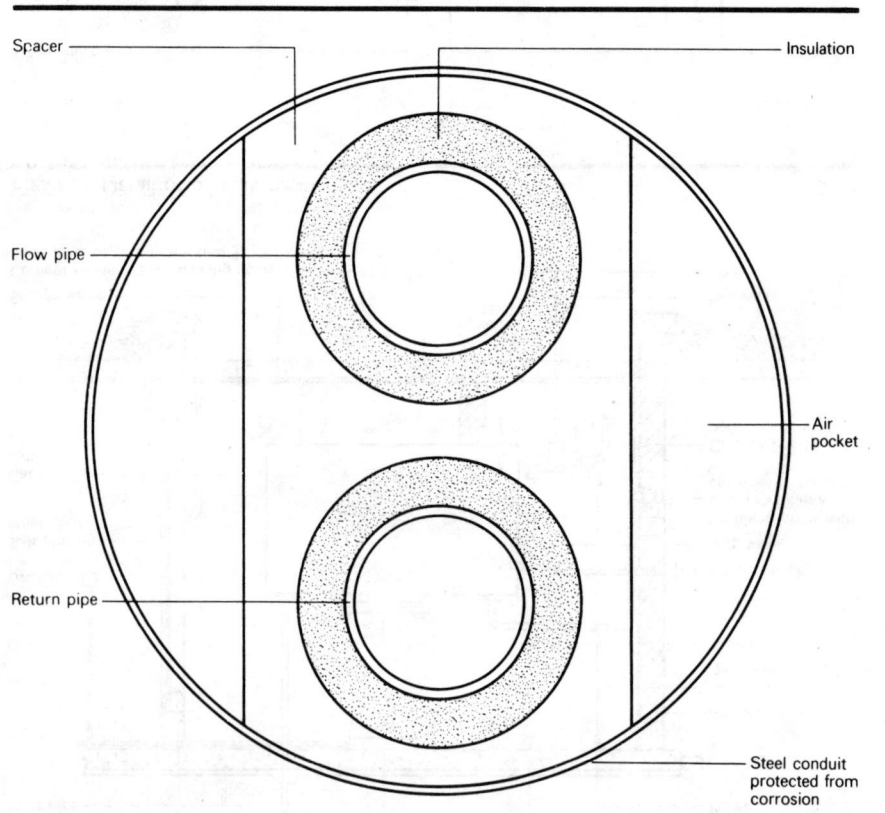

Spacer

Insulation

Flow pipe

Air
pocket

Return pipe

Steel conduit
protected from
corrosion

Fig. 7.33 Steel pipe conduit

distribution mains, the heating pipes can be fitted inside a watertight pipe conduit casing.

Figure 7.33 shows a section through a pipe conduit and the heating mains. The conduit and heating pipes are manufactured as complete units in 12 m lengths. The normal method of installation is to weld three lengths together above the trench and hydraulically testing their line. This section of the pipework is then lowered into the trench and this practice is continued along the full length of the line. The conduit joints are given a protective coating before the trench is backfilled. This method of pipe laying ensures a quick installation, with trenches open for minimum periods, thus reducing labour costs and the inconvenience of open trenches.

The pipe conduit casing is protected from corrosion by layers of coal tar, reinforced with glass fibre and an outer layer of coal tar-saturated asbestos felt. The conduit should be provided with accessible drain plugs for emergency use in the event of a leak on the pipework.

Chapter 8

Drainage below ground

Principles of drainage below ground

The efficient disposal of foul and surface water from a building is of great importance to public health and is an essential part of the construction of a building. If a drain is unsound and leaks, the escaping water may contaminate the water supply or air. The escaping water may also wash away the soil below the foundation and cause a risk of settlement of part of the building. The Building Regulations 1985 deals with drainage below ground and the student is strongly advised to become familiar with the various requirements.

Terms used

The following definitions are laid down by section 343 of the Public Health Act 1936:

Drain
Means a drain used for the drainage of one building, or of any buildings or yards, appurtenant to buildings within the same curtilage.

Sewer
Does not include a drain as defined in this section but, save as aforesaid, includes all sewers and drains used for the drainage of buildings and yard appurtenant to buildings.

Private sewer
Means a sewer which is not a public sewer. In general terms, drain, sewer and private sewer may be described as follows:

Drain: a system of pipes used for the drainage of one or more buildings within a private boundary. The owner of the building, or buildings, is responsible for the maintenance of a drain.

Private sewer: a system of pipes used for the drainage of two or more buildings belonging to separate owners. The pipes are inside a private boundary and their maintenance is shared jointly by the separate owners of the buildings.

Public sewer: a system of pipes belonging to the local authority and maintained by the authority. The pipes are outside the private boundary, usually under the roadway. It is the duty of every local authority to provide public sewers and the provision, by means of sewage disposal works, for the effectual disposal of the contents of public sewers. If the public sewer is within 30 m of a building site, the local authority may require the building drainage to connect to the public sewer.

Figure 8.1 shows a drainage system for three houses having separate owners, with each house connected to the public sewer by means of a private sewer, which may be run either at the back or front of the houses. Figure 8.2 shows how the houses may be connected to the public sewer by an individual drain from each house.

The first method has the advantage of reducing the number of connections required to the public sewer in the roadway and also the reduction in pipe length and number of manholes. The second method has the advantage of preventing a dispute in the event of a repair or blockage. If, for example, a blockage occurred in the private sewer at A, the owner of the house nearest to the public sewer may refuse to pay towards the work of clearing the blockage because he is not affected. The shared responsibility of maintenance of the private sewer is often included in the title deeds of each house, but this does not always prevent a dispute occurring. The owner will be eventually compelled to pay towards the cost of the maintenance of the private sewer, but a good deal of ill-feeling between the neighbours will have been caused. If a blockage occurs in the public sewer, the local authority would be responsible for its clearance.

Systems of drainage

The type of system used depends upon the local authority regulation. There are three types:

Separate system: in which the foul water discharges from W.C.s, basins, sinks, baths, etc., are conveyed by foul water drains to a foul water sewer, or private sewage disposal plant; the rainwater or surface water from roofs and paved areas are conveyed by surface water drains to a public surface water sewer or soakaway. Figure 8.3 shows a separate system for a medium-sized building.

Combined system: in which the foul water from sanitary appliances and surface water from roofs and paved areas are conveyed by a single drain to a combined sewer. The system saves on drainage costs, but the cost of sewage disposal is increased. Figure 8.4 shows a combined system for the same medium-sized building.

Partially separate system: in which most of the surface water is conveyed by a

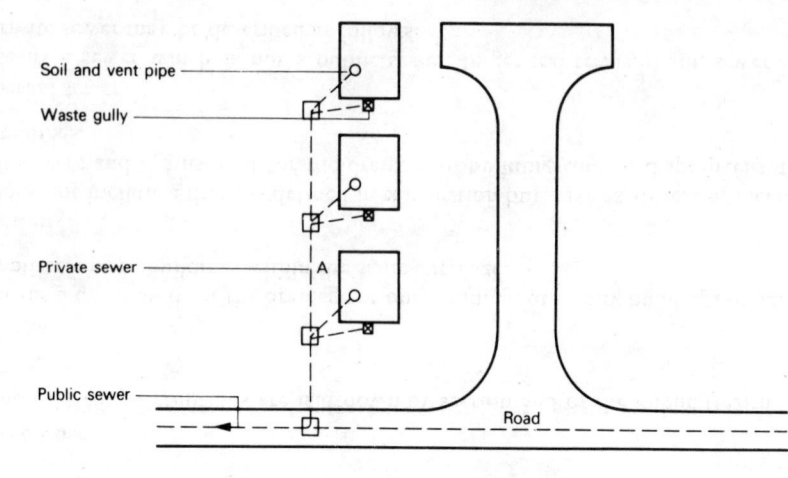

Soil and vent pipe

Waste gully

Private sewer — A

Public sewer

Road

Fig. 8.1 Use of private sewer

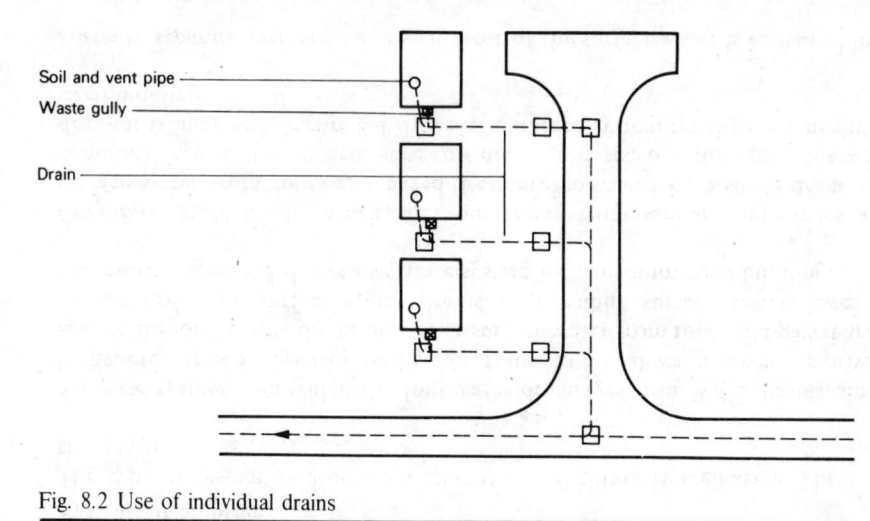

Soil and vent pipe

Waste gully

Drain

Fig. 8.2 Use of individual drains

- - - Foul water drain   ▨ Trapless surface water gully

-·- Surface water drain   ○ Rodding eye

▨ Trapped waste gully   ▢ Foul water manhole

○ Soil and vent pipe   ◯ Surface water manhole

⬡▢ Rainwater shoe

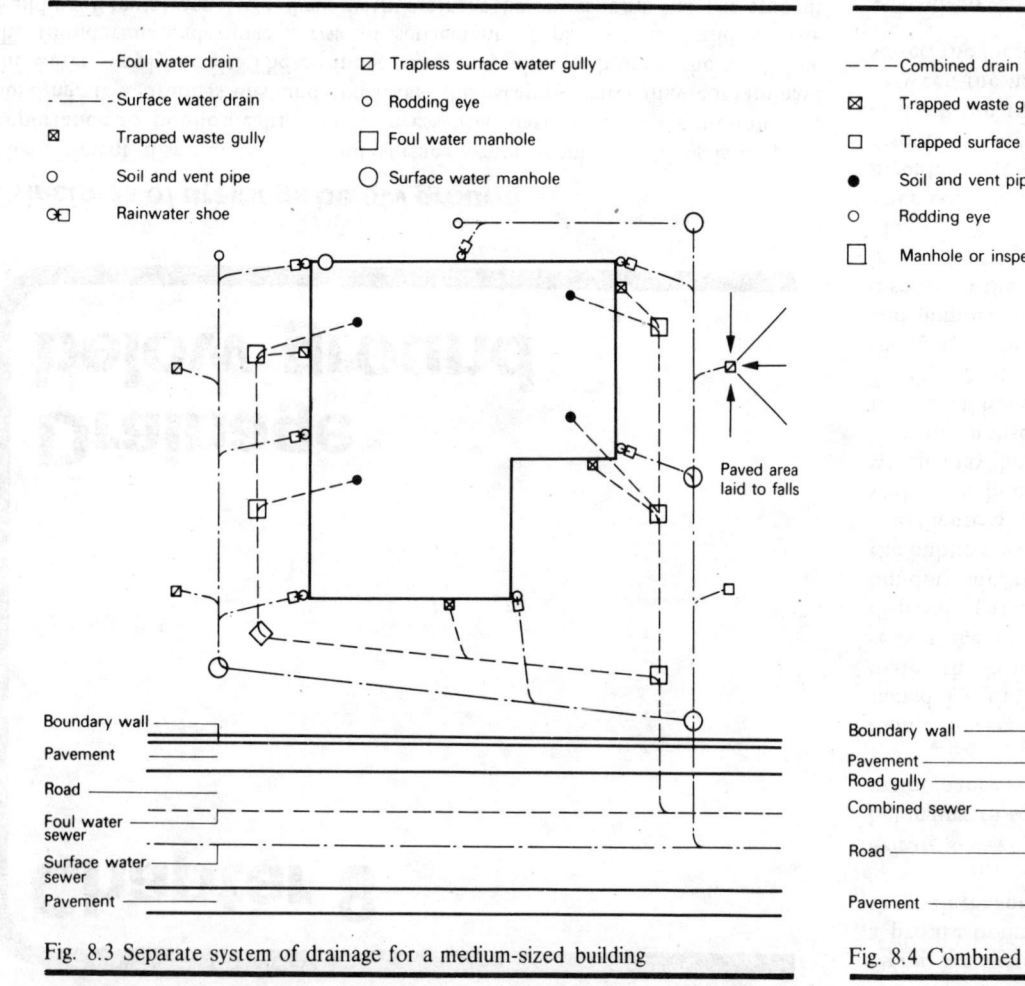

Paved area laid to falls

Boundary wall

Pavement

Road

Foul water sewer

Surface water sewer

Pavement

Fig. 8.3 Separate system of drainage for a medium-sized building

- - Combined drain

▨ Trapped waste gully

▢ Trapped surface water gully

● Soil and vent pipe

○ Rodding eye

▢ Manhole or inspection chamber

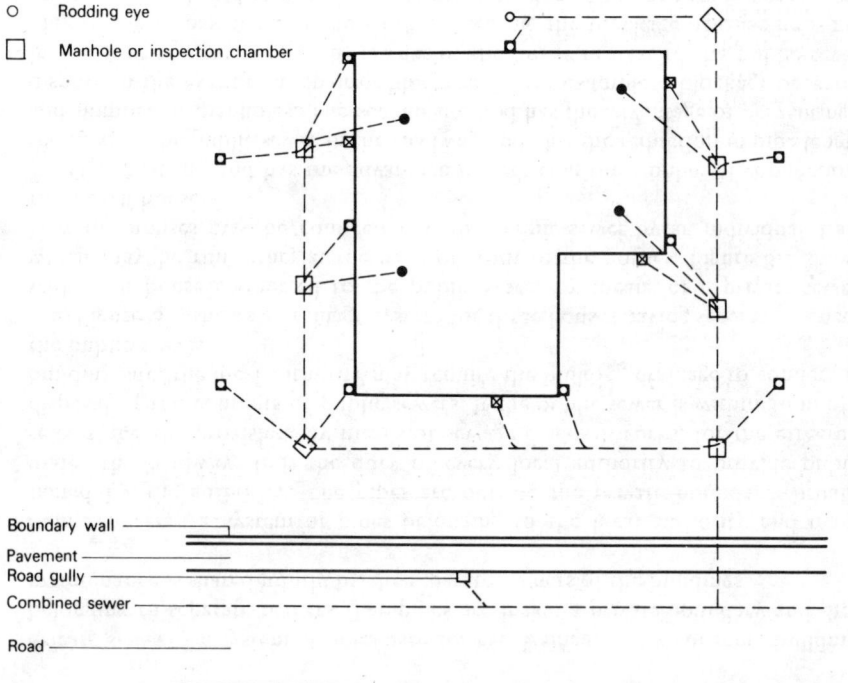

Boundary wall

Pavement

Road gully

Combined sewer

Road

Pavement

Fig. 8.4 Combined system of drainage for a medium-sized building

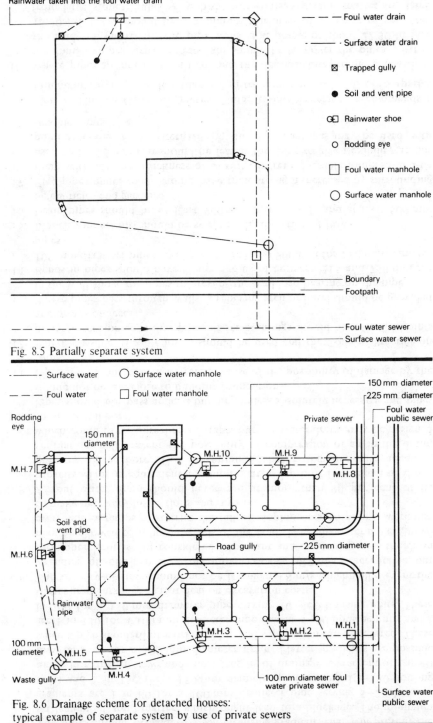

Rainwater taken into the foul water drain

— — Foul water drain

— · Surface water drain

⊠ Trapped gully

● Soil and vent pipe

Œ Rainwater shoe

○ Rodding eye

□ Foul water manhole

○ Surface water manhole

Boundary
Footpath

Foul water sewer
Surface water sewer

Fig. 8.5 Partially separate system

Surface water   ○ Surface water manhole

Foul water   □ Foul water manhole

Rodding eye

150 mm diameter

M.H.7

150 mm diameter

225 mm diameter

Foul water public sewer

Private sewer

M.H.10   M.H.9

M.H.8

Soil and vent pipe

M.H.6

Road gully   225 mm diameter

Rainwater pipe

100 mm diameter

M.H.5

M.H.3   M.H.2   M.H.1

M.H.4

Waste gully

100 mm diameter foul water private sewer

Surface water public sewer

Fig. 8.6 Drainage scheme for detached houses:
typical example of separate system by use of private sewers

surface water drain to a surface water sewer or soakaway, but some of the rainwater is connected to the foul water drain. This is done when the rainwater can be conveniently connected to the foul water drain, usually at the rear of the building.

This arrangement, when compared to the separate system, saves on drainage costs and the rainwater will also flush the foul water drain. Figure 8.5 shows a partially separate system for a detached house and Fig. 8.6 shows a drainage scheme for a small private development of detached houses. Table 8.1 gives a comparison of the separate and combined systems.

Table 8.1   Summary of the comparisons between the separate and combined drainage system

Separate system	Combined system
1. Two sets of drains; increases the cost of building drainage	1. Only one drain for both foul and surface water; reduces the cost of building drainage
2. There is a risk of a wrong connection, e.g., a foul water branch drain may be wrongly connected to a surface water drain	2. There is no risk of making a wrong connection
3. The foul water drain is not thoroughly flushed by rainwater	3. Foul water is flushed through the drain by the surface water
4. There is no risk of foul air passing through an unsealed rainwater gully trap	4. The loss of a trap seal in a rainwater gully allows the foul gas from the drain to pass into the open air around the building
5. The size of the sewage disposal plant is much smaller	5. The size of the sewage disposal plant is greater
6. The cost of sewage purification is less	6. The cost of the sewage disposal is greater
7. If the sewage is pumped to the sewage disposal works there is a reduction in the cost of pumping. The surface water may flow by gravity to a nearby river	7. Possibly greater pumping costs, due to both surface water and foul water to have to reach the sewage disposal works

The following points should be considered in the design of drainage systems.

1. The layout of the system should be as simple and direct as possible and the number of bends, traps and manholes kept to a minimum.
2. The pipes should be laid in straight lines, from point to point.
3. The pipes should be non-absorbent, durable, smooth in bore and of adequate strength.
4. The pipes should be adequately supported without restricting movement.
5. Foul water drains should be well ventilated, to prevent the accumulation of foul gases and fluctuation of air pressure within the pipe, which could lead to the unsealing of gully or W.C. traps.
6. All the parts of the drainage system should be accessible for inspection and cleaning.

7. The pipes should be laid to a self-cleansing gradient, that will prevent the settlement of solid matter, which might lead to a blockage. The minimum gradients are 1 in 80 for a 100 mm diameter pipe serving 5—20 housing units and 1 in 150 for a 150 mm diameter pipe serving 10—150 housing units. For smaller flows and short lengths of drainage, a gradient of 1 in 40 may be used for a 100 mm diameter drain. Flatter gradients are possible when a high standard of workmanship and supervision can be assured. These are 1 in 130 for a 100 mm diameter pipe serving 5—20 housing units and 1 in 200 for a 150 mm diameter pipe serving 10—150 housing units. These flatter gradients save a great deal on excavation costs.

8. The velocity of flow should not be less than 0.8 m/s which will prevent the stranding of solid matter. A maximum velocity of 2 m/s is acceptable, but the upper limit is not considered important and on sloping sites the drain may be allowed to follow the fall of the land. Large diameter sewers, however, require ramps to restrict the fall on sloping sites, to permit workmen to make necessary inspections and repairs.

9. A foul water drain should never run at more than 90 per cent of its capacity. This is equivalent to running at a depth of flow equal to three-quarters of the bore. This maximum discharge, together with adequate ventilation, will prevent the possibility of compression of air in the drain, which could cause unsealing of traps. Surface water drains may be designed to run at full bore.

10. Pipes should not pass under a building unless absolutely necessary and pipes should not be laid close to building foundations.

11. Pipes should not pass near trees because of the possibility of damage by the roots.

12. Where possible, flexible joints should be used and the Code of Practice on drainage recommends that pipes under buildings should have flexible joints and means of access.

13. Where pipes pass through walls, a relieving arch or lintel should be provided in the wall above the pipes to prevent the wall load bearing on the pipe.

14. Bends in pipes should have a large radius of between 215 and 750 mm for 100 mm diameter pipes and between 225 and 900 mm for 150 mm diameter pipes.

15. Branch connections should be swept in the direction of flow.

16. Drain pipes should be at least 900 mm below roads and at least 600 mm below fields and gardens.

17. Clay pipes under roads should have their strength increased by surrounding them with 150 mm thickness of *in situ* concrete. Flexible joints should be used and a 25 mm gap should be left at the joint to give flexibility at this point (see Fig. 8.7). Alternatively, ductile iron pipes may be used with flexible joints.

Note: The BS Code of Practice 2005, 1968, 'Sewerage', recommends a minimum cover of not less than 1.2 m for sewers under roads or footpaths.

18. Where pipes are not under a road and the depth below ground is less than 600 mm, two pre-cast concrete slabs should be laid over the pipes, so that the load transmitted to the pipe walls is at points of about 22.30 hr and 13.30 hr. There should be a minimum thickness of 150 mm of soil, free from large stones, building rubbish, tree roots, vegetable matter and large lumps of clay. Alternatively, broken stone or gravel 10 mm nominal single size may be placed between the concrete slabs and the top of the pipe.

Connections to drains

The provision of a water-sealed trap to each connection to foul water drainage is essential and the seal should be maintained under working conditions. The trap prevents the entry of foul air into the building, or into the open air. Soil, vent and waste stacks do not require traps at their bases, because the sanitary fittings connected to the stacks are provided with traps. Other connections to foul water drains such as rainwater pipes in a combined system should be provided with gully traps, as shown in Fig. 8.8.

In order to reduce the number of gully traps from rainwater connections in a combined system, a master gully trap may be used, as shown in Fig. 8.9. Each rainwater pipe is then connected to a rainwater shoe, before discharging into the master gully. This method also reduces the length of drainage and the number of connections to the inspection chamber. Figure 8.10 shows a detail of a rainwater shoe and Fig. 8.11 shows a detail of a master gully trap.

Use of intercepting traps

In the past, intercepting traps were fitted between the drain and a sewer, with the object of preventing sewer gases entering the drainage system (see Fig. 8.12). Modern drainage systems do not include an interceptor and this has led to better ventilation of the sewer, fewer blockages and saving in costs (see Fig. 8.13). The local authority, however, may require in interceptor trap where a new connection is to be made to an old sewer in poor condition, or in a district where the trap has been fitted to each drainage system. Figure 8.14 shows a section through various types of intercepting traps.

Ventilation

The ventilation of the drain is provided by means of a vent stack at the head of the foul water drain. A soil or waste stack may be used as a vent pipe, providing it is of suitable diameter. The vent stack should be terminated at least 900 mm above any window, within a horizontal distance of 3 m from the stack and a wire balloon should be fitted to the top of the pipe. Foul water drain pipes exceeding 6.4 m long are usually required to be vented.

Surface water from paved areas

When a combined system is used a yard gully having a water seal of not less than 50 mm in depth must be fitted on the surface water branch drain before it connects to the combined foul water drain (see Fig. 8.15).

When a separate system is used, a trapless yard gully may be fitted and the branch pipe from the gully may be connected directly into the surface water drain (see Fig. 8.16).

Drain diameters

Under the Building Regulations, the minimum diameters of surface water and foul water drains are 75 and 100 mm respectively. The size of a surface water drain should be sufficient to carry away the usual maximum rainfall intensity for the district, with an allowance for the impermeability of the surface. For

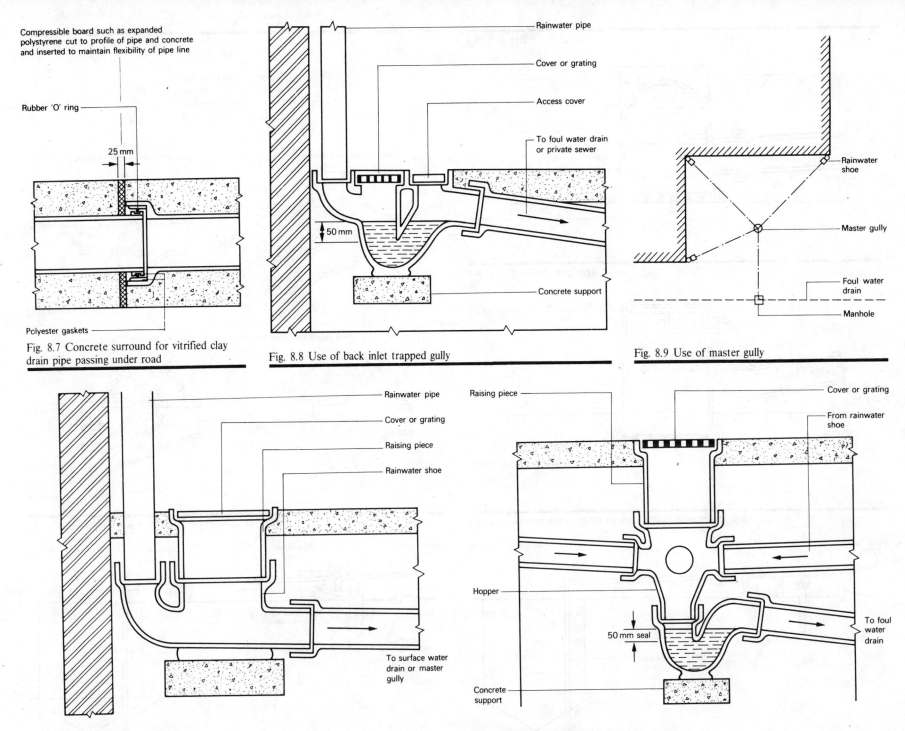

Compressible board such as expanded polystyrene cut to profile of pipe and concrete and inserted to maintain flexibility of pipe line

Rubber 'O' ring

25 mm

Polyester gaskets

Fig. 8.7 Concrete surround for vitrified clay drain pipe passing under road

Rainwater pipe

Cover or grating

Access cover

To foul water drain or private sewer

50 mm

Concrete support

Fig. 8.8 Use of back inlet trapped gully

Rainwater shoe

Master gully

Foul water drain

Manhole

Fig. 8.9 Use of master gully

Rainwater pipe

Cover or grating

Raising piece

Rainwater shoe

To surface water drain or master gully

Fig. 8.10 Use of rainwater shoe

Raising piece

Cover or grating

From rainwater shoe

Hopper

50 mm seal

Concrete support

To foul water drain

Fig. 8.11 Detail of master gully

84

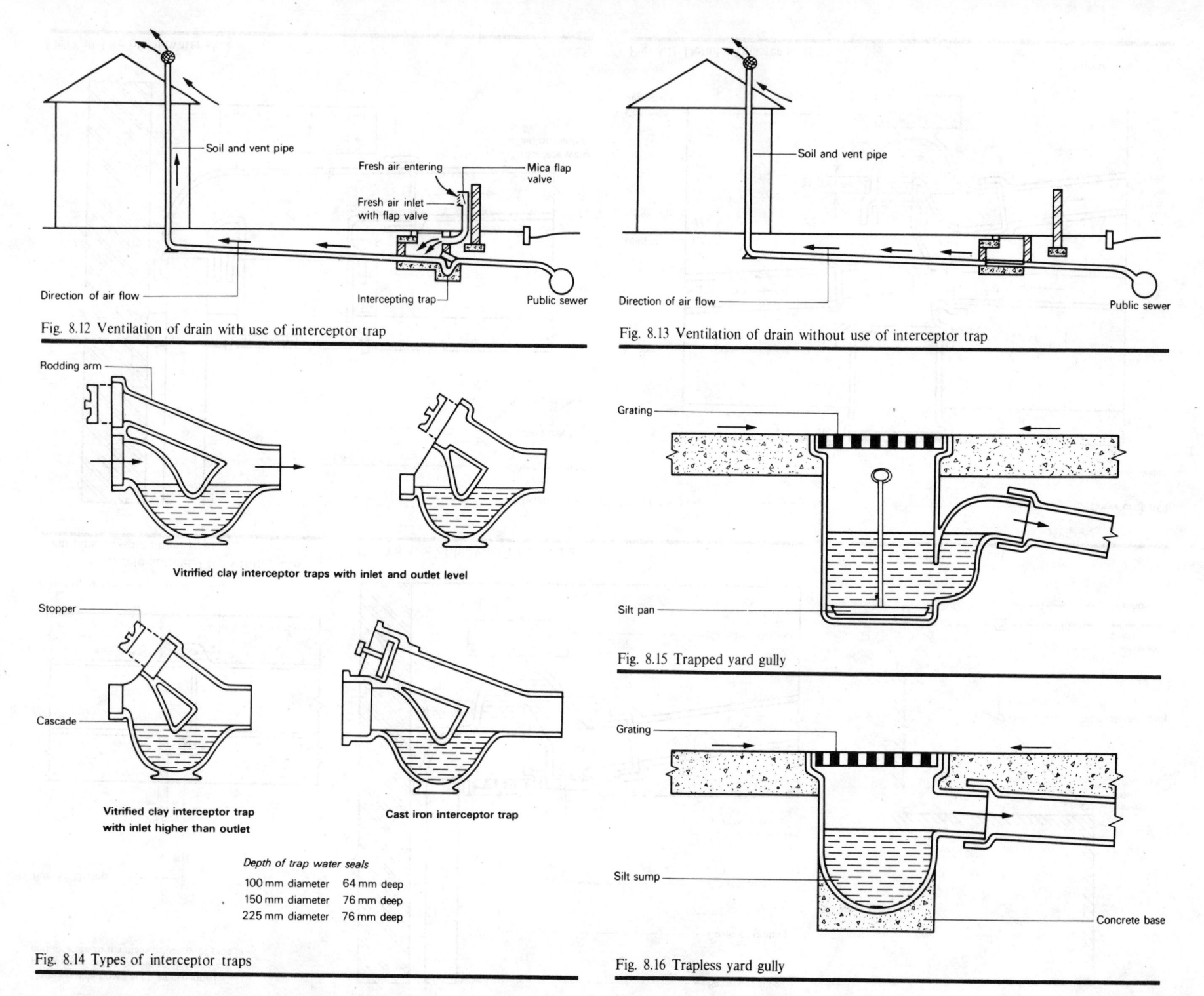

Fig. 8.12 Ventilation of drain with use of interceptor trap

Fig. 8.12 labels:
Soil and vent pipe
Fresh air entering
Mica flap valve
Fresh air inlet with flap valve
Direction of air flow
Intercepting trap
Public sewer

Fig. 8.13 Ventilation of drain without use of interceptor trap

Fig. 8.13 labels:
Soil and vent pipe
Direction of air flow
Public sewer

Fig. 8.14 labels:
Rodding arm

Vitrified clay interceptor traps with inlet and outlet level

Stopper
Cascade

Vitrified clay interceptor trap with inlet higher than outlet

Cast iron interceptor trap

Depth of trap water seals
100 mm diameter   64 mm deep
150 mm diameter   76 mm deep
225 mm diameter   76 mm deep

Fig. 8.14 Types of interceptor traps

Fig. 8.15 labels:
Grating
Silt pan

Fig. 8.15 Trapped yard gully

Fig. 8.16 labels:
Grating
Silt sump
Concrete base

Fig. 8.16 Trapless yard gully

building drainage, a rainfall intensity of 50 mm per hour is usually allowed. Table 8.5 gives the impermeability factors for the various types of surfaces.

Table 8.2

Type of surface	Impermeable factor
Watertight roof surfaces	0.70 to 0.95
Asphalt pavements in good order	0.85 to 0.90
Closely jointed wood and stone pavements	0.80 to 0.85
Macadam roadways	0.25 to 0.45
Black pavements with wide joints	0.50 to 0.70
Lawns and gardens	0.05 to 0.25
Wooded areas	0.01 to 0.20

If a drain is oversized, the depth of water will be reduced which may not then be sufficient to carry away the solids. An undersized drain, on the other hand, will lead to a surcharge and water will back up through the gully traps and cause flooding. An allowance must be considered if there is a possibility of future extension to the building. The discharge capacity of a drain depends upon the gradient at which it is laid, the diameter and smoothness of the bore and the accurancy of pipelaying.

Various tables and charts may be used to find the diameters of both foul and surface water drains. Alternatively, the diameters may be found by calculations.

Example 8.1 (surface water drain). *The total surface area of footpaths and roadways of a building development scheme is found to be 4000 m². Calculate the diameter of the main surface water drain, using the following data:*

1. Rainfall intensity 50 mm/hr
2. Impermeable factor of surface 0.9
3. Full bore discharge
4. Velocity of flow required 0.8 m/s

Using the formula

$$Q = V.A.$$

Where

Q = Volume of flow m³/s

V = Velocity of flow m/s

A = Area of pipe m²

$$Q = \frac{\text{Area to be drained (m}^2) \times \text{Rainfall intensity in (m/hr)} \times \text{Impermeable factor}}{60 \text{ (to bring to minutes)} \times 60 \text{ (to bring to seconds)}}$$

$$Q = \frac{4000 \times 0.05 \times 0.9}{60 \times 60}$$

$$Q = 0.05 \text{ m}^3/\text{s}$$

$$A = \frac{\Pi D^2}{4}$$

therefore

$$Q = \frac{V \Pi D^2}{4}$$

$$D = \sqrt{\frac{4Q}{V \Pi}}$$

$$D = \sqrt{\frac{4 \times 0.05}{0.8 \times 3.142}}$$

$$D = \sqrt{\frac{0.2}{2.514}}$$

$$D = 0.282 \text{ m}$$

$$D = 282 \text{ m}$$

Nearest pipe size = <u>300 mm diameter</u>

The diameter of a foul water drain will depend upon the discharge from the sanitary fittings during the peak demand period. The amount of water used in the building will have to be discharged to the drain and the drain must be sized, therefore, to carry away this discharge.

The following stages may be followed:

1. Establish the possible maximum number of occupants in the building.
2. Establish the consumption of water per day by the occupants.
3. From 1 and 2 find the average flow of water passing into the drain, during a 6 hour period (it may be assumed that half the daily flow will take place during a 6 hour period).
4. Find the average flow rate m³/s.
5. Find the maximum flow rate m³/s (it may be assumed that the maximum flow rate will be between four to six times the average flow rate).
6. Calculate the pipe diameter, using the formula $Q = V.A.$

Example 8.2 (foul water sewer). *Calculate the diameter of a foul water private sewer suitable for fifty houses. The average number of occupants is assumed to be four per house and the water consumption 225 litres per head, per day. The sewer is to be sized, so as to run half-full bore, at a velocity of 0.8 m/s during the peak demand period.*

Flow rate per day	= 225 × 50 × 4
Flow rate per day	= 45 000 litre
Average flow rate during a 6-hour period, assumed half the daily flow	= 22 500 litre
Average flow rate per hour	= $\dfrac{22\,500}{6}$

Average flow rate per hour $= 3750$ litre

Average flow rate per second $= \dfrac{3750}{60 \times 60}$

Average flow rate per second $= 1.041$ litre

Maximum flow rate per second assumed four times the average $= 1.041 \times 4$

Therefore maximum flow rate per second $= 4.164$ litre

Maximum flow rate per second $= 0.004$ m^3/s

$Q = V.A.$

where

Q = Volume of flow m^3/s

V = Velocity of flow m/s

A = Area of pipe m^2

Area of pipe for half-full bore $= \dfrac{\Pi D^2}{8}$

$Q = \dfrac{V \Pi D^2}{8}$

therefore

$D = \sqrt{\dfrac{8 Q}{V \Pi}}$

$D = \sqrt{\dfrac{8 \times 0.004}{0.8 \times 3.142}}$

$D = \sqrt{\dfrac{0.032}{2.514}}$

$D = 0.1128$ m

$D = 113$ mm

Nearest pipe size = 125 mm diameter

When the number and type of sanitary appliances are known, the diameter and gradient of a foul water drain may be found by a discharge unit method. Each sanitary appliance is given a discharge unit value, which represents the rate of discharge, capacity and frequency of use of the appliance. The sum of all the discharge units of the sanitary appliances connected to the drain is found and the British Standard 5572, 1978, gives a list of discharge units and the number that may be connected to 100, 125 and 150 mm nominal bore foul water drains, laid at gradients of 1 in 96, 1 in 48 and 1 in 24. Table 8.3 gives the discharge unit values and Table 8.4 the number of discharge units allowed on vertical stacks and horizontal branches or drains.

Table 8.3 Discharge unit values B.S. 5572 1978

Type of appliance	Frequency of use (minutes)	Discharge unit value
Spray tap (basin)	Add 0.06 litres per second per tap	—
W.C. (9-litre cistern)	20 10 5	7 14 28
Sink	20 10 5	6 14 27
Wash basin	20 10 5	1 3 6
Bath	75 (domestic) 30 (commercial and congested)	7 18
Shower (per head)	Add 0.1 litre per second per spray	—
Urinal (per stall or bowl)	20 (commercial and congested)	0.3
One group consisting of one WC one bath and 1 or 2 basins, sink		14
Washing machine (automatic)	250	4

Table 8.4(a) Maximum number of discharge units to be allowed on vertical stacks

Nominal internal diameter (mm)	Discharge units	
50	10	
65	60	
75	200	(not more
90	350	than 1 W.C.)
100	750	
125	2500	
150	5500	

Table 8.4(b) Maximum number of discharge units to be allowed on horizontal branches

Internal diameter of pipe (mm)	Gradient		
	½° 9 mm/m	1¼° 22 mm/m	2½° 45 mm/m
32	—	1	1
40	—	2	8
50	—	10	26
65	—	35	95
75	—	100	230
90	120	230	460
100	230	430	1050
125	780	1500	3000
150	2000	3500	7500

Example 8.3. *By the use of the discharge unit value method, find the diameter and gradient of a drain to serve a factory containing 20 W.C.s, 25 wash basins and 4 sinks. The W.C.s are to be provided with 9 litre flushing cisterns.*

W.C.s   20 × 28 DUs = 560

Basins   25 × 6 DUs = 150

Sinks   4 × 27 DUs = 108

Total   818

With reference to Table 8.4, a 150 mm diameter drain having a gradient of 9 mm/m or a 125 mm diameter drain having a gradient of 22 mm/m would be suitable.

Gradient

Various formulae and tables may be used to find the gradient or fall of a drain. One of the best known formulae which may be used for pipes and channels, is known as 'Chezy's' expressed as follows:

$$V = C\sqrt{(mi)}$$

where

C = Chezy constant

V = velocity of flow in m/s

m = hydraulic mean depth

i = inclination or fall

Chezy constant may be found from the following formula:

$$C = \sqrt{\left(\frac{2g}{f}\right)}$$

where

g = acceleration due to gravity (9.81)

f = coefficient of friction

The average coefficient of friction may be taken as 0.0064 and therefore C would be given by:

$$C = \sqrt{\frac{2 \times 9.81}{0.0064}}$$

$C = 55$

The hydraulic mean depth is found from:

$$m = \frac{\text{wetted area}}{\text{wetted perimeter}}$$

For half or full bore discharge, the hydraulic mean depth is equal to $D/4$ which can be shown as follows:

Full bore

$$m = \frac{\dfrac{\Pi D^2}{4}}{\Pi D}$$

$$m = \frac{1}{4} \times \frac{D^2}{1} \times \frac{1}{\Pi} \times \frac{1}{D}$$

by cancellation

$$m = \frac{D}{4}$$

Half-full bore

$$m = \frac{\dfrac{\Pi D^2}{8}}{\dfrac{\Pi D}{2}}$$

$$m = \frac{1}{8} \times \frac{D^2}{1} \times \frac{8}{1} \times \frac{1}{\Pi} \times \frac{1}{D}$$

$$m = \frac{D}{4}$$

The inclination or fall is equal to H/L, where (L) is the length of drain in metres with a head (H) of 1 m.

Example 8.4. *Calculate the gradient required for a private sewer flowing half-full bore at a velocity of 0.8 m/s (Chezy constant = 55).*

Putting the above values into the formula as follows:

$$V = C\sqrt{\left(\frac{D}{4} \times \frac{1}{L}\right)}$$

By transposition

$$\left[\frac{V}{C}\right]^2 = \frac{D}{4} \times \frac{1}{L}$$

$$\frac{1}{L} = \left[\frac{V}{C}\right]^2 \times \frac{4}{D}$$

$$L = \left[\frac{C}{V}\right]^2 \times \frac{D}{4}$$

$$L = \left[\frac{55}{0.8}\right]^2 \times \frac{0.150}{4}$$

$$L = 68.75^2 \times 0.375$$

$$L = 177 \text{ (approx.)}$$

Gradient = 1 in 177

Siting of access points

The Building Regulations 1985 requires access to drains at the following points:

1. At a bend or change of direction.
2. At a junction, unless each run can be cleared from an access point.
   Note: some junctions can only be rodded through from one direction.
3. On or near the head of each drain run.
4. On long runs.
5. At a change of pipe size.

Figures 8.17 to 8.21 show the positions of access points. Distances marked A depend upon the type of access; see Table 8.6.

Manhole

The Building Regulations state that a manhole shall be capable of sustaining the loads which may be imposed on it, exclude subsoil water and be watertight. The size of the changer should be sufficient to permit ready access to the drain or private sewer for inspection, cleansing and rodding and have a removable and non-ventilating cover of adequate strength, constructed of suitable and durable material. Where the depth of the chamber so requires step irons, ladders or other fittings, should be provided to ensure safe access to the level of the drain or private sewer. If the inspection chamber contains an open channel, benching should be provided having a smooth finish and formed so as to allow the foul matter to flow towards the pipe and also to ensure a safe foothold.

An inspection chamber or manhole may be constructed of engineering brick or well-burnt common brick, bedded in cement mortar. The brickwork should be finished in English bond, with the internal joints finished flush. Cement and sand rendering to the inside of the chamber is not required and may lead to blockage of the drain, due to the rendering flaking off and falling into the channel at the bottom of the chamber. Where a high water table is evident, the chamber will require waterproofing externally, to prevent infiltration of ground water.

Figure 8.22 shows a detail of a shallow brick manhole. Pre-cast concrete manholes are formed with successive sections, jointed together with cement mortar, built up from a pre-cast concrete base unit. Figure 8.23 shows a section through a shallow pre-cast concrete manhole.

The vertical sections are usually circular in plan, although rectangular sections may be obtained. Deep pre-cast concrete manholes should be backfilled with 150 mm thickness of concrete. For chambers exceeding 1 m deep malleable iron steps should be provided, and for manholes exceeding 4.5 m deep a wrought iron ladder is preferable to step irons. The channel at the base of a brick manhole is formed with vitrified clay, which may also be used for a pre-cast concrete manhole. A half round channel is used for through the chamber and a half round channel may also be used for the branch connections to the main channel. Three-quarter section channel bends may be used for the branch connections and these will prevent any risk of solid matter being washed up on the benching. Concrete benching is formed on each side of the channel and is trowelled to a smooth surface (see Fig. 8.24).

If a cast iron drain is used, a cast iron inspection junction which has branches cast on at an angle of 135° is fitted at the bottom of the manhole. The inspection junction has a bolted cover, jointed with a greased felt gasket (see Fig. 8.25).

Unplasticised polyvinyl chloride (upvc) access units are obtainable and one type, manufactured by Mascar Limited of Weybridge in Surrey, consists of a bowl measuring 114 mm to 485 mm in diameter. The bowl has an outlet at its base and side entries for the branch drains. The outlet pipe and branch pipes are of unplasticised polyvinyl chloride, which are connected to the bowl by means of solvent cement joints. The bowl replaces the traditional brick or concrete manhole and has the advantage of reducing the excavation, drainage costs and the drain is accessible from ground level. Figure 8.26 shows a section through the 'Mascar' system.

A plastic manhole base unit manufactured by Osma Plastics Limited has four branch inlets swept into the main channel and most types of brickwork may be connected to the base unit. The unit obviates conventional channel laying and also provides smooth contours (see Fig. 8.27).

Sizes of access points

The size of an access point depends upon its depth to the invert of the drain, and in the cases of inspection chambers and manholes the number of branch drains. Table 8.5 gives the minimum dimensions of access points (Building Regulations 1985). Table 8.6 gives the maximum spacing of access points for drains up to and including 300 mm in diameter (Building Regulations 1985).

Manhole covers

Manhole covers and frames are generally made of cast iron. Covers are also made of steel which are unbreakable and therefore safer. Concrete covers may be used when a complete airtight seal is not required between the cover and the frame. BS 497 specifies three grades of manhole cover and frame.

Grade C. Light duty: which are suitable for housing, or in situations where they will not have to withstand wheeled traffic.

Grade B. Medium duty: which are suitable for footpaths, carriage drives and cycle tracks.

Grade A. Heavy duty: for roadways where they must be capable of withstanding constant vehicle traffic.

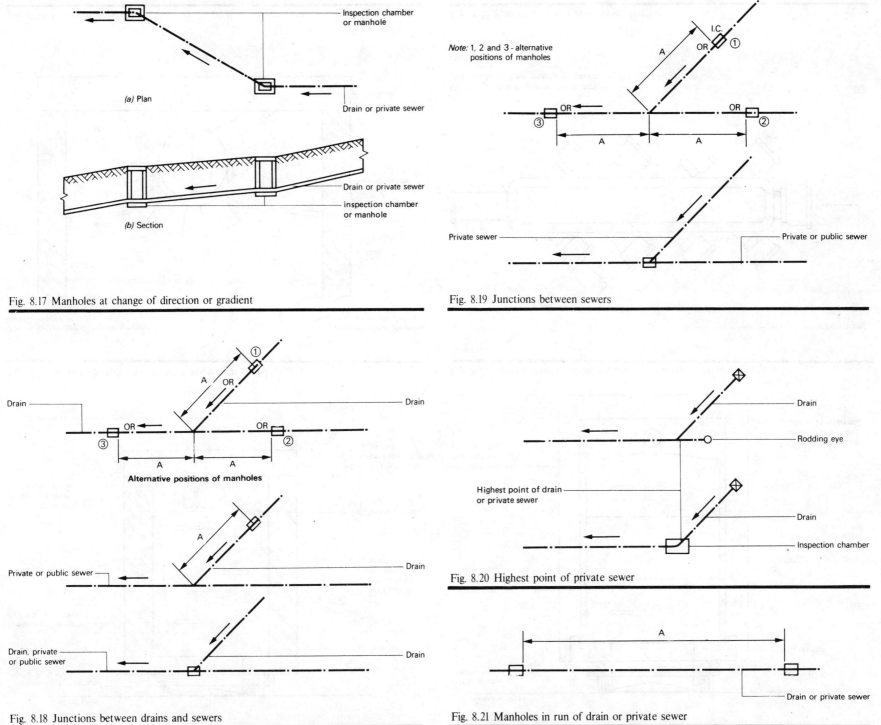

Fig. 8.17 Manholes at change of direction or gradient

(a) Plan

Inspection chamber or manhole

Drain or private sewer

Drain or private sewer

inspection chamber or manhole

(b) Section

Fig. 8.18 Junctions between drains and sewers

Drain

Alternative positions of manholes

Drain

Private or public sewer

Drain

Drain, private or public sewer

Drain

Fig. 8.19 Junctions between sewers

Note: 1, 2 and 3 - alternative positions of manholes

I.C.

OR

Private sewer

Private or public sewer

Fig. 8.20 Highest point of private sewer

Drain

Rodding eye

Highest point of drain or private sewer

Drain

Inspection chamber

Fig. 8.21 Manholes in run of drain or private sewer

Drain or private sewer

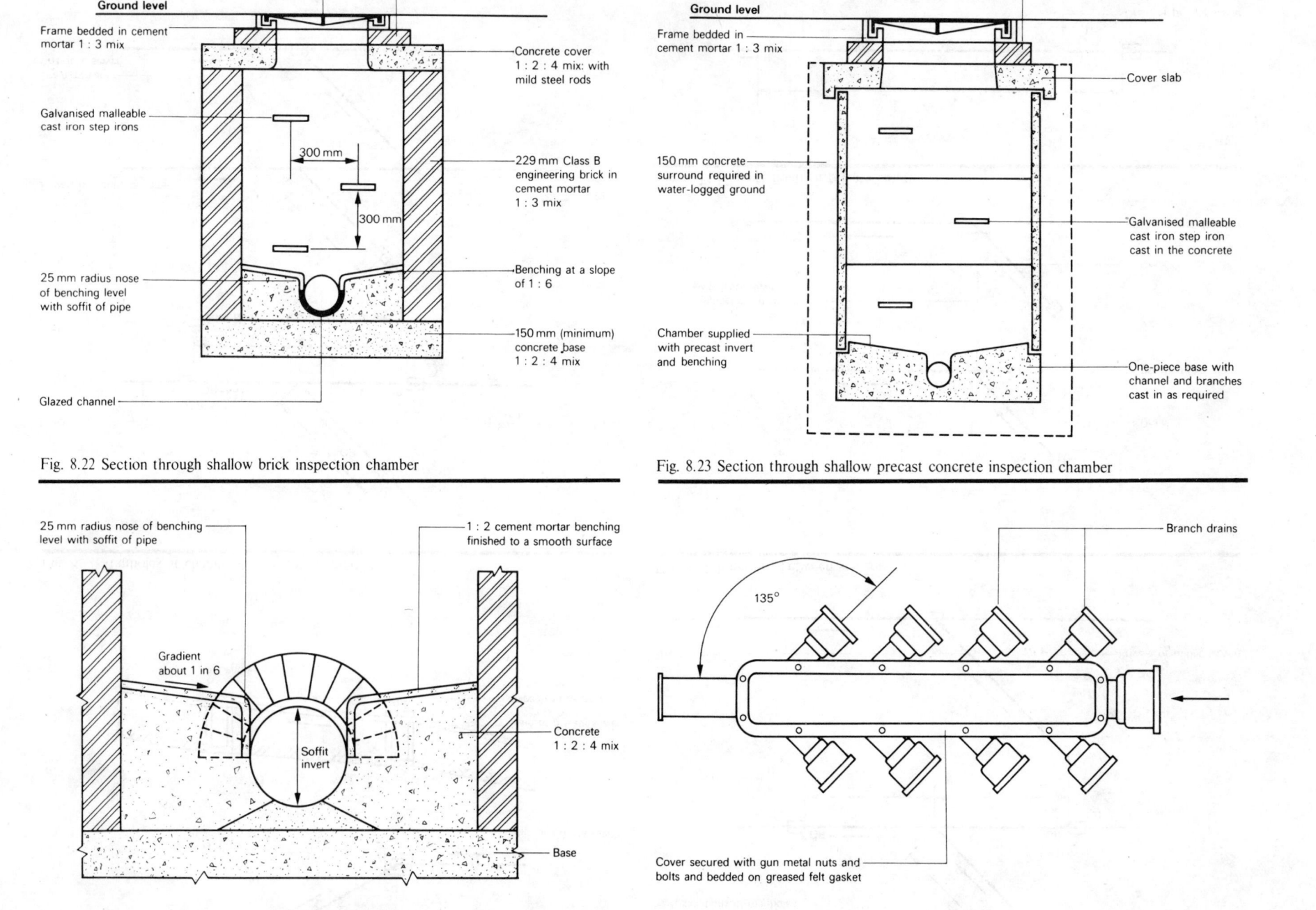

Light duty single seal cover to BS 497

Brick levelling coarse

Ground level

Frame bedded in cement mortar 1 : 3 mix

Concrete cover 1 : 2 : 4 mix: with mild steel rods

Galvanised malleable cast iron step irons

300 mm

300 mm

229 mm Class B engineering brick in cement mortar 1 : 3 mix

25 mm radius nose of benching level with soffit of pipe

Benching at a slope of 1 : 6

150 mm (minimum) concrete base 1 : 2 : 4 mix

Glazed channel

Fig. 8.22 Section through shallow brick inspection chamber

Light duty single seal manhole cover to BS 497

Brick levelling course

Ground level

Frame bedded in cement mortar 1 : 3 mix

Cover slab

150 mm concrete surround required in water-logged ground

Galvanised malleable cast iron step iron cast in the concrete

Chamber supplied with precast invert and benching

One-piece base with channel and branches cast in as required

Fig. 8.23 Section through shallow precast concrete inspection chamber

25 mm radius nose of benching level with soffit of pipe

1 : 2 cement mortar benching finished to a smooth surface

Gradient about 1 in 6

Soffit invert

Concrete 1 : 2 : 4 mix

Base

Fig. 8.24 Detail of benching

Branch drains

135°

Cover secured with gun metal nuts and bolts and bedded on greased felt gasket

Fig. 8.25 Plan of cast iron inspection chamber

Table 8.5 Minimum dimensions for access points and chambers (Building Regulations 1985)

Type	Depth to invert m	Internal sizes		Cover sizes	
		Length X width mm X mm	Circular mm	Length X width mm X mm	Circular mm
Rodding eye	–	As drain but min 100			
Access fitting					
small	0.6 or less	150 x 100	150	250 x 100	150
large		225 x 100	–	225 x 100	–
Inspection chamber	0.6 or less	–	190*	–	190*
	1.0 or less	450 x 450	450	450 x 450	450†
Manhole	1.5 or less	1200 x 750	1050	600 x 600	600
	over 1.5	1200 x 750	1200	600 x 600	600
Shaft	over 2.7	900 x 840	900	600 x 600	600

*Drains up to 150 mm
†For clayware or plastics may be reduced to 430 mm in order to provide support for cover and frame.

Table 8.6 Maximum spacing of access points for drains up to and including 300 mm diameter (Building Regulations 1985)

From	To	Access Fitting			Inspection chamber	Manhole
		Small	Large	Junction		
Start of external drain*		12	12	–	22	45
Rodding eye		22	22	22	45	45
Access fitting						
small 150 diam						
150 x 100		–	–	12	22	22
large 225 x 100		–	–	22	45	45
Inspection chamber		22	45	22	45	45
Manhole		22	45	45	45	90

* Connection from ground floor appliances or stack.

Light and medium duty covers are available with single and double seals, whilst heavy duty covers are single seal type only. A double seal cover is required over an open channel type manhole inside a building. Figure 8.28 shows single and double seals for manhole covers. Covers may be rectangular, triangular or circular on plan. Triangular and circular covers have three points of support and have the advantage of being non-rocking. Frames are solidly bedded in 1 in 3 cement and sand mortar on top of the manhole. Covers are bedded in the frames with grease to ensure an airtight joint.

Rodding points

A rodding point may be used in the following systems:

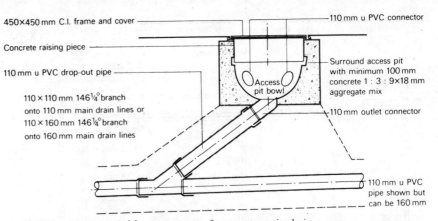

Fig. 8.26 Section through Marscar system of access to main drain

450×450 mm C.I. frame and cover
Concrete raising piece
110 mm u PVC drop-out pipe
110 × 110 mm 146¼° branch onto 110 mm main drain lines or 110 × 160 mm 146¼° branch onto 160 mm main drain lines
110 mm u PVC connector
Surround access pit with minimum 100 mm concrete 1 : 3 : 9×18 mm aggregate mix
Access pit bowl
110 mm outlet connector
110 mm u PVC pipe shown but can be 160 mm

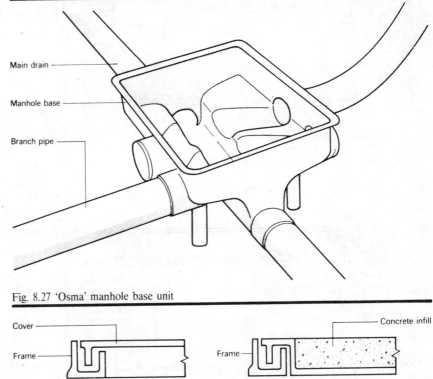

Fig. 8.27 'Osma' manhole base unit

Main drain
Manhole base
Branch pipe

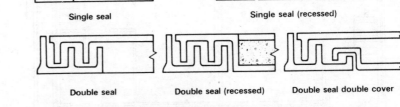
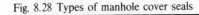

Fig. 8.28 Types of manhole cover seals

Cover
Frame
Concrete infill
Single seal
Single seal (recessed)
Double seal
Double seal (recessed)
Double seal double cover

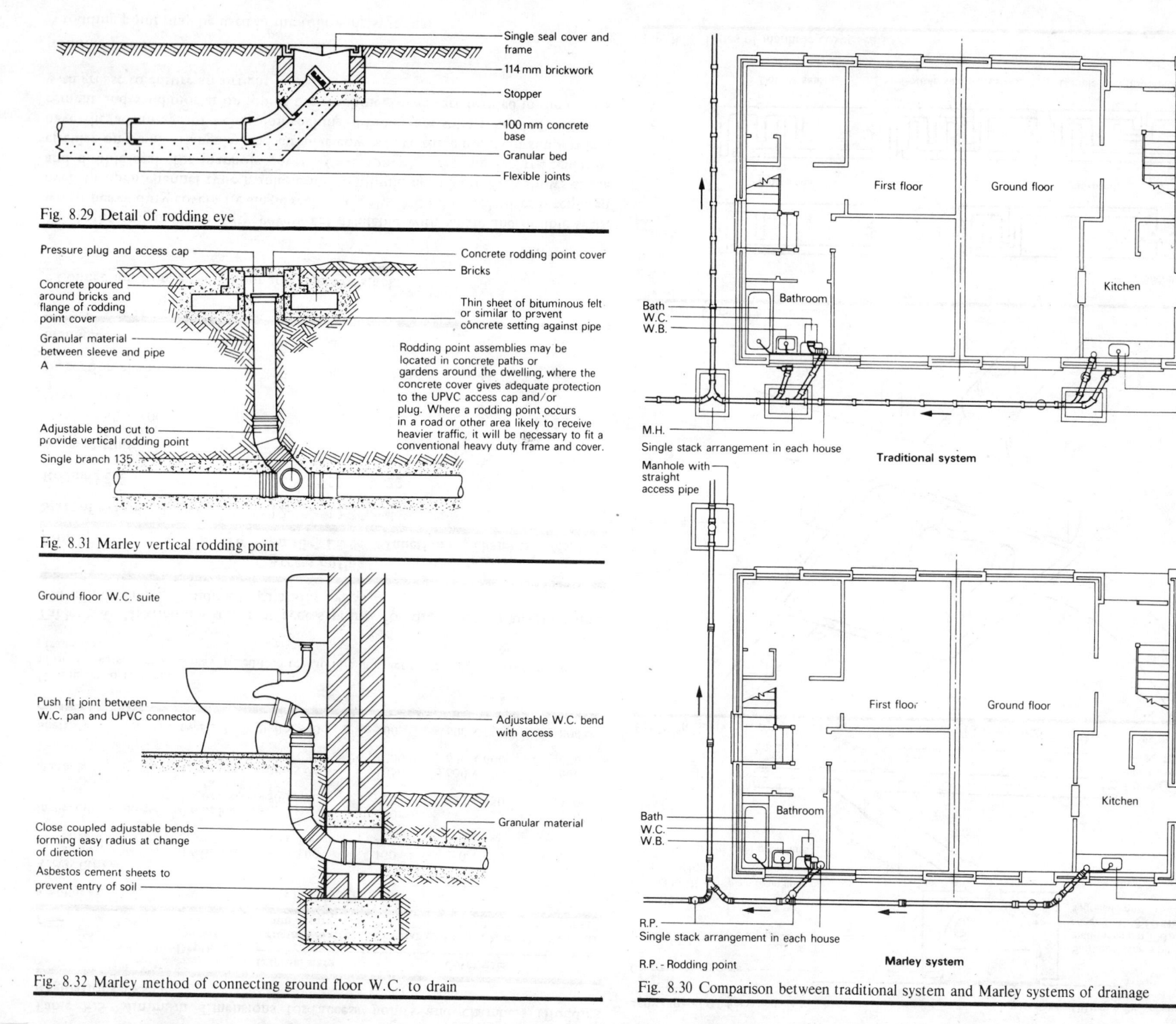

Single seal cover and frame

114 mm brickwork

Stopper

100 mm concrete base

Granular bed

Flexible joints

Fig. 8.29 Detail of rodding eye

Pressure plug and access cap

Concrete rodding point cover

Bricks

Concrete poured around bricks and flange of rodding point cover

Thin sheet of bituminous felt or similar to prevent concrete setting against pipe

Granular material between sleeve and pipe A

Rodding point assemblies may be located in concrete paths or gardens around the dwelling, where the concrete cover gives adequate protection to the UPVC access cap and/or plug. Where a rodding point occurs in a road or other area likely to receive heavier traffic, it will be necessary to fit a conventional heavy duty frame and cover.

Adjustable bend cut to provide vertical rodding point

Single branch 135

Fig. 8.31 Marley vertical rodding point

Ground floor W.C. suite

Push fit joint between W.C. pan and UPVC connector

Adjustable W.C. bend with access

Close coupled adjustable bends forming easy radius at change of direction

Granular material

Asbestos cement sheets to prevent entry of soil

Fig. 8.32 Marley method of connecting ground floor W.C. to drain

First floor

Ground floor

Kitchen

Bathroom

Bath
W.C.
W.B.

Sink

M.H.

M.H.

Single stack arrangement in each house

Traditional system

Manhole with straight access pipe

First floor

Ground floor

Kitchen

Bathroom

Bath
W.C.
W.B.

Sink

R.P.

R.P.

Single stack arrangement in each house

Marley system

R.P. - Rodding point

Fig. 8.30 Comparison between traditional system and Marley systems of drainage

1. In a traditional system, where manholes are used at the various positions required under the Building Regulations and the rodding point is permitted at the head of a private sewer. Figure 8.29 shows a detail of a rodding point at the head of a private sewer, or drain.

2. In a closed drainage system, known as the Marley Rodding Point system, which considerably reduces the number of manholes required for the traditional drainage systems, and there is, therefore, a reduction in the cost of drain laying. The local authority however will have to relax the Building Regulations before the system is adopted. The system is covered by Agreement Certificates Nos 70/72 and 71/99 and consists of unplasticised polyvinyl chloride pipes and fittings. A full range of fittings is available in 82.4, 110 and 160 mm outside diameter sizes.

One of the most important items is a patented adjustable bend, which may be adopted on site to meet the various angles as they occur at changes of direction in the drainage pipework. The bend is also used in combination with a basic 135° junction, to form vertical rodding points.

Figure 8.30 shows a drainage system for housing, using the Marley system, and Fig. 8.31 shows a vertical rodding point used in the system. Figure 8.32 shows the Marley method of connecting a ground floor W.C. to the drain.

Connection to public sewer

Wherever possible, a new drain should be connected to an existing manhole, as this will save breaking up the road and connecting the drain to a public sewer. If this is not possible, the work of breaking into an existing public sewer and forming a connection between the new drainage system and the sewer is usually carried out by the local authority employees.

With a vitrified clay sewer of large diameter, breaking into the sewer should be affected by the continuous enlargement of a small hole made into the sewer. The connection is made by a suitable saddle, bedded on to the sewer by cement mortar and completely surrounding the joint with 1:2:4 concrete 150 mm thick. Every precaution should be taken to prevent any jointing or excavating material entering the sewer. After the completion of a new connection, the excavation around the joint requires careful reinstatement. Figure 8.33 shows a saddle connection to a sewer.

When the sewer is too small for the opening to use a saddle, it is necessary to insert a 135° junction by one of the following methods:

1. Three pipes are taken out of the sewer line and 135° junction and two plain pipes are inserted, as shown in Fig. 8.34.

2. Two pipes are taken out of the sewer line and a collar slipped over the end of the spigot end of the sewer pipeline. A 135° junction and a double spigot pipe are inserted and the collar then slipped into position over the two spigot ends of the pipeline, as shown in Fig. 8.35.

Backdrop connections

Where there is a considerable drop in level between the incoming drain and the manhole bottom, or public sewer, backdrop connections are required. The backdrop is required to reduce the cost of excavation (see Fig. 8.36). In order to save space inside the manhole, drains above 150 mm bore have the backdrop usually

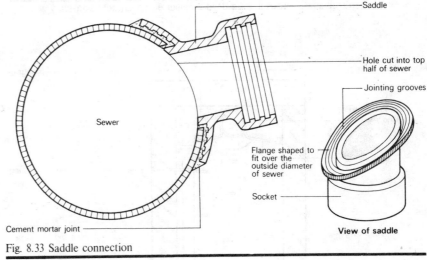

Fig. 8.33 Saddle connection

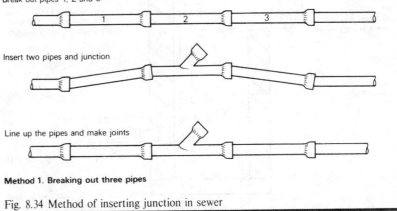

Break out pipes 1, 2 and 3

Insert two pipes and junction

Line up the pipes and make joints

Method 1. Breaking out three pipes

Fig. 8.34 Method of inserting junction in sewer

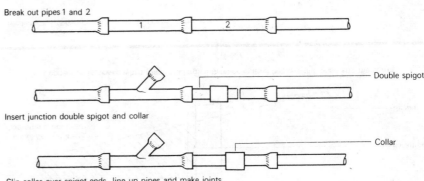

Break out pipes 1 and 2

Insert junction double spigot and collar

Slip collar over spigot ends - line up pipes and make joints

Fig. 8.35 Method of inserting junction in sewer

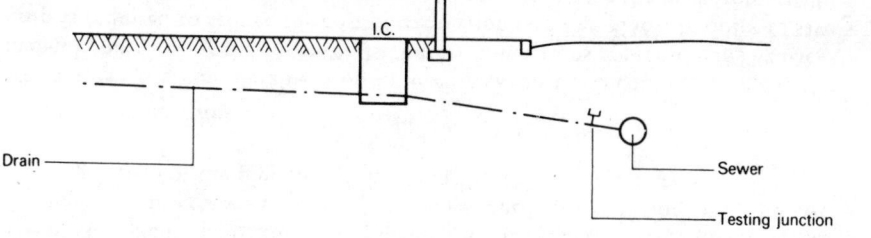

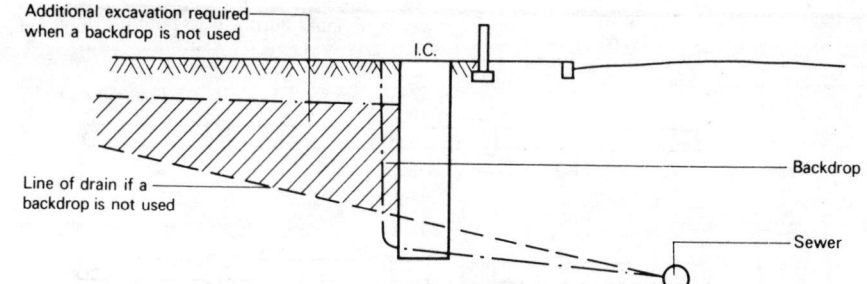

I.C.
Drain
Sewer
Testing junction
Connection to sewer when there is little difference in level between the drain and the sewer
Additional excavation required
when a backdrop is not used
I.C.
Backdrop
Line of drain if a
backdrop is not used
Sewer
Connection to sewer when there is a large difference in level between the drain and the sewer

Fig. 8.36 Use of back drop

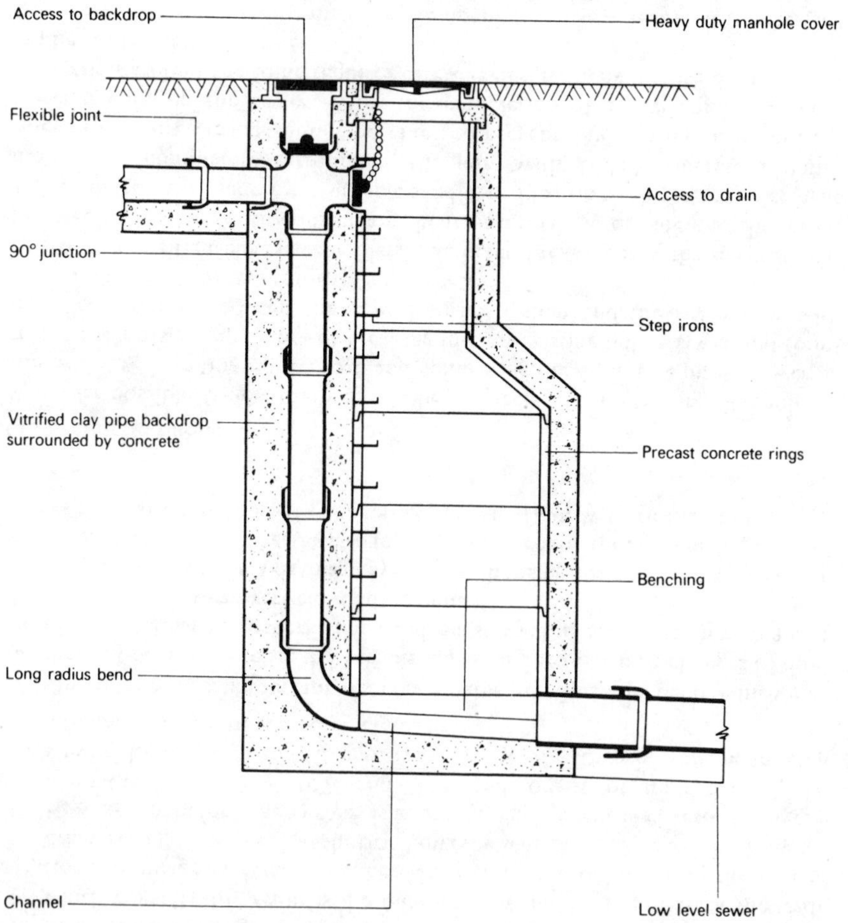

Access to backdrop
Heavy duty manhole cover
Flexible joint
Access to drain
90° junction
Step irons
Vitrified clay pipe backdrop
surrounded by concrete
Precast concrete rings
Benching
Long radius bend
Channel
Low level sewer

Fig. 8.37 Vitrified clay pipe backdrop outside a deep-precast concrete inspection chamber

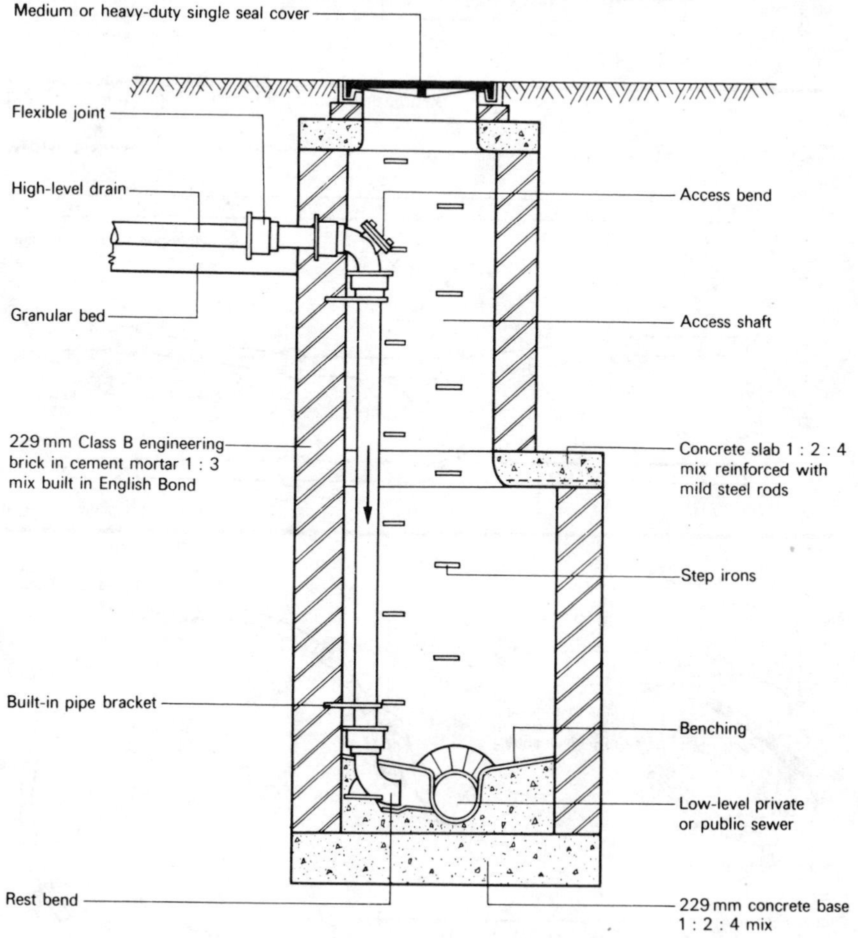

Medium or heavy-duty single seal cover
Flexible joint
High-level drain
Access bend
Granular bed
Access shaft
229 mm Class B engineering
brick in cement mortar 1 : 3
mix built in English Bond
Concrete slab 1 : 2 : 4
mix reinforced with
mild steel rods
Step irons
Built-in pipe bracket
Benching
Low-level private
or public sewer
Rest bend
229 mm concrete base
1 : 2 : 4 mix

Fig. 8.38 Cast-iron pipe backdrop inside a deep-brick inspection chamber

constructed outside the manhole, using vitrified clay pipe, surrounded with 150 mm of 1:2:4 concrete. Figure 8.37 shows a backdrop in vitrified clay pipe outside a pre-cast concrete manhole. For drains up to 150 mm bore, the backdrop may be constructed inside the manhole, providing there is sufficient space left for access. The backdrop may be made by using cast iron, pitch fibre or unplasticised polyvinyl chloride pipe and the pipe should be secured to the manhole face by means of pipe brackets. Figure 8.38 shows a backdrop in cast iron pipe, inside a deep brick manhole. At one time backdrops were used to reduce the gradient of a drain laid below sloping sites, but in modern drainage practice the drain is laid at the same slope of the ground and this method saves a great deal of drainage costs. It has been found that with steeper gradients the tendency for solids to remain behind due to the water running away from the solids does not occur in practice and a steeper gradient, in fact, reduces the risk of blockages. It was also thought that with high velocities grit would cause abrasion to the inside of the pipe, but the high velocity of flow actually reduces scour by carrying the solids in suspension, which at low velocities would travel in contact with the surface of the pipe.

Sewage pumping

Wherever possible drainage schemes should be designed so that the liquid gravitates to the public sewer, or sewage-disposal plant. Cases arise, however, when the drain is below the sewer level, or where the site levels make it necessary to raise the liquid to a higher level before it can discharge by gravity to the sewer, or sewage-disposal plant. The raising of the liquid can be achieved by the use of either a pneumatic ejector or a centrifugal pump.

Sewage ejectors

The pneumatic ejector is an efficient, reliable method of pumping sewage. It has very few working parts, which require little maintenance and the sewage is contained in a closed cylinder. The installation usually requires an automatic self-starting air compressor, an air-storage cylinder and, if possible, two ejectors, so that one of the ejectors may be used whilst the other is being repaired. Figure 8.39 shows the method of duplicating ejectors.

One central air compressor station may be used to supply air to several ejectors and the station may be situated at any convenient position, inside or outside the building, providing that the frictional resistance inside the air pipeline is not excessive. Where a breakdown would cause serious inconvenience, two air compressors should be installed, so that one can be used for standby purposes.

The size of the ejector depends upon the flow rate entering the cylinder. Cylinder sizes are between 500 and 900 mm inside diameter and between 1 and 1.30 m in height. The discharge from an ejector is from 1.13 litre/s to 15 litre/s. Figure 8.40 shows a section through an ejector, which operates as follows:

1. The liquid to be raised gravitates through the inlet pipe into the cylinder and the non-return valve prevents the liquid from flowing back.
2. The liquid gradually fills the cylinder and as the level of the liquid rises the float slides along the float rod until it reaches to top collar.

3. The float rod is then raised, which lifts the rocking weight, and when the weight passes the vertical position it suddenly falls by gravity, allowing the valve to close and the compressed air inlet valve to open.
4. Compressed air enters the cylinder and forces out the liquid through the outlet pipe until the cylinder is empty.
5. The float falls, thereby actuating the automatic valve which cuts off the compressed air supply and opens the exhaust valve, and the ejector is again ready for the next operation. Figure 8.41 shows the installation of the ejector below ground.

Unit ejector

A self-contained unit ejector is operated by means of a small electric motor and compressor, mounted directly on top of the ejector, which eliminates the need for a compressor room, air-receiving cylinder and pipelines.

It operates as follows:

1. Sewage enters through the inlet non-return valve and gradually fills the cylinder, until the liquid reaches the tip of a short electrode.
2. The electric current is switches on, which brings the motor and air compressor into operation.
3. Compressed air enters the ejector and forces out the liquid through the outlet pipe, until the level of liquid falls below a second, longer electrode.
4. The compressor is switched off and the compressed air is exhausted to the atmosphere through an automatic valve.

Figure 8.42 shows the installation of the Simplex self-contained unit ejector. In order to save the cost of excavation and the construction of an injector chamber below ground, a lift and force ejector may be installed. With this type of ejector the whole of the mechanical and electrical equipment is housed above ground, thus providing easy access. Figure 8.43 shows a detail of the installation of the lift and force ejector.

Operation

1. When the liquid reaches the high level, the float switches on the compressor and the suction created at its outlet creates a partial vacuum inside the ejector.
2. The sewage is forced into the ejector by the atmospheric pressure acting on top of the liquid.
3. When the ejector is full the automatic valve closes the suction valve and opens the compressed air valve.
4. The compressed air enters the ejector and forces out the liquid.

Connection to sewer

Before the liquid discharged from an ejector or a pump enters the sewer it should pass through a manhole so that the liquid gravitates to the sewer. This method prevents a surcharge in the sewer, backflooding of the drain and provides access for inspection and rodding of the drain. Figure 8.44 shows the type of manhole for connecting the discharge from an ejector, or pump to the sewer.

Centrifugal pumps

A centrifugal pump can lift a large volume of liquid at a greater speed than the ejector and is therefore suitable for the pumping of sewage or surface water discharges from large building schemes. The pump should be wear resisting and sited so that it is accessible for easy maintenance and, wherever possible, two pumps should be installed, so that one pump can be repaired whilst the other is in use. In cases where a failure of the electrical supply would cause serious inconvenience, such as hospitals, consideration must be given to the installation of a standby petrol or diesel electric generator, so that this may be used to provide electrical power for pumping in the event of the electrical supply from the mains failing due to a power cut or a fault.

Sewage pumps used for building drainage are usually of the following types:

1. Vertical spindle centrifugal.
2. Horizontal spindle centrifugal.
3. Self-priming centrifugal.

The vertical and horizontal centrifugal pumps have full-way impellers, made from special grade cast iron, to ensure maximum life and the impellers are designed to give minimum resistance to the flow of liquid, whilst enabling any solids entering the suction pipe to pass freely through the pump.

Vertical spindle pump

The vertical spindle pump is usually designed for installation in a dry well, constructed below the level of the incoming drain or sewer, discharging into a wet well adjacent to the dry well. In this position the pump is primed under the gravitational head from the liquid inside the wet well and thus avoids the necessity of providing automatic self-starting apparatus. Figure 8.45 shows the installation of a vertical spindle pump. The power required to drive the pump is provided from a vertical spindle electric motor installed in a motor room above the pump chamber. Figure 8.46 shows a section through a vertical centrifugal pump and Fig. 8.47 a section through a horizontal spindle centrifugal pump. It is sometimes necessary to pump sewage from sanitary appliances installed in the basement or sub-basements of buildings. Figure 8.48 shows the installation of pumping equipment required when the basement drain is below the sewer.

Horizontal spindle pumps

In order to reduce the cost of installation, a horizontal spindle pump which has the electric motor fixed close to the pump may be used. This method does not require a motor room above ground, but the electrical equipment below ground is more exposed to dampness and is not as accessible as when the motor room is above ground. Figure 8.49 shows the installation of a horizontal spindle centrifugal pump, which eliminates the construction of a motor room above ground.

Self-priming centrifugal pump

This type of pump has the advantage of being sited above ground, which provides protection from dampness for the pump and electrical apparatus. Figure 8.50 shows the installation of a self-priming centrifugal pump, which operates as follows:

1. When the pump is started, liquid is drawn from the wet well by a partial vacuum, caused by the displacement of a body of water trapped in the lower half of a 'two-storey' priming chamber, to which the suction line is connected.
2. This priming water is delivered by the pump into an upper section of the priming chamber, where it is retained under normal discharge pressure, whilst the liquid from the pump continues to enter the pump and is discharged to the pumping main by centrifugal force in the usual manner.
3. When the pump is stopped by the action of the float switch, the contents of the upper portion of the priming chamber gravitates back through the pump into the lower half, thus repriming the pump ready for the next operation.

The Code of Practice Building Drainage, gives recommendations for pumping installations, some of which are as follows:

1. The velocity of flow through the pumping main should be between 0.6 m/s and 1.2 m/s. Velocities below 0.6 m/s are likely to permit the accumulation of solids in the main, whereas velocities above 1.2 m/s are generally considered to be uneconomical due to the resultant high friction head and possible increased surge problems.
2. The layout of pipework should provide for isolating sluice valves on each side of each pump, with a reflux valve between the pump and the sluice valve on the delivery side. The reflux valve should be fitted in a horizontal run of pipe, between the pump and the rising main to avoid sedimentation of solids which occurs in the vertical pipework.
3. The dry well to house vertical spindle pumps or ejectors should be watertight and should provide ample space for easy maintenance of the equipment installed.
4. Pump starters should incorporate in each phase a magnetic over load release, controlled by a hydraulic relay dash pot and they should be of the pattern rated for forty starts per hour.
5. The working capacity of a wet well should be designed in conjunction with the selected pump size to ensure reasonable frequency of operation and reasonable pumping period. On a separate system of drainage and with a pump output of six times the 24 hour average, a well capacity of 5 minute pump output may give a reasonable compromise between frequency of starts and duration of pumping periods. To provide against breakdown, the capacity of the wet well below overflow level and above the pump working level, should not be less than 1 hour's average flow, or more if practicable.

Sump drainage

It is sometimes necessary to drain basements and boiler rooms below ground of sub-soil or surface water, which seeps through the structure, or is used for washing down purposes. A concrete or brick sump measuring 460 x 460 mm is constructed and a centrifugal sump pump fitted inside the chamber, as shown in Fig. 8.51. The sump pump shown is fully automatic, but non-automatic sets may also be obtained. Sump pumps are not designed for sewage and a suction strainer is fitted to prevent the entry of large particles which would affect the pump operation. The electric motor is totally enclosed, but if the pump units are required to operate in hazardous atmospheres a flameproof motor should be fitted. The pump will usually lift water to a height of up to 11 m and discharge about 3.2 litre/s.

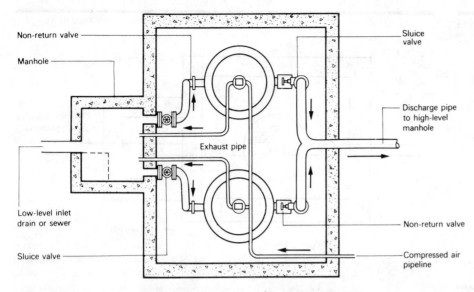

Fig. 8.39 Plan of ejector chamber showing method of duplication of the ejectors

Sizes of ejector chambers

Discharge in litres per second

	1.13	2.3	4.5	6.8	9.5	15
Length (m)	2.1	2.4	2.4	2.4	2.8	3.0
Width (m)	2.1	2.4	2.4	2.4	2.8	3.0

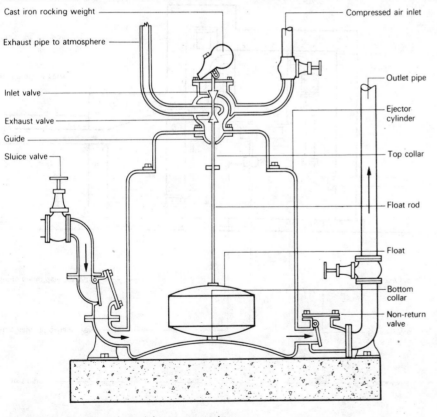

Fig. 8.40 Section through an Adams sewage ejector

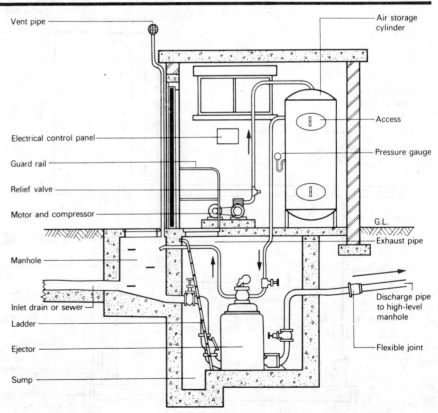

Fig. 8.41 Installation of pneumatic ejector for lifting sewage

98

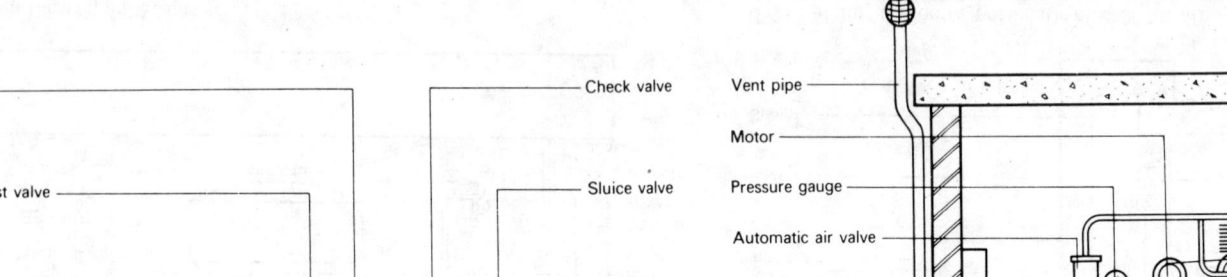

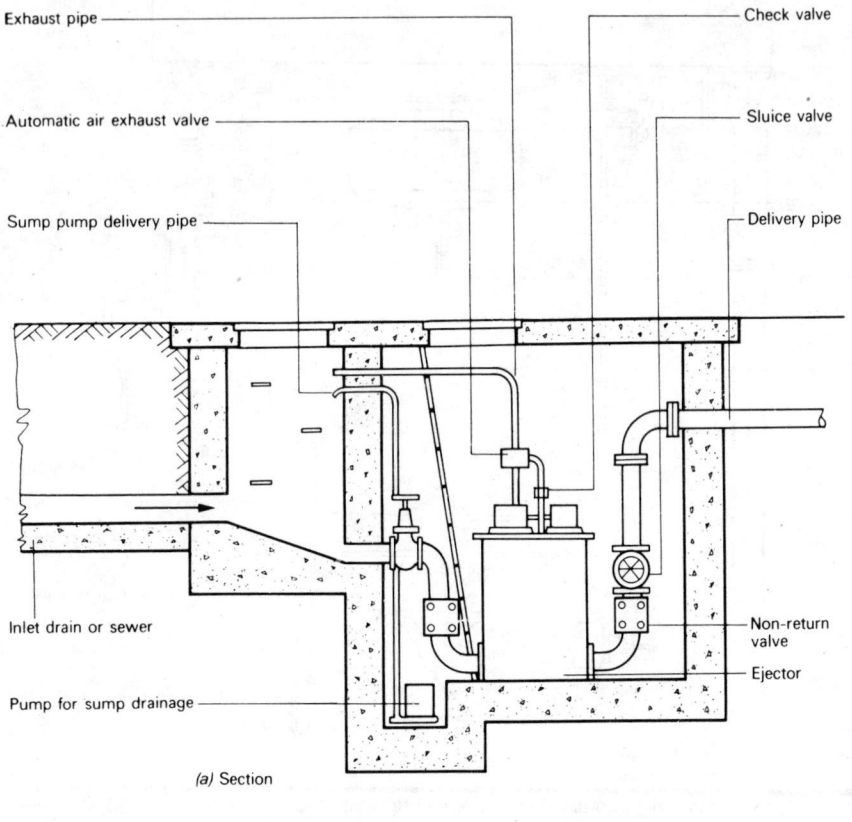

(a) Section

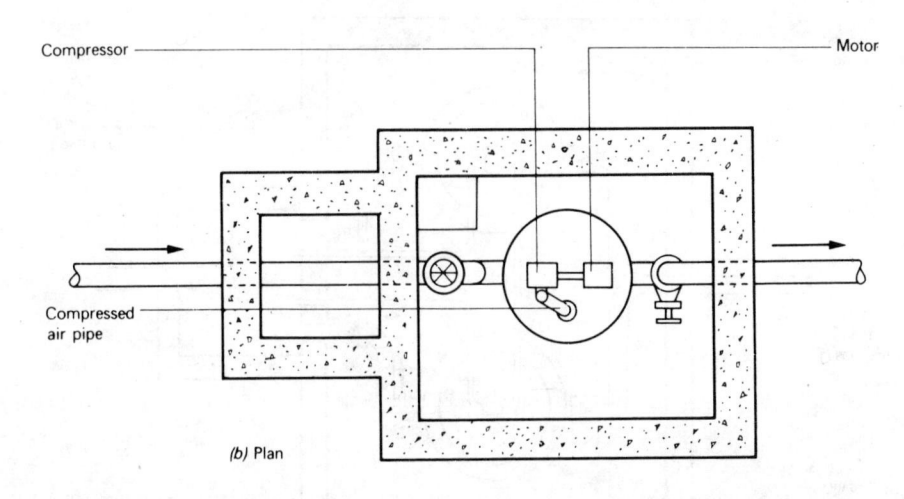

(b) Plan

Fig. 8.42 Installation of the Simplex self-contained ejector

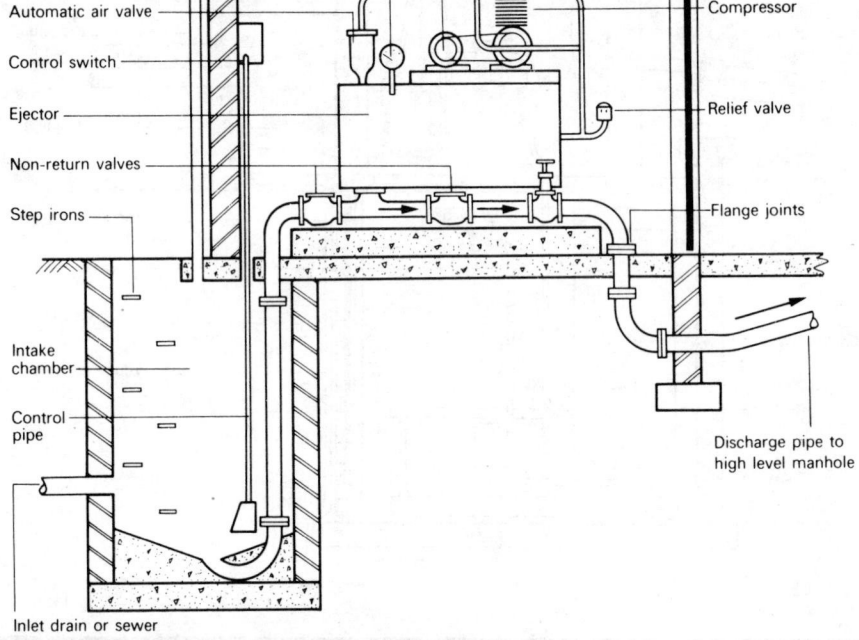

Fig. 8.43 Installation of lift and force sewage ejector

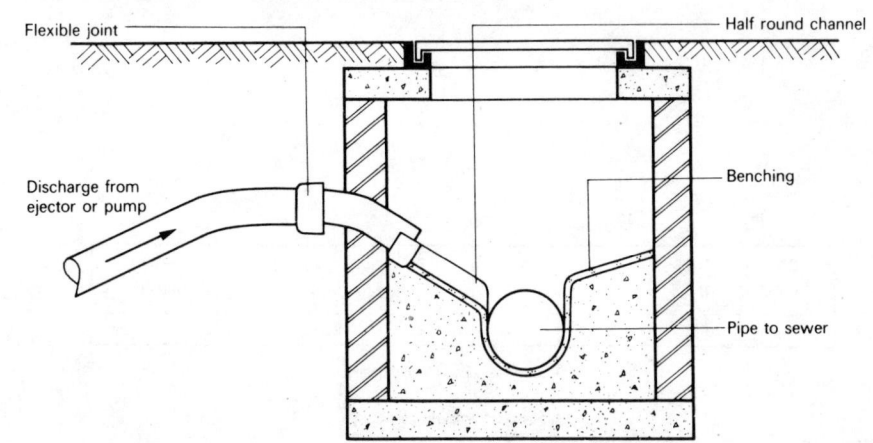

Fig. 8.44 Detail of high-level manhole

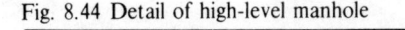

Mutrator pump

A small-packaged pumping unit consisting of a glass fibre collecting chamber, macerator and a self-priming centrifugal pump may be installed for the pumping of sewage from small or medium sized buildings. The sewage is macerated and then pumped through a 38 mm diameter pipe to the sewer or small sewage-disposal plant.

Drainage pipe materials

Clay

There are four British Standards.

1. BS 65 and 540:1971, clay drain and sewer pipes, including surface water pipes.
2. BS 539:1968, dimensions of fittings for use with clay drain.
3. BS 1143:1955, salt-glazed pipes with chemically resistant properties.
4. BS 1196:1971, clay field drain pipes.

All pipes covered by these standards are often described as salt glazed, but the use of salt glazing is diminishing and unglazed pipes are no less suitable for usage than the older glazed pipe. Clay pipes are resistant to attack by a wide range of effluents, including acids and alkalis, but some flexible jointing material may be affected by certain substances contained in trade wastes.

There are two classes of pipes, British Standard for foul and or surface water, and British Standard Surface Water, for surface water only. When pipes have been tested in accordance with clause 10 of BS 65 they should bear the marking 'Extra Strength', and when tested hydraulically in accordance with clause 12 they should be marked 'Tested'. Pipes may also be marked with the maker's name. Pipes are available in a wide range of sizes, from 75 mm bore to 900 mm bore, and may be obtained in lengths from 300 mm to 8.5 m, either with or without sockets for use with flexible or rigid joints. Figure 8.52 shows a typical BS 65 clay drain pipe.

Methods of jointing

The traditional method of jointing of clay pipes is by means of cement and sand mortar, as shown in Fig. 8.53. The joint may still be used for certain trade wastes, but is liable to fracture due to thermal expansion, or ground movement, and in order to prevent fracture various types of flexible telescopic joints are obtainable. These joints are quicker to make, allow up to 5° in lateral movement and allow telescopic movement of up to 18 mm. Figures 8.54 and 8.55 show two types of flexible telescopic joints on clay pipe. With both types of flexible joints shown it is essential that the mating surfaces are clean and that the lubricant supplied by the manufacturer is used.

With the O-ring joint, the O-ring is placed into the groove on the spigot moulding of one pipe and the socket of the other pipe is lubricated. The spigot is then placed up to the socket of the mating pipe and pushed in with a slight side-to side action. The O-ring is compressed against the polyester moulds, thus making a flexible telescopic watertight joint. The sleeve joint is made by lubricating the end of the pipe and pushing the sleeve on to it. The plain end of the other pipe is then lubricated and pushed into the sleeve.

Unplasticised polyvinyl chloride (upvc)

BS 4660:1971, unplasticised polyvinyl chloride underground drain pipes and fittings, covers the requirements for pipes of nominal 100 mm internal bore, but some manufacturers produce sizes of 89, 110, 160 and 200 mm external diameter pipes, in lengths of 1, 2, 3 and 6 m. BS 4660 pipe is coloured brown, so as to distinguish it from the thinner walled pipe used for drainage above ground. The pipe resists attack by normal domestic wastes and a wide range of acids, alkalis and sulphates. It may be attacked by certain organic solvents, but exposure to small quantities of such substances is unlikely to have significant effect on the pipe. Where large quantities are discharged, however, advice should be sought from the manufacturer.

If the pipe is to be used for prolonged discharges of very hot liquids, there is a risk of distortion of the pipe and it is necessary to cool the liquid before it enters the drain. Before the pipe is to be used for laundries, factories or hospitals, it is essential to ascertain the maximum temperatures to which it is to be subjected and to obtain advice from the manufacturers where necessary. The pipe is light in weight, very quick to lay, permits thermal and ground movement and has a smooth internal bore. Figure 8.56 shows the type of joint used for unplasticised polyvinyl chloride drain pipe.

Pitch fibre

Pitch fibre pipe is made by impregnating wood-fibre with pitch. The pipe is suitable for normal drainage use, but is unsuitable for prolonged discharges of very hot liquids and the liquid should be cooled before it is discharged into the drain. Where there is a high concentration of certain trade effluents, such as, oils, fats and organic solvents, the manufacturer's advice should be sought before using pitch fibre pipes.

British Standard 2760:1967 covers pitch fibre pipes and fittings for drainage below and above ground. Standard internal diameters of the pipes are 50, 75, 100, 125, 150 and 200 mm, in lengths of 2.5 and 3 m. When these pipes were first used, they were jointed by means of a coupling, tapered internally at each end at 2° and the pipe ends were machined externally to the same taper. The couplings were driven on to the pipe and a joint made, without the need of a jointing compound. This type of joint however was inflexible and tended to leak in use due to expansion and contraction of the pipe or ground movement; also sometimes the coupling was over driven and split. A flexible joint is now used, consisting of a polypropylene sleeve containing a rubber 'D' ring, as shown in Fig. 8.57. The joint is made by placing the 'D' ring on the end of the pipe and pushing the end firmly into the socket of the coupling. This movement will cause the 'D' ring to rotate through 360°, snapping into position and creating the seal. Figure 8.58 shows the taper joint for pitch fibre pipe. Connections between pitch fibre and other pipe materials can be made with special adaptors.

Cast iron

There are a number of British Standards for cast iron drain pipes:

BS 78:Part 1, 1961, spigot and socketed pipes, vertically cast, covers pipes for the conveyance of water, gas or sewage. Pipes are available in internal diameters of between 75 and 300 mm and in lengths of 2.74 to 3.66 m.

Shaft bearings

Packing gland

Pressure gauge tapping

Two-blade impeller

Pressure gauge tapping

Handhole cover

Fig 8.46 Verticle spindle centrifugal pump

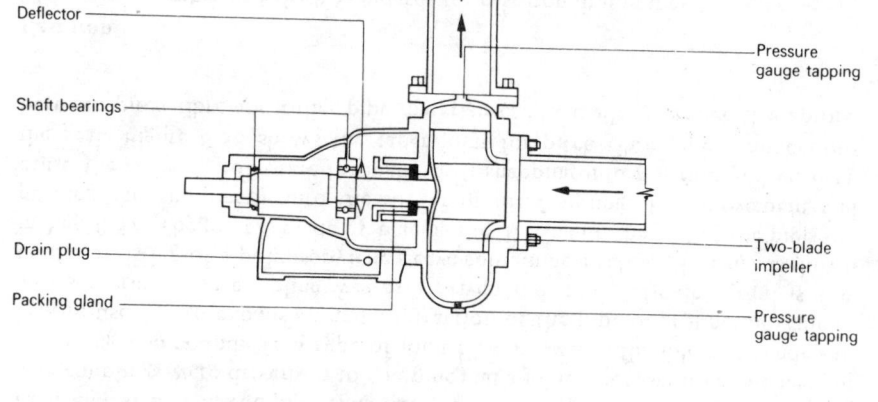

Deflector

Shaft bearings

Drain plug

Packing gland

Pressure gauge tapping

Two-blade impeller

Pressure gauge tapping

Fig. 8.47 Horizontal spindle centrifugal pump

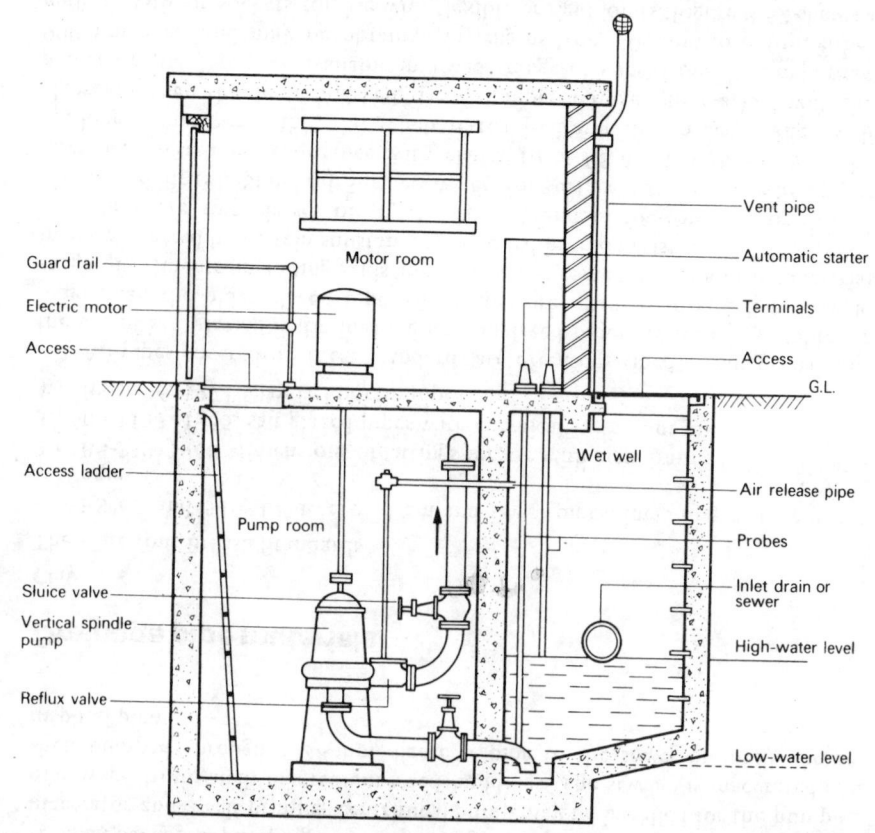

Vent pipe

Automatic starter

Terminals

Access

G.L.

Guard rail

Motor room

Electric motor

Access

Access ladder

Wet well

Air release pipe

Pump room

Probes

Sluice valve

Inlet drain or sewer

Vertical spindle pump

High-water level

Reflux valve

Low-water level

Fig. 8.45 Installation of centrifugal sewage pump with motor room above ground level

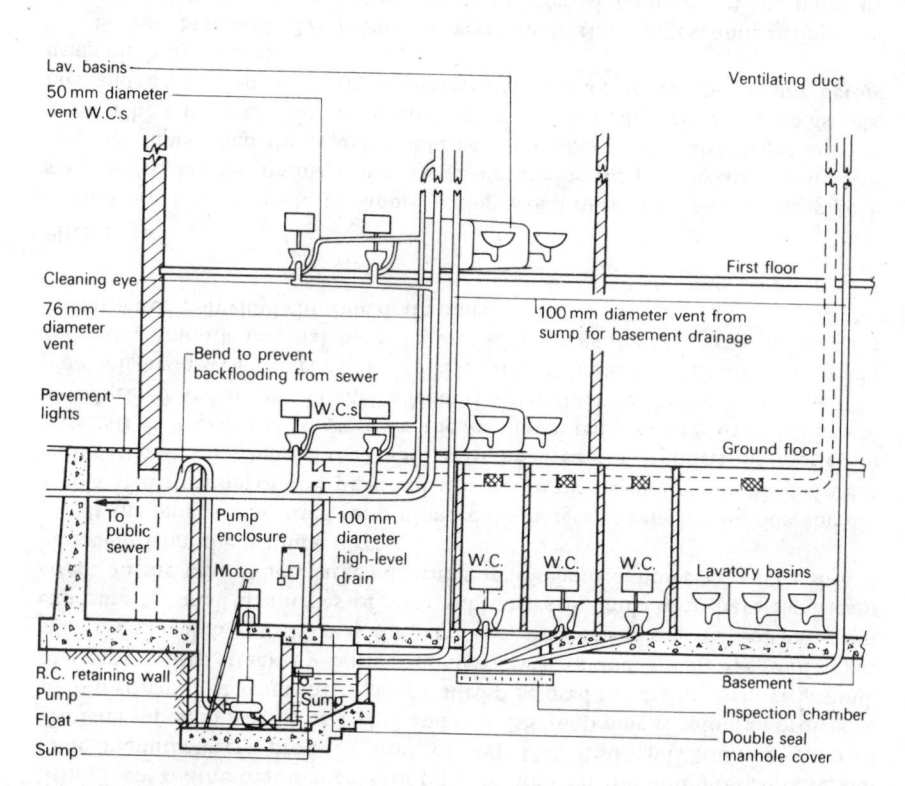

Lav. basins

50 mm diameter vent W.C.s

Ventilating duct

Cleaning eye

First floor

76 mm diameter vent

100 mm diameter vent from sump for basement drainage

Bend to prevent backflooding from sewer

Pavement lights

W.C.s

Ground floor

To public sewer

Pump enclosure

100 mm diameter high-level drain

Motor

W.C.

W.C.

W.C.

Lavatory basins

R.C. retaining wall

Pump

Float

Sump

Sump

Basement

Inspection chamber

Double seal manhole cover

Fig. 8.48 Pumping of drainage from a basement

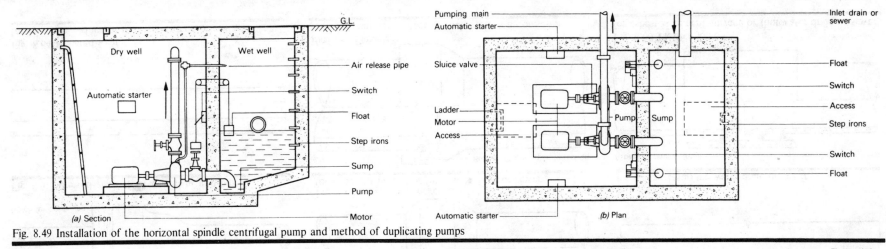

Fig. 8.49 Installation of the horizontal spindle centrifugal pump and method of duplicating pumps

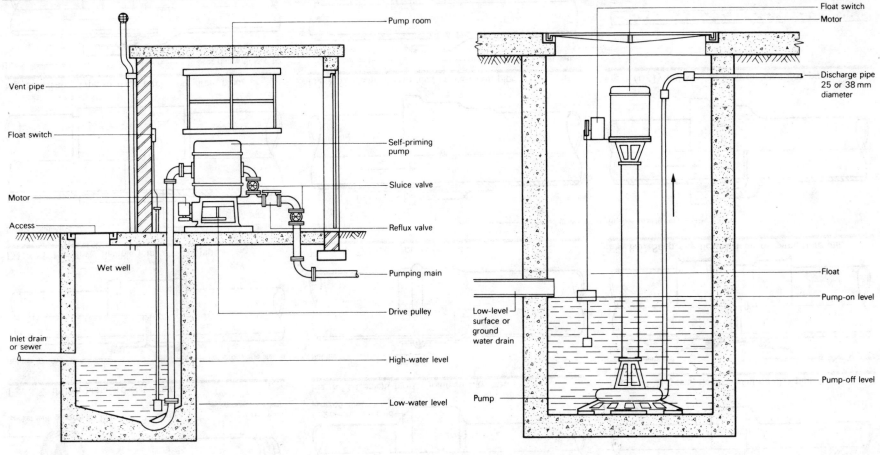

Fig. 8.50 Installation of self-priming centrifugal pump

Fig. 8.51 Sump drainage

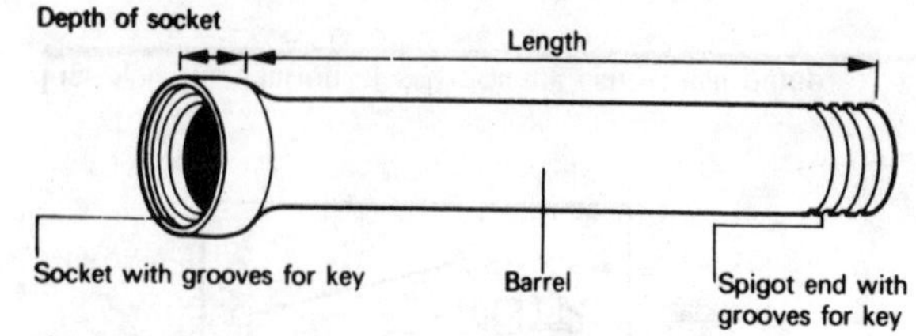

102

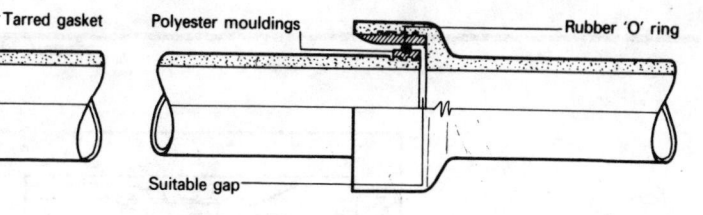

Fig. 8.52 British Standard clay drain pipe

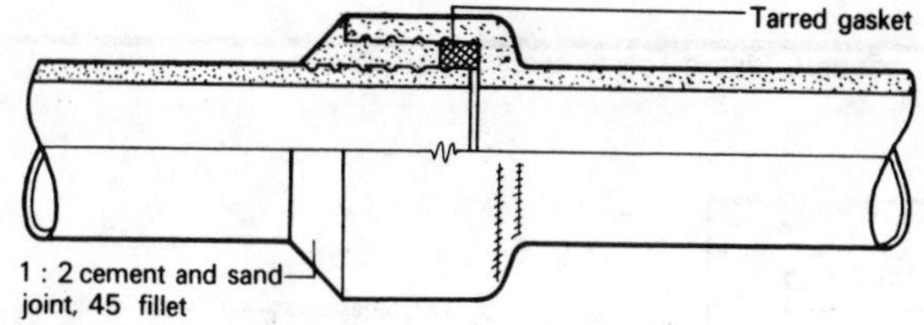

Labels: Depth of socket; Length; Socket with grooves for key; Barrel; Spigot end with grooves for key

Fig. 8.53 Cement mortar joint on clay pipe

Labels: Tarred gasket; 1 : 2 cement and sand joint, 45 fillet

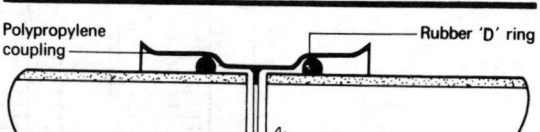

Fig. 8.54 'Hepseal' flexible joint on clay pipe

Labels: Polyester mouldings; Rubber 'O' ring; Suitable gap

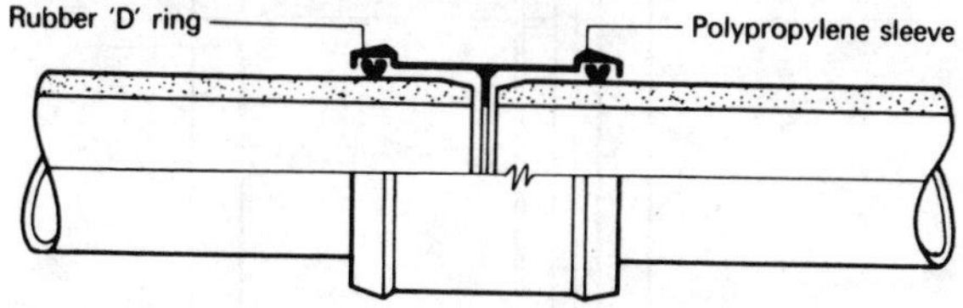

Fig. 8.55 'Hepsleve' flexible joint on clay pipe

Labels: Rubber 'D' ring; Polypropylene sleeve

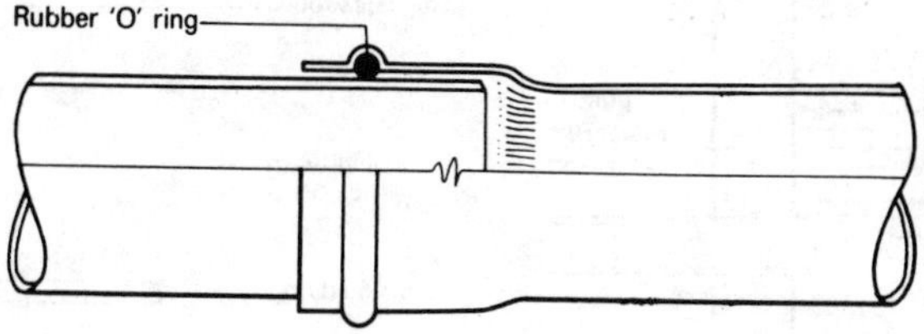

Fig. 8.56 Flexible joint on upvc drain pipe

Labels: Rubber 'O' ring

Fig. 8.57 Snap ring flexible joint on pitch fibre joint

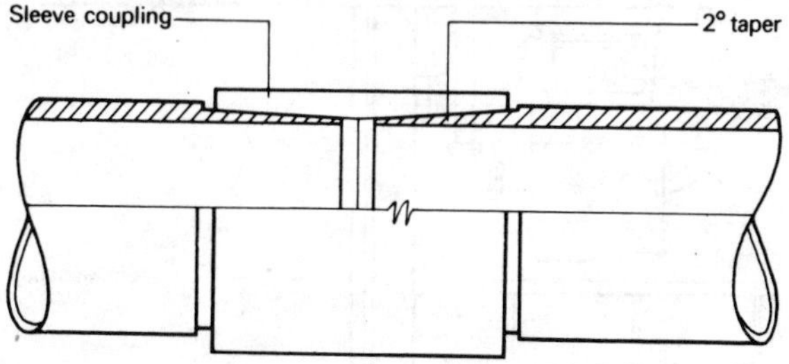

Labels: Polypropylene coupling; Rubber 'D' ring

Fig. 8.58 Taper type joint on pitch fibre drain pipe

Labels: Sleeve coupling; 2° taper

Fig. 8.59 Caulked lead joint on cast iron pipe

Labels: Caulked lead; Tarred gasket

Fig. 8.60 'Tyton' flexible joint on spun iron pipe

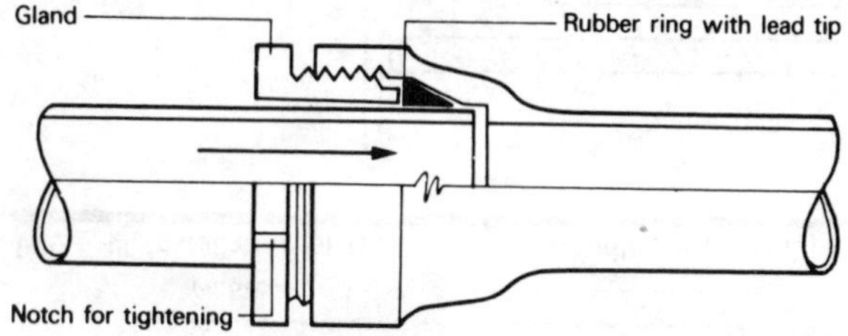

Labels: Rubber sealing ring

Fig. 8.61 Screwed gland flexible joint in iron pipe

Labels: Gland; Rubber ring with lead tip; Notch for tightening

Fig. 8.62 Spigot and socket joint for concrete surface or foul water drain or sewer

Labels: 1 : 2 cement and sand 45° fillet; Tarred gasket

Fig. 8.63 Ogee joint for concrete surface water drain or sewer

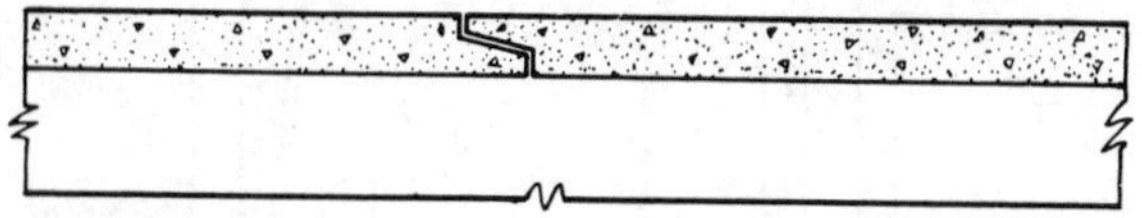

Fig. 8.64 Rebated joint for concrete surface or foul water drain or sewer

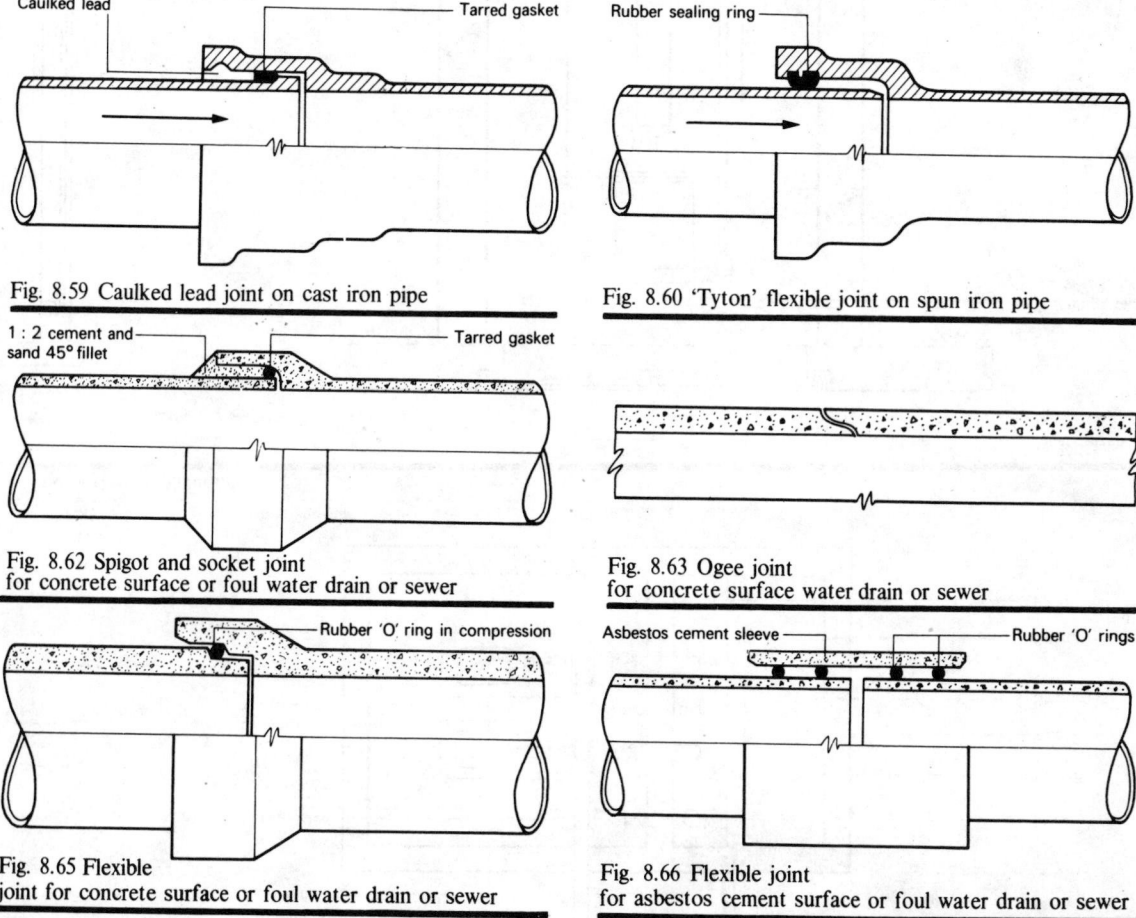

Fig. 8.65 Flexible joint for concrete surface or foul water drain or sewer

Labels: Rubber 'O' ring in compression

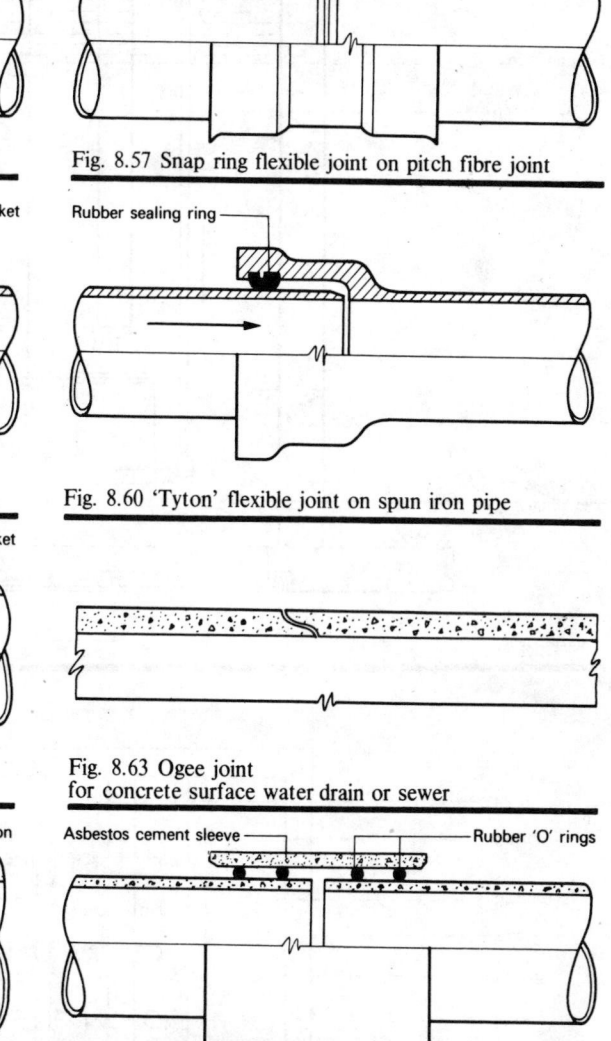

Fig. 8.66 Flexible joint for asbestos cement surface or foul water drain or sewer

Labels: Asbestos cement sleeve; Rubber 'O' rings

BS 437:Part 1, 1970, cast iron pipes are suitable for both surface water and foul water drainage. The pipes are made, socket and spigot, with internal diameters of 48, 74, 100, 150 and 225 mm and in lengths of 1.8, 2.74 and 3.66 m. The pipes are the ones most commonly used for drainage.

BS 1211: centrifugal spun iron pressure pipes are suitable for water, gas and sewage, where higher internal pipe pressures are expected. They are available in internal diameters from 75 to 900 mm and in lengths of 3.66, 4, 4.88 and 5.5 m. The pipes can be jointed by the traditional caulked lead, or by a flexible joint.

BS 4622:1970, grey cast iron pipes and fittings are suitable for drainage and are available with internal diameters of 80 to 700 mm and a variety of lengths. A variety of cast iron drainage fittings are also available. The pipes and fittings are protected from corrosion by a bituminous coating, both inside and outside the pipe, but care should be taken to prevent damage to the coating as it is affected by acidic effluents, or sulphates and acids in the soil. The pipe is very strong and is often used suspended above ground, on brick and concrete piers, under roads and buildings. Figure 8.59 shows a detail of a traditional caulked lead joint in case iron and Figs 8.60 and 8.61 show flexible telescopic joints, for cast iron and spun iron pipes.

Concrete

BS 556:Part 2, 1972, concrete pipes and fittings, lists pipes with internal diameters from 150 mm to 1.8 m, in lengths from 900 mm to 5 m. The pipes are suitable for normal drainage effluents, but may be liable to attack by acids. Pipes of 225 mm diameter and upwards are available with an external wrapping of glass fibre laminate, which reinforces the pipes and protects them from external attack. Various proprietory protective coatings are also available for internal and external application. Pipes may also be obtained manufactured from sulphate-resisting cement, which should be used in soil containing sulphates, but this type of pipe should not be used for acidic effluents, without first consulting the manufacturers. The pipes may be jointed by cement and sand mortar, which is similar to the clay pipe joint, or with a rubber O-ring.

Figure 8.62 shows a rigid spigot and socket joint for concrete surface, or foul water drain or sewer and Figs 8.63 and 8.64 show ogee and rebated joints for concrete surface water, drain or sewer. The flexible O-ring joint for concrete surface, or foul water drain, or sewer (shown in Fig. 8.65) is made by placing the O-ring on the spigot and then, by means of a ratchet pulling mechanism, forcing the pipe into the socket of the mating pipe, which allows the O-ring to roll into the position shown.

Asbestos-cement

BS 3656:1963, asbestos-cement pipes and fittings for sewage and drainage, covers pipes and fittings from 100 to 900 mm internal diameters, in lengths from 1 to 5 m. Substances that attack concrete may also attack asbestos-cement and where exposure to these substances is likely the manufacturer's advice should be sought. Bitumen-dipped pipes are available which gives increased resistance to attack, both internally and externally. There are three classes of pipe strength which can be obtained, depending upon the load which the pipe has to carry. The pipes are normally obtained with flexible O-ring joints as shown in Fig. 8.66.

Drainlaying

Pipes under roads should be laid with from 900 mm to 1.2 m of cover and pipes under fields and gardens with at least 600 mm of cover. For lesser depths, particularly where there is likelihood of heavy traffic, special protection of the pipes may be necessary. The drain trench should be as narrow as possible, so as to reduce the backfill load on the pipe to the minimum. Local soft spots in the trench bottom should be stabilised by tamping in granular material and large boulders and tree roots should be removed and replaced, by tamped granular material.

The drainage pipeline should be laid so as to provide flexibility and flexible joints are normally preferable to rigid joints, for the following reasons:

1. A minimum of skill is required in laying and is quicker, more reliable and cheaper.
2. Because of increased speed of laying, the time the trench is kept open is reduced to the minimum, with a possible reduction in pumping and less risk of the trench bottom becoming muddy.
3. The flexible joints reduce the risk of fracture of the pipeline, due to the ground movement, backfill or superimposed loads.
4. The pipeline may be tested immediately after laying.
5. Rectifying faults is quick and easy.
6. There is less delay in laying, due to wet or freezing site conditions.

Types of bedding

Class A: this type of bedding is used for rigid pipes such as vitrified clay and asbestos-cement. It is used where additional supporting strength of the pipe is required under roads, or where there is a risk of disturbing the pipeline after laying, such as when excavations have to be made alongside it at a later date. Figure 8.67 shows class A bedding, which is carried out as follows:

1. A layer of concrete at least 50 mm thick should be spread along the prepared trench bottom.
2. The pipes should be supported clear of this concrete by means of blocks or cradles placed under the pipes and a piece of resilient material about 14 mm thick placed between the pipes and the supporting blocks. The total clearance under the pipes should not be less than 100 mm.
3. The pipes should be tested and a piece of compressed board cut to the profile of the pipe, placed at the face of each pipe joint, as shown in Fig. 8.67.
4. A 1:2:4 concrete, using 14 mm maximum size aggregate, giving a minimum 28 days cube strength of 20 M N/m^2 should be carefully placed under the pipes between the profiles and extended up to the barrel of the pipes to the required height.
5. The trench should be backfilled by first placing three layers of selected soil, free from hard objects, to a depth of 100 mm each and carefully compacting each layer by hand tamping separately. The trench can now be backfilled, but a mechanical rammer should not be used until there is a minimum cover over the pipe of 600 mm.

Class B: this is the usual method of bedding, of both rigid and flexible pipes. Figure 8.68 shows class B bedding, which is carried out as follows:

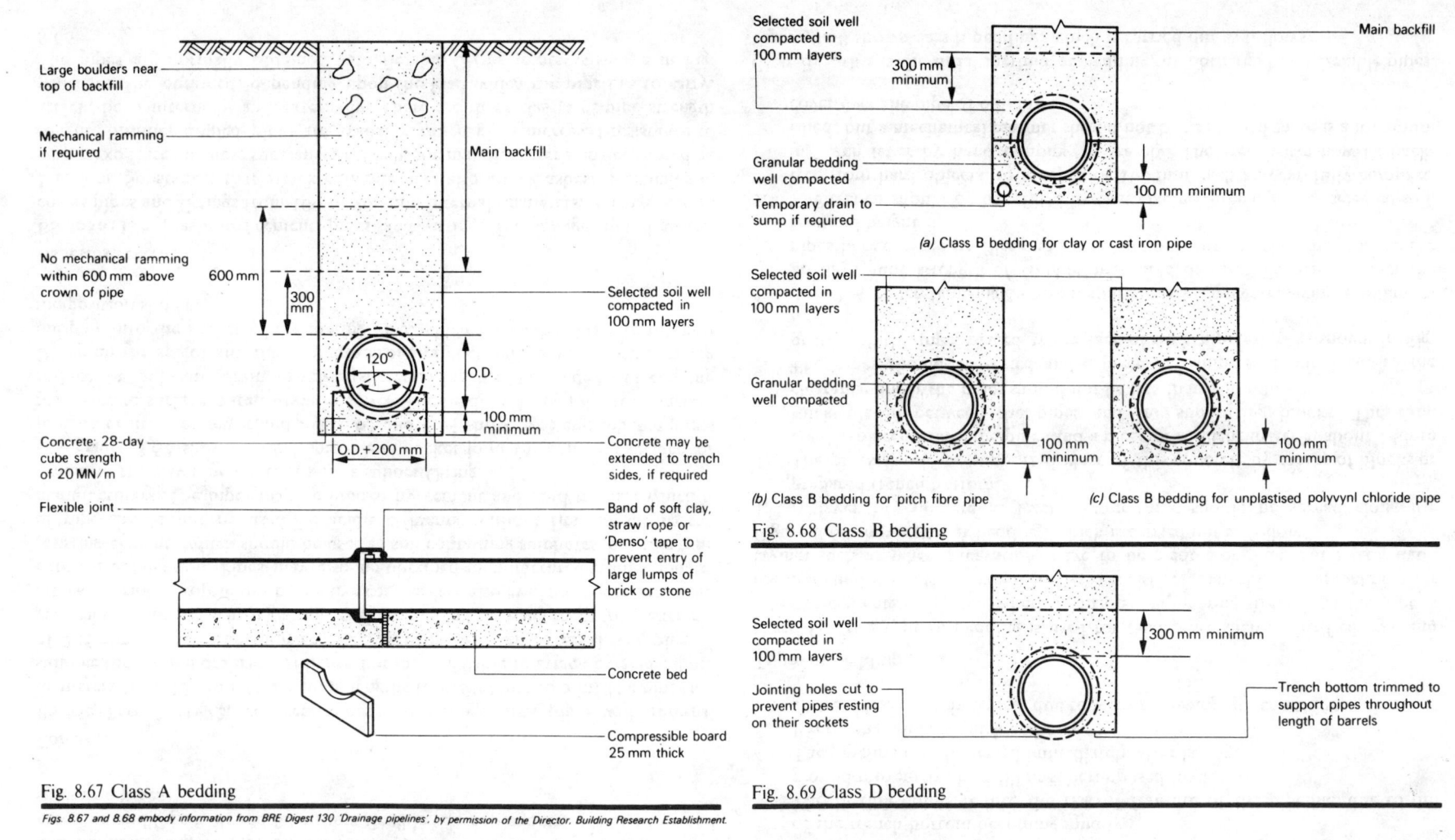

Fig. 8.67 Class A bedding

Fig. 8.68 Class B bedding

Fig. 8.69 Class D bedding

Figs. 8.67 and 8.68 embody information from BRE Digest 130 'Drainage pipelines', by permission of the Director, Building Research Establishment.

1. The trench bottom should be prepared and granular material consisting of broken stone or gravel 5 to 10 mm in size should be spread along the prepared trench bottom, to a depth of at least 100 mm; after compacting, the top should be levelled off.
2. Socket holes should be formed where necessary to allow the pipeline to rest uniformly on top of the granular material.
3. The pipeline should be laid and tested and granular material placed at either side of rigid pipes, to a depth of half the diameter of the pipe. For pitch fibre pipes the granular material should be brought up to the top of the pipe, and for unplasticised polyvinyl chloride pipe the granular material should be carried up to a minimum height of 100 mm above the crown of the pipe. For all types of pipes the granular material should be carefully compacted by hand tamping and the main backfill should be placed as described for class A bedding.

Class D: where the soil is reasonably dry, soft and fine grained, the pipes may be laid directly on the trench bottom. Figure 8.69 shows class D bedding, which is carried out as follows:

1. The trench bottom should be prepared by accurately trimming by hand and socket holes cut out, so that the soil will be in contact with the pipe barrel over the whole length of the pipeline. If too much soil is removed, it must be replaced to the correct level and thoroughly rammed.
2. The pipes should be laid and tested and the trench backfilled in 100 mm layers up to a level of 300 mm above the pipes. The main backfill should then be placed as described for class A bedding.

Pipes laid under buildings

The placing of drains under buildings should wherever possible be avoided, but if there is no other alternative the following points should be observed:

1. The drain should be laid in a straight line and at one gradient.
2. Access for cleansing should be provided to all parts of the drain.
3. Manholes placed inside the building should have double-sealed covers.
4. Flexible joints should be used and the pipe surrounded by 150 mm of granular material, or concrete, with a piece of compressed board 25 mm thick at the face of each pipe joint to maintain flexibility.

Trenches for drains and private sewers

The Building Regulations 1985 states that where any drain or private sewer is constructed adjacent to a loadbearing part of a building such precautions should be taken as may be necessary to ensure that the trench in which the drain or private sewer is laid in no way impairs the stability of the building. Except where the nature of the ground makes it unnecessary, where any drain or private sewer is adjacent to a wall and the bottom of the trench is lower than the foundation of the wall, the trench should be filled in with concrete, to a level which is not lower than the bottom of the foundation of the wall, by more than the distance from the foundation to the near side of the trench, less 150 mm. Provided that where the trench is within 1 m of the foundation of the wall the trench should be filled in with concrete to the level of the underside of the foundation. The concrete filling required by the foregoing paragraph should have such expansion joints as are necessary to ensure that no continuous length of filling exceeds 9 m. Figure 8.70 shows the method of ensuring that the drain or sewer trench is prevented from weakening the stability of the building.

Use of soakaways

In the absence of a surface water sewer and if it is unnecessary to store rainwater, it is often advantageous to dispose of surface water from impermeable areas to an underground soakaway. The use of a soakaway will reduce the flow of rainwater to a river or stream and thus help to prevent flooding of land during a storm or continuous heavy rainfall. It may also be used to dispose of the final effluent from a small sewage purification plant.

Types of soakaways

A soakaway consists of a pit dug into impermeable sub-soil, from which the water may percolate into the surrounding ground.

There are two types:

1. *Filled:* for small soakaways, the pit can be filled with coarse granular material such as broken bricks, crushed sound stone or river gravel, with a size range of 10 to 150 mm. The end of the inlet pipe should be surrounded with large pieces, to ensure that the rainwater can flow freely into the granular material. This type of soakaway is cheap to construct, but its water-holding capacity is greatly reduced and there is also a risk that leaves and silt may enter and reduce the water-holding capacity still further. Figure 8.71 shows a section through a filled soakaway.

2. *Unfilled:* for larger soakaways, the excavated pit may be prevented from collapsing by lining with brickwork or stone laid dry, or cement-jointed honey-combed brickwork. Alternatively, the lining may consist of perforated concrete rings laid dry, surrounded by granular material to lead the water into the soil. The top can be covered with a standard reinforced concrete manhole top and fitted with an access cover, so that it is possible to clean out leaves and silt. Figures 8.72 and 8.73 show small and large unfilled soakaways.

Design

A soakaway can only be used in pervious subsoil such as gravel, sand, chalk or fissured rock and its base must be above the ground water table. In order to design a soakaway, the following points must be taken into account:

1. The local authority drainage regulations.
2. Area of impermeable surface, such as, roofs and paved areas.
3. Type of subsoil.
4. The rainfall intensity for the district during a storm, or continuous heavy rainfall.
5. The height of the ground water table at all times of the year.
6. Whether it is to be filled or unfilled.

For large impermeable areas it may be more economical to provide two or more soakaways and this saves on the lengths of the surface water drains.

Siting

Soakaways should be at least 3 m from a building and built on land lower than, or sloping away from, the building. In this position there is less risk that the building foundations are not weakened by the percolation of water from the soakaways.

Size

The British Standard, Code of Practice Building Drainage, suggests that a common method of sizing a soakaway is to provide a water storage capacity equal to at least 13 mm of rainfall over the impermeable area. The Building Research Establishment Digest 151 describes the method of testing the permeability of the soil and provides a graph from which the size of a soakaway may be found from the permeability test time and the drained area. The digest also suggests that a soakaway may be sized on the basis of a rainfall intensity of 15 mm per hour. The capacity of a soakaway for an impermeable surface of 100 m^2 would be:

$$100 \text{ m}^2 \times 0.015 \text{ m} = 1.5 \text{ m}^3$$

Garage drainage

The Public Health Act 1936 makes it illegal to discharge spirit into a public sewer. Inflammable vapours given off by petrol would cause considerable danger if ignited and manhole covers in the roads or footpaths could be blown off by an explosion, thus endangering the public. These vapours may also cause danger to workmen carrying out maintenance work in the sewers. To prevent the entry of petrol into the sewer, the floor washings of large garages, petrol stations and even small garages should be intercepted. For a large garage a petrol interceptor

should be built and the garage drain connected to it before being connected to the public sewer.

Figure 8.74 shows a section of a petrol interceptor housing three compartments, each provided with a ventilating pipe so that the petrol vapours may pass to safety. Petrol floats upon the water of each compartment, and as the water flows through the interceptor the petrol will evaporate before the water discharges to the public sewer. The outlet pipe in the first chamber is set higher than those in the second and third chambers, to allow for sediment from the washing-down areas. The chambers should be emptied by a pump, flushed and refilled with clean water periodically.

Floor drainage

Water from the garage floor should be drained into a trapped garage gully, or gullies, before discharging into the interceptor. Figure 8.75 shows a section through a garage gully, which is provided with a perforated galvanised steel bucket, and if cleaned out regularly will prevent most of the silt from passing into the interceptor. The gully would be sufficient for a small garage without an interceptor.

Drainage layout

The petrol interceptor should only be used for water from the garage floor and surface water from the roof and external paved areas should be excluded. Figure 8.76 shows the layout of a drainage system for a garage floor. The minimum slope of the floor to each gully should be 1 in 80 and the area of surface to be drained by each gully should be about 50 m².

Drain testing and inspection

After the drain has been laid and before backfilling, or placing concrete, or granular material round the pipes, it should be tested by either water or air. If any leak occurs, the defective pipe or joint should be rectified and the drain again tested. Wherever possible, testing should be carried out between the manholes and short branch drains tested along with the main drainage system. Long branch drains and manholes should be tested separately. The test before backfilling should be carried out as soon as is practicable and the pipe should be strutted, to prevent any movement of the drain during the test. A temporary bend and stand pipe should be fitted at the head of the drain and a stopper fitted at the lower end. Alternatively, the test may be applied by means of a rubber tube connected to a vessel and the drain stopper. Figure 8.77 shows the method of carrying out the water test.

Building Regulations and Code of Practice tests

The Building Regulations 1985 and the Code of Practice B.S. 8301, Building Drainage 1985 give the following procedures for carrying out tests on gravity drains and private sewers up to 300 mm in diameter.

Water test

1. The drain should be filled with water to give a test pressure equal to 1.5 m head of water above the soffit of the drain at the high end but not more than 4 m head of water above the soffit fo the drain at the low end. Steeply

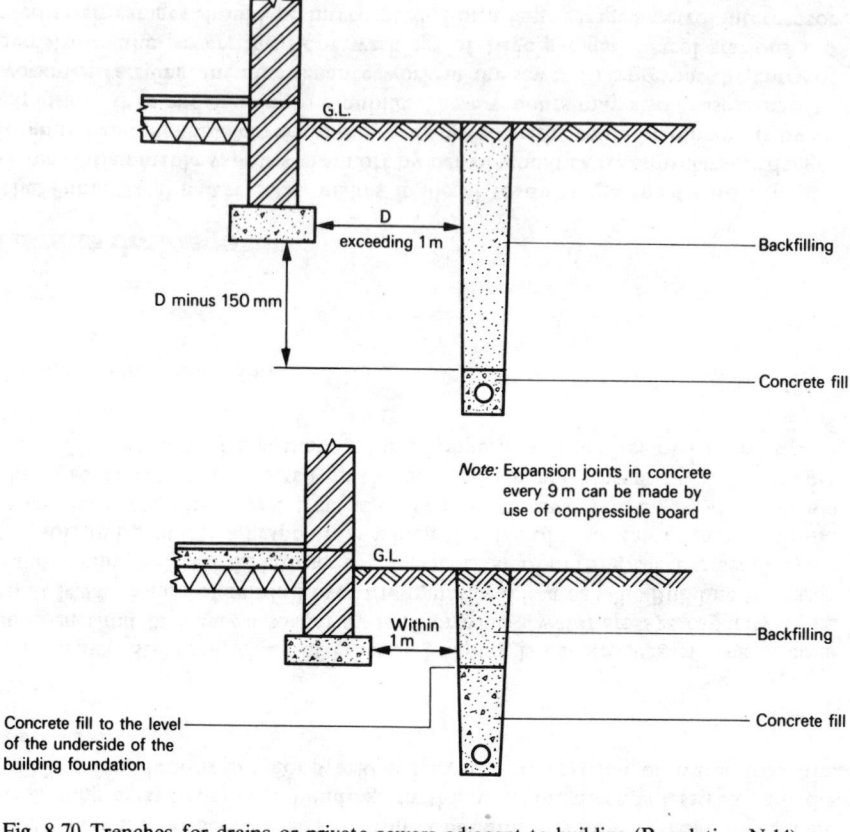

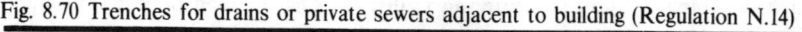

Fig. 8.70 Trenches for drains or private sewers adjacent to building (Regulation N.14)

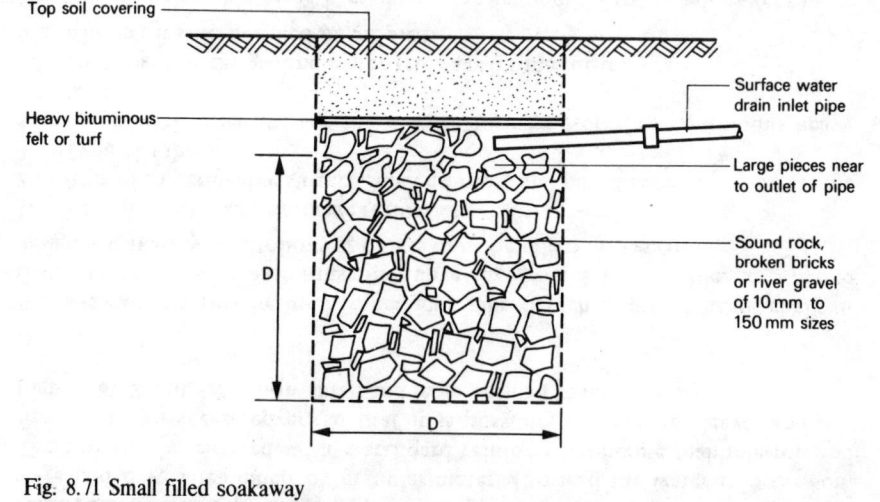

Fig. 8.71 Small filled soakaway

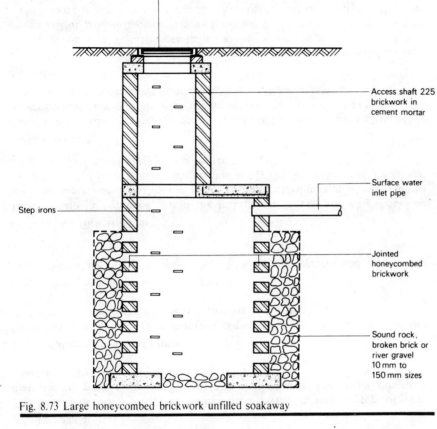

Fig. 8.73 Large honeycombed brickwork unfilled soakaway

Access cover and frame

Access shaft 225 brickwork in cement mortar

Step irons

Surface water inlet pipe

Jointed honeycombed brickwork

Sound rock, broken brick or river gravel 10 mm to 150 mm sizes

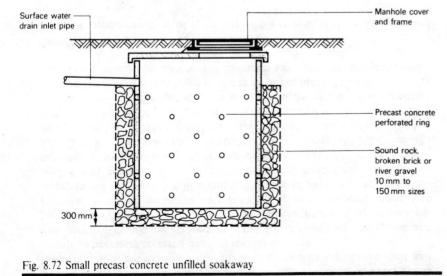

Fig. 8.72 Small precast concrete unfilled soakaway

Surface water drain inlet pipe

Manhole cover and frame

Precast concrete perforated ring

Sound rock, broken brick or river gravel 10 mm to 150 mm sizes

300 mm

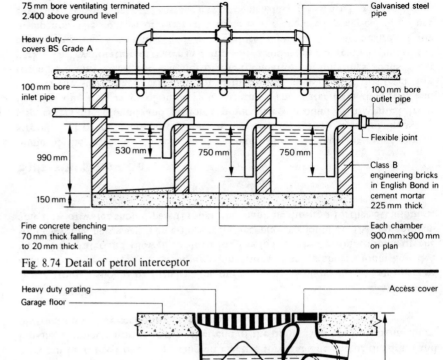

Fig. 8.74 Detail of petrol interceptor

75 mm bore ventilating terminated 2.400 above ground level

Galvanised steel pipe

Heavy duty covers BS Grade A

100 mm bore inlet pipe

100 mm bore outlet pipe

Flexible joint

990 mm

530 mm   750 mm   750 mm

150 mm

Class B engineering bricks in English Bond in cement mortar 225 mm thick

Fine concrete benching 70 mm thick falling to 20 mm thick

Each chamber 900 mm × 900 mm on plan

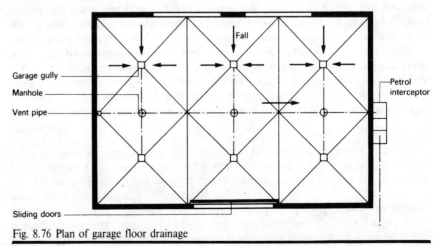

Fig. 8.75 Section through garage gully

Heavy duty grating

Garage floor

Access cover

Galvanised steel perforated sediment pan

600 mm

Fig. 8.76 Plan of garage floor drainage

Fall

Garage gully

Manhole

Vent pipe

Petrol interceptor

Sliding doors

graded drains should be tested in stages so that the head of water at the lower end does not exceed 4 m. This is to prevent damage to the drain and it may be necessary to test a drain in several sections.

2. The pipeline should be allowed to stand for 2 hours for absorption and topped up with water as necessary. After 2 hours the loss of water from the pipeline should be measured by noting the quantity of water needed to maintain the test head for 30 minutes. The rate of water loss should not exceed 1 litre/hour per metre diameter, per metre run of pipe. For various pipe diameters the rate of loss per metre run during the 30 minute period is 0.05 litre for 100 mm pipe; 0.8 litre for 150 mm pipe; 0.12 litre for 225 mm pipe and 0.15 litre for 300 mm pipe. Figure 8.77 shows the method of carry out the water test using a rubber tube connected to a testing vessel.

Alternatively, the test may be carried out by means of a temporary bend and stand-pipe connected to the head of the drain. The fall of water in the vessel or stand-pipe may be due to one or more of the following causes:

(a) Absorption by pipes or joints

The initial absorption may be of the order of 5 per cent of the total weight of the pipeline.

(b) Trapped air

This usually occurs at the joints and the amount will vary with the type of joint, diameter of pipe, number of joints and the gradient. Eventually the air is absorbed by the water, but this can take some time.

(c) Sweating of pipes and joints

Occasionally pipes and joints under water pressure may sweat slightly, but this should not be considered a cause for rejection of the pipeline.

(d) Leakage from defective pipes or joints

The defective pipe or joint should be taken out and replaced and the pipeline re-tested.

(e) Leakage from stoppers

These should be tightened, but if the leakage continues a new plug should be inserted. Sometimes a leakage can occur on the threaded portion of the plug. All equipment used for testing should be thoroughly checked before being used and the rubber surfaces of plugs should be free from grit.

Final water test

There is a risk that the drain or private sewer can be damaged during backfilling or surrounding the pipeline with concrete. The final test therefore should be carried out after the backfilling or concreting has been completed.

Air test

The length of drain to be tested should be plugged and air pumped into the pipe until a pressure of slightly more than 100 mm water gauge is obtained. Where gullies and/or ground-floor appliances are connected the test pressure should be slightly more than 50 mm water gauge. A change in air temperature will affect the test pressure and therefore 5 minutes should be allowed for pressure stabi-

lisation. The air pressure should be adjusted to 100 mm or 50 mm water gauge as necessary. Without further pumping, the head of water during a period of 5 minutes should not fall more than 25 mm and 12 mm for a 100 mm and 50 mm water gauge test pressures respectively.

Figure 8.78 shows the method of carrying out an air test. The test is carried out by fixing a stopper, sealed with a cap, at one end of the drain and pumping in air at the other end until the U gauge shows the required head of water.

Smoke test

The use of smoke cartridges for this test is not to be recommended, due to the possibility of the build up of high pressure inside the drain. The test by use of a smoke machine is usually applied to existing drains with the purpose of locating a leak. In order to ensure that the drain is filled with smoke, a rubber tube should be passed through the water seal of gully traps, so that air contained inside the drain may escape. Figure 8.79 shows the method of carrying out a smoke test. The test is carried out by placing a stopper, sealed with a cap, at one end of the drain and pumping in smoke at the other. The dome rises with the action of the bellows and if the drain is sound is maintained in this elevated position. Figure 8.80 shows the smoke machine during a test.

Types of stoppers

The drain is plugged or stopped by either an expanding drain plug (shown in Fig. 8.81) or by an inflatable bag (shown in Fig. 8.82).

Tests for straightness and obstruction

This can be carried out by placing an electric torch at one end of the drain and looking through a mirror at the other end of the drain. Any fault in alignment or obstruction will be seen through the mirror.

Testing manholes

A bag stopper should be fitted in the outlet of the manhole and bag stoppers or expanding plugs fitted at all other connections. The manhole should be filled with clean water and allowed to stand for 8 hours for absorption, topping up the water level as necessary. The criterion for acceptance should be, that no appreciable flow of water should pass through the manhole during a further 30 minutes.

Sub-soil drainage

Some of the water which reaches the ground either as snow or rain flows to streams and rivers or into the surface water drainage system. Some of the water also enters the ground and at some point below ground is prevented from flowing, due to a layer of impervious stratum. The water level thus rises and reaches the sub-surface level, known as the 'water table', and if this level rises too high it may affect the stability of the foundations of buildings, roads or car parks. The evaporation of water in damp soils also lowers the temperature of the air surrounding a building and is therefore injurious to the health of the occupants. The control of the water table is also required in agriculture, since excessive moisture near the surface excludes air which is essential to plant life.

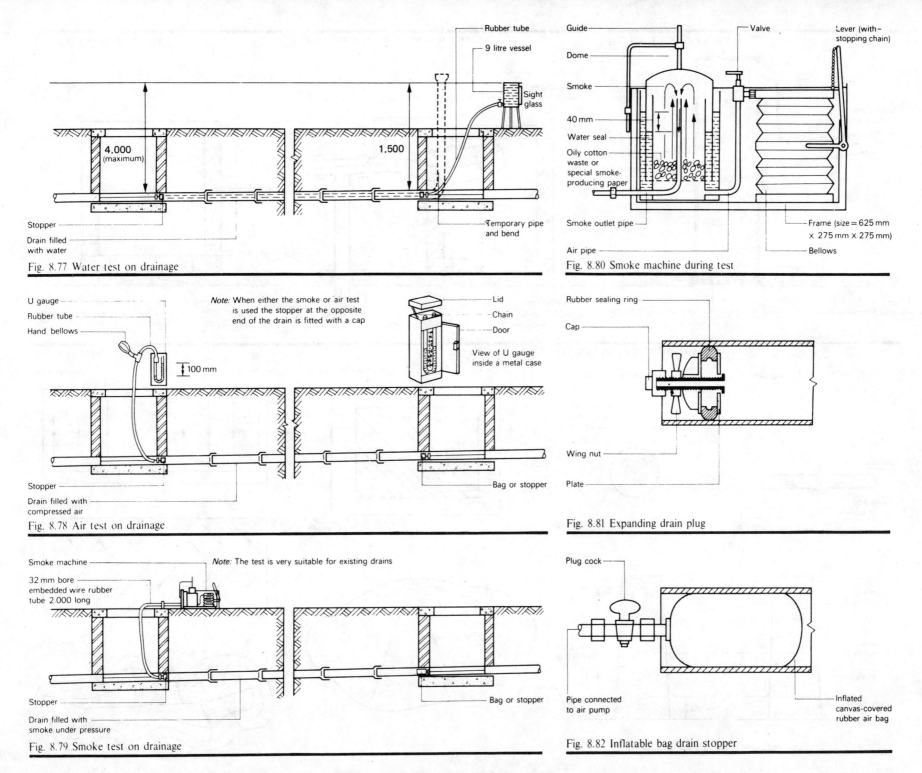

Rubber tube
9 litre vessel
Sight glass
4,000 (maximum)
1,500
Stopper
Drain filled with water
Temporary pipe and bend

Fig. 8.77 Water test on drainage

Guide
Valve
Lever (with–stopping chain)
Dome
Smoke
40 mm
Water seal
Oily cotton waste or special smoke-producing paper
Smoke outlet pipe
Air pipe
Frame (size = 625 mm × 275 mm × 275 mm)
Bellows

Fig. 8.80 Smoke machine during test

U gauge
Rubber tube
Hand bellows
100 mm
Stopper
Drain filled with compressed air

Note: When either the smoke or air test is used the stopper at the opposite end of the drain is fitted with a cap

Lid
Chain
Door
View of U gauge inside a metal case

Bag or stopper

Fig. 8.78 Air test on drainage

Rubber sealing ring
Cap
Wing nut
Plate

Fig. 8.81 Expanding drain plug

Smoke machine
32 mm bore embedded wire rubber tube 2.000 long
Stopper
Drain filled with smoke under pressure

Note: The test is very suitable for existing drains

Bag or stopper

Fig. 8.79 Smoke test on drainage

Plug cock
Pipe connected to air pump
Inflated canvas-covered rubber air bag

Fig. 8.82 Inflatable bag drain stopper

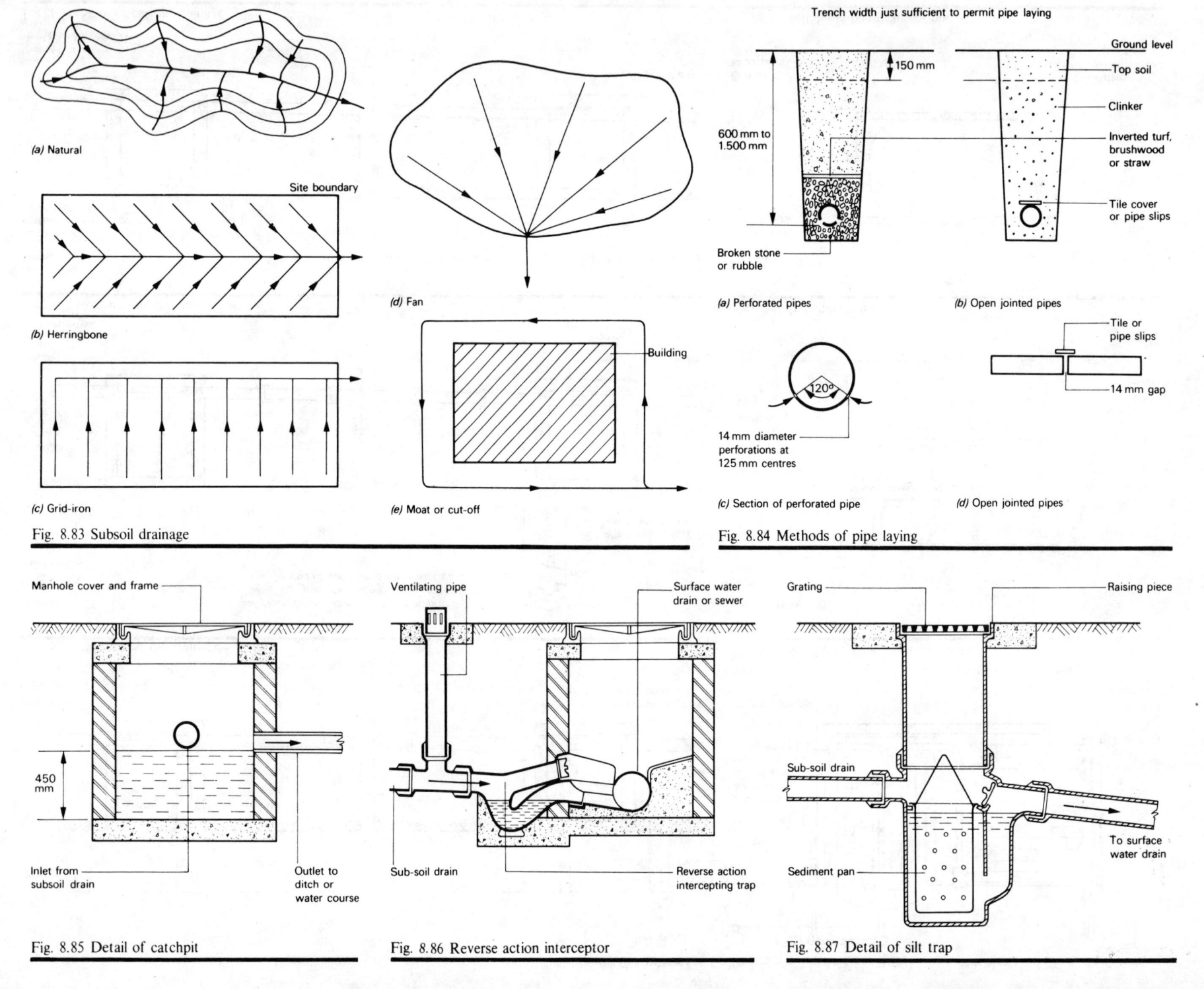

(a) Natural

(b) Herringbone

Site boundary

(c) Grid-iron

Fig. 8.83 Subsoil drainage

(d) Fan

Building

(e) Moat or cut-off

Trench width just sufficient to permit pipe laying

Ground level

150 mm

Top soil

600 mm to
1.500 mm

Clinker

Inverted turf,
brushwood
or straw

Tile cover
or pipe slips

Broken stone
or rubble

(a) Perforated pipes

(b) Open jointed pipes

Tile or
pipe slips

14 mm gap

120°

14 mm diameter
perforations at
125 mm centres

(c) Section of perforated pipe

(d) Open jointed pipes

Fig. 8.84 Methods of pipe laying

Manhole cover and frame

450
mm

Inlet from
subsoil drain

Outlet to
ditch or
water course

Fig. 8.85 Detail of catchpit

Ventilating pipe

Surface water
drain or sewer

Sub-soil drain

Reverse action
intercepting trap

Fig. 8.86 Reverse action interceptor

Grating

Raising piece

Sub-soil drain

Sediment pan

To surface
water drain

Fig. 8.87 Detail of silt trap

The Building Regulations 1985 states: 'Wherever the dampness or position of the site renders it necessary, the sub-soil of the site shall be effectively drained, or such other steps shall be taken as will effectively protect the building against damage by moisture.'

Systems of sub-soil drainage

The artificial control of the water table is achieved by a system of sub-soil drainage, using open jointed, porous or perforated pipes and the following systems may be used (see Fig. 8.83).

1. *Natural:* the pipes are laid to follow the natural depressions or valleys of the site; branches discharge into the main as tributaries do into a river.

2. *Herringbone:* the system consists of a number of drains into which discharges from both sides smaller subsidiary branch drains parallel to each other, but at an angle to the mains forming a series of herringbone patterns. Normally these branch drains should not exceed 30 m in length.

3. *Grid iron:* a main or mains drain is laid near to the boundaries if the site into which subsidiary branches discharge from one side only.

4. *Fan:* the drains are laid so as to converge on a single outlet at one point on the boundary of a site without the use of a main collecting drain.

5. *Moat or cut off:* this system consists of drains laid on one or more sides of a building to intercept the flow of sub-soil water and carry it away, so protecting the foundations of a building.

Table 8.6 gives the spacing of branch drains for the herringbone and grid iron systems.

Pipe laying

The main pipes should be either 75 or 100 mm nominal bore and the branch pipes 65 or 75 mm nominal bore. The pipes should be laid at between 600 and 900 mm in heavy soils, and deeper in light soils and the gradients rather by the fall of land than by consideration of self-cleansing velocity. Figure 8.84 shows the methods of laying the pipes.

Discharge of water

The ground water may discharge into a soakaway, or through a catch pit into the nearest ditch or water course (see Fig. 8.85). Alternatively, the water may discharge into a surface water drain, through either a silt trap or a reverse action intercepting trap (see Figs 8.86 and 8.87).

Types of pipes

1. Clayware field drain pipes BS 1196.
2. Concrete pipes BS 4101.
3. Perforated clay pipes BS 65 and 540.
4. Perforated pitch fibre pipes BS 2760.
5. Porous concrete pipes BS 1194.
6. Surface water clay pipes BS 65 and 540.
7. Plastic pipes BS 3506.

Note: The use of concrete pipes may be unsuitable where sub-soil water carries

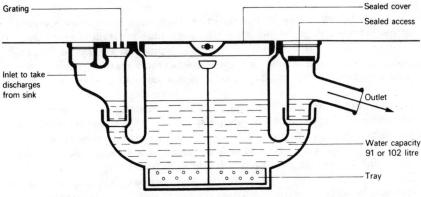

Dimensions: 600 mm × 450 mm × 135 mm and 915 mm × 450 mm × 135 mm

Fig. 8.88 Details Detail of grease trap

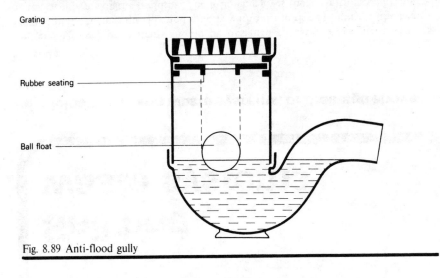

Fig. 8.89 Anti-flood gully

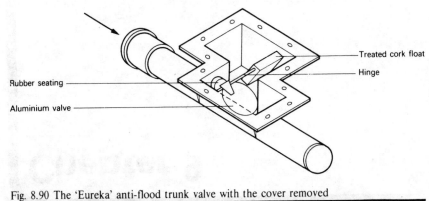

Fig. 8.90 The 'Eureka' anti-flood trunk valve with the cover removed

sulphates, or is acid due to the presence of peat. In other circumstances, porous concrete pipes prevent the entry of fine particles of silt into the pipes. In order to prevent the soil from entering open jointed, or perforated pipes, they should be surrounded by a filter composed of clinker, broken stone or rubble.

Table 8.7 Spacing of branch drains. British Standard Code of Practice BS 8301: 1985 Building Drainage.

Soil	Distance between groundwater drains for various depths to invert of main drains	
	Mains 0.8 to 1.0 m deep	Mains 1.0 to 1.5 m deep
Sand	–	45–90
Sandy loam	–	30–45
Loam	18–20	20–30
Clay loam	12–16	15–20
Sandy clay	6–12	–
Clay	2–6	–

Grease traps

Where grease may enter the drainage system from canteen sinks, a grease trap should be fixed between the sinks and the drain. Figure 8.88 shows a detail of a grease trap which contains a large volume of water. Grease entering the trap is congealed by the water and is periodically removed by lifting the perforated tray.

Anti-flooding devices

On a drainage system which is liable to back flooding from a surcharged sewer, it is sometimes necessary to provide an anti-flooding device. Figure 8.89 shows an anti-flood gully which contains a copper or plastic ball. When back-flooding occurs, the water rising in the gully forces the ball float against a rubber seating and so prevents water escaping through the top of the gully. Figure 8.90 shows the 'Eureka' anti-flooding trunk valve. Under normal conditions the hinged valve remains open, but when back-flooding occurs the cork float rises and closes the valve against the rubber seating.

Chapter 9

Soil and waste systems

Principles of soil and waste systems, or drainage above ground

General principles

Soil and waste pipe systems should be designed to carry away the discharges from sanitary fittings quickly and quietly without the risk of injury to the health of the occupants of the building. The following gives the definitions and requirements for the installation of soil and waste systems:

Definitions

1. *Soil appliances:* includes a water closet or urinal receptacle, bed-pan washer, bed-pan sink and slop sink.
2. *Soil pipe:* means a pipe (not being a drain) which conveys soil water either alone or together only with waste water, or rainwater, or both.
3. *Ventilating pipe:* means a pipe (not being a drain) open to the external air at its highest point, which ventilates a drainage system, either by connection to a drain, or to a soil pipe, or waste pipe and does not convey any soil water, waste water or rainwater.
4. *Waste appliance:* includes a slipper bath, lavatory basin, bidet, domestic sink, cleaner's bucket sink, drinking fountain, shower tray, wash fountain, washing trough and wash-tub.
5. *Waste pipe:* means a pipe (not being a drain, or overflow pipe) which conveys waste water, either alone or together only with rainwater.

6. *Waste water*: means used water not contaminated by soil water or trade effluent.

Requirements

1. *Water seals in traps*

Such provision shall be made in the drainage system of a building, either above or below the ground as may be necessary to prevent the destruction under working conditions of the water seal in any trap in the system, or in any appliance which discharges into the system.

2. *Sizes of pipes*

The internal diameter of any soil, waste pipe or ventilating pipe shall not be less than the internal diameter of any pipe, or the outlet of any appliance which discharges into it. The internal diameter of a soil pipe shall not be less than: (*a*) 50 mm, if it exclusively serves one or more urinals, or (*b*) 75 mm, in any other case. The internal diameter of a waste pipe shall not be less than 32 mm if it serves a lavatory basin.

3. *Materials, fixing and joints*

Any soil pipe, waste pipe or ventilating pipe shall be composed of suitable materials of adequate strength and durability. Bends in pipes should have the largest practicable radius of curvature and there should be no change in the cross-section of the pipe throughout the bend. Pipes should be adequately supported throughout their lengths, without restricting thermal movement, and any fitting which gives such support should be securely attached to the building.

A soil or waste pipe should be fixed inside the building, except a waste pipe from a waste fitting fixed on the ground floor, providing that the pipe discharges into a trap fitted with a suitable grating, so that the water may discharge below the grating and above the level of water in the trap (see Fig. 9.1). All pipes should be placed so that they are reasonably accessible for maintenance and repair, and be provided with such means of access for internal cleansing. Joints should be made so that they remain air-tight and any dissimilar materials used for jointing should not cause electrolytic corrosion. *Note:* The Building Regulations 1985 permit a soil and vent stack sited externally for buildings up to three storey in height.

4. *Traps*

A trap should be fitted close to a waste or soil appliance, unless the appliance has an integral trap. Alternatively:

(*a*) a waste pipe serving two or more lavatory basins, may discharge into a sealed gully which is provided with means of access for cleaning. Bottle traps with anti-siphon valves may be used (see Figs. 9.2 and 9.27).

(*b*) a waste pipe serving a number of lavatory basins, or shower trays, or both fixed in a range, may have one trap fitted at the end of the pipe, providing the length of the pipe does not exceed 5 m in length (see Fig. 9.3). A trap should have an adequate diameter water seal and means of access for internal cleansing. Figure 9.4 shows three basins having individual traps, connected to a common waste pipe.

5. *Tests*

A soil and waste system should be capable of withstanding a smoke or air test for a minimum period of 3 minutes, at a pressure equivalent to a head not less than 38 mm head of water (see Fig. 9.5). The Building Regulations 1985 require that during the 3 minutes every trap should maintain a water seal of at least 25 mm.

Loss of water seal in traps

1. *Self-siphonage*

This is usually greater in wash basins and other small appliances, due to their curved shape, and is caused by a moving plug of water in the waste pipe. A partial vacuum is created at the outlet of the trap, thus causing siphonage of the water. Air bubbles are drawn through the trap during discharge of the water, which assist the siphonage by their pumping action.

The diameter, length and slope of the waste pipe affect the hydraulic behaviour of the plug of water. Experimental work at the Building Research Establishment has set limiting values within which it is safe to say that the seal of a trap will not be lost by self-siphonage. If the diameter, length and slope of the waste pipe fall outside these limits set for unvented wastes, then either a vent pipe must be provided, or a resealing trap used. Figure 9.6 shows how self-siphonage occurs.

2. *Induced siphonage*

This is caused by the discharge of water from another sanitary fitting connected to the same pipe. In either a vertical, or a horizontal main waste pipe, water passing the connection of a branch waste pipe may draw air from it, thus creating a partial vacuum at the outlet side of the trap and causing siphonage of the water. Figure 9.7 shows how induced siphonage occurs.

3. *Compression or back pressure*

This is caused by a build up of air pressure near the bend at the foot of the stack. A waste pipe connected to the stack in the pressure zone may have the seal of the trap lost by the compressed air forcing out the water. Detergent foam increases the risk of compression. Figure 9.8 shows the loss of the seal of a trap by compression.

4. *Capillary attraction*

This is caused by a piece of porous material being caught at the outlet of the trap, so that one end is in the water and the other end is hanging over the outlet. The water may be drawn out of the trap by capillary attraction (see Fig. 9.9).

5. *Wavering out*

This is caused by high-velocity gusts of wind passing over the vent pipe, which draw some of the air out of the pipe, thus creating a partial vacuum on the outlet side of the trap. The gusts of wind cause the water in the trap to oscillate, until the water seal in the trap is broken (see Fig. 9.10).

6. *Evaporation*

When the trap is not in use, the rate of evaporation of the water will depend

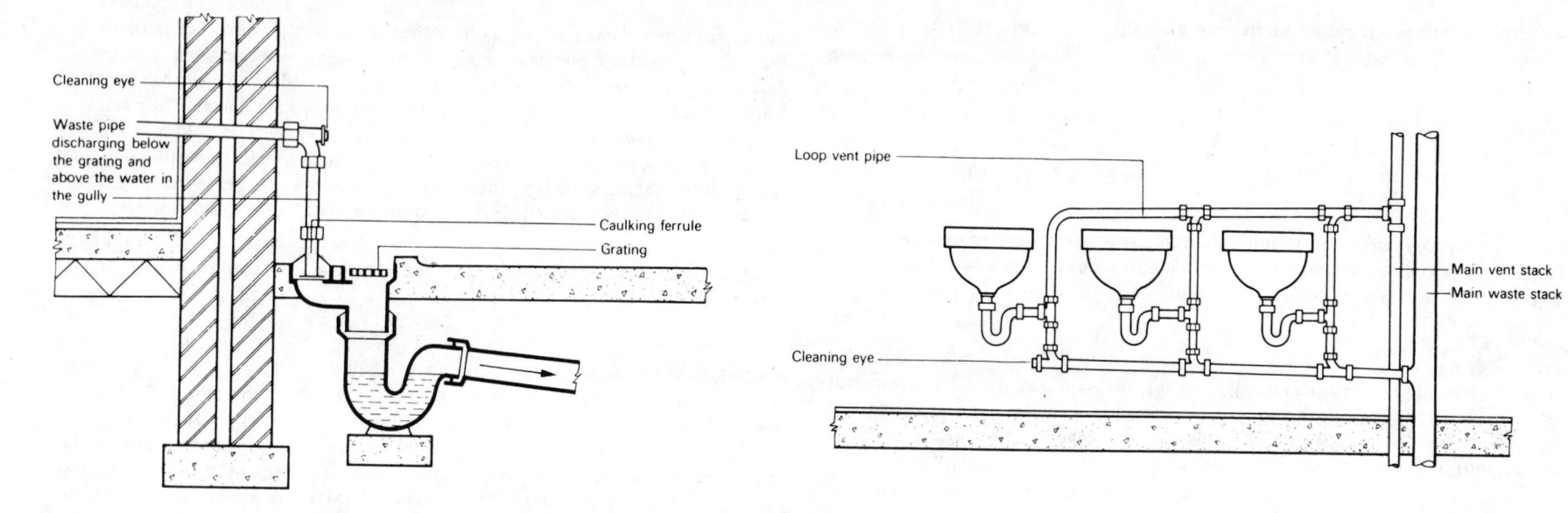

Fig. 9.1 Waste pipe from an appliance situated on the ground floor

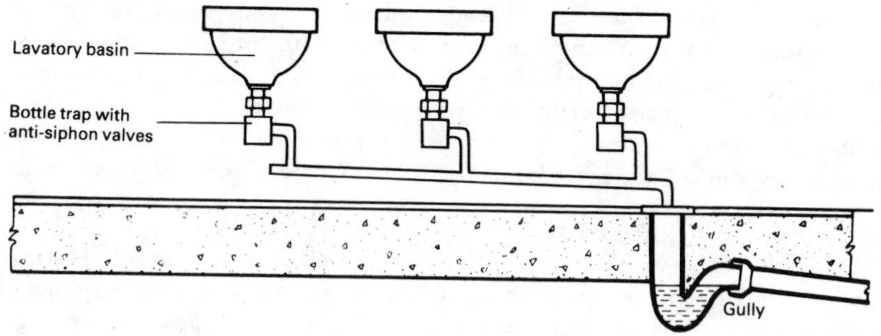

Fig. 9.4 Individual traps connected to a common waste pipe

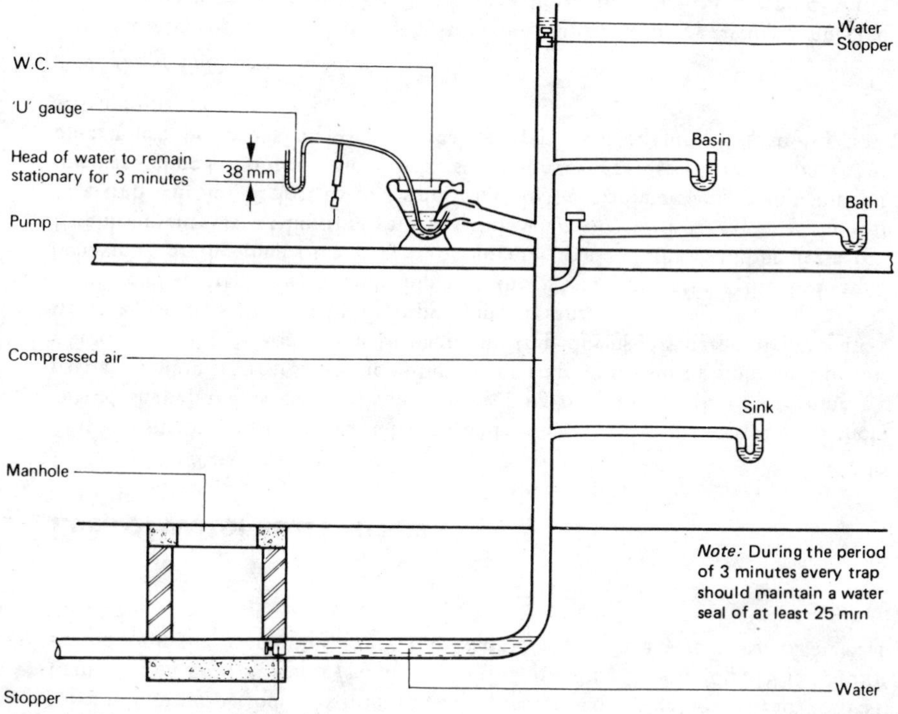

Fig. 9.2 Use of bottle traps with anti-siphon valves

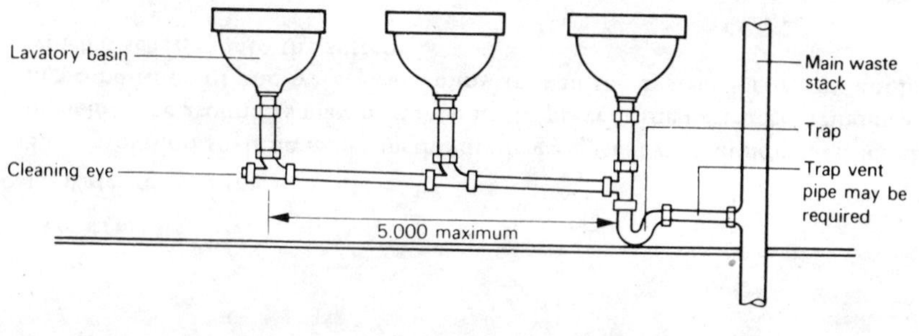

Fig. 9.3 Use of common waste and trap

Fig. 9.5 Air test on soil and waste pipes

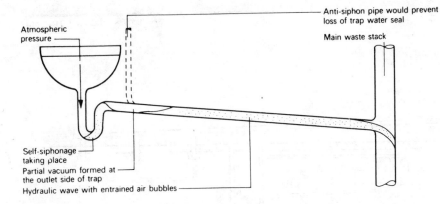

Fig. 9.6 Self siphonage

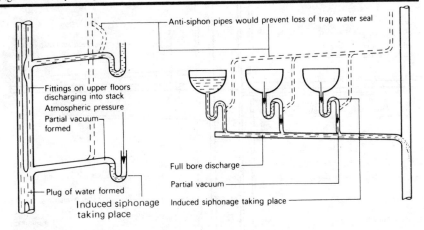

Fig. 9.7 Induced siphonages

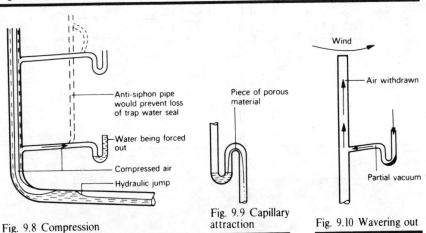

Fig. 9.8 Compression

Fig. 9.9 Capillary attraction

Fig. 9.10 Wavering out

upon the relative humidity of the air in the room. The rate is approximately 2.5 mm per week, so a 25 mm seal would last 10 weeks.

7. *Momentum*

This is caused by a sudden discharge of water into a fitting and the force of the water may be sufficient to unseal the trap. The discharge of a bucket-full of water into a W.C. is the most usual cause.

8. *Leakage*

This is usually due to a faulty joint on the cleaning eye, or a crack in the trap due to expansion and contraction of the material.

The two- or dual-pipe system

This is the traditional method of removing the discharges from sanitary fittings and is used when waste fittings are at some distance from the soil fittings. In buildings such as schools, factories and hospitals, wash basins and sinks may be installed in rooms which are at some distance from the main soil stack and it is therefore often cheaper to install a separate vertical waste stack to the drain than to run a long horizontal waste to the main vertical soil stack for these fittings. In the system, the sanitary fittings are divided into two groups, namely:

1. *Soil fittings*

These are for the disposal of foul matter and include W.C.s, urinals, slop sinks and bed-pan washers. These fittings are connected to a vertical soil stack, which is connected to the drain by means of a rest bend at the foot of the vertical stack.

2. *Waste or ablution fittings*

These are for the disposal of mainly soapy water and include wash basins, showers, baths, sinks, bidets, wash-tubs, drinking fountains and island washing fountains. These fittings are connected to a vertical waste stack, which is connected to the drains by means of a rest bend at its base.

Note: Some local authorities may require a back inlet gully at the foot of the waste stack. This was the traditional method of connecting the vertical waste stack to the underground drain.

Depth of trap water-seals

If the internal diameter of the trap is 64 mm or more it should have a 50 mm water-seal. If the trap has an internal diameter less than 64 mm it should have a 38 mm water-seal.

Soil stack

The internal diameter should not be less than the internal diameter of any soil fitting discharging into it and in any case not less than 75 mm.

Waste stack

The internal diameter will depend upon the type and number of fittings discharging into it, but should not be less than 50 mm.

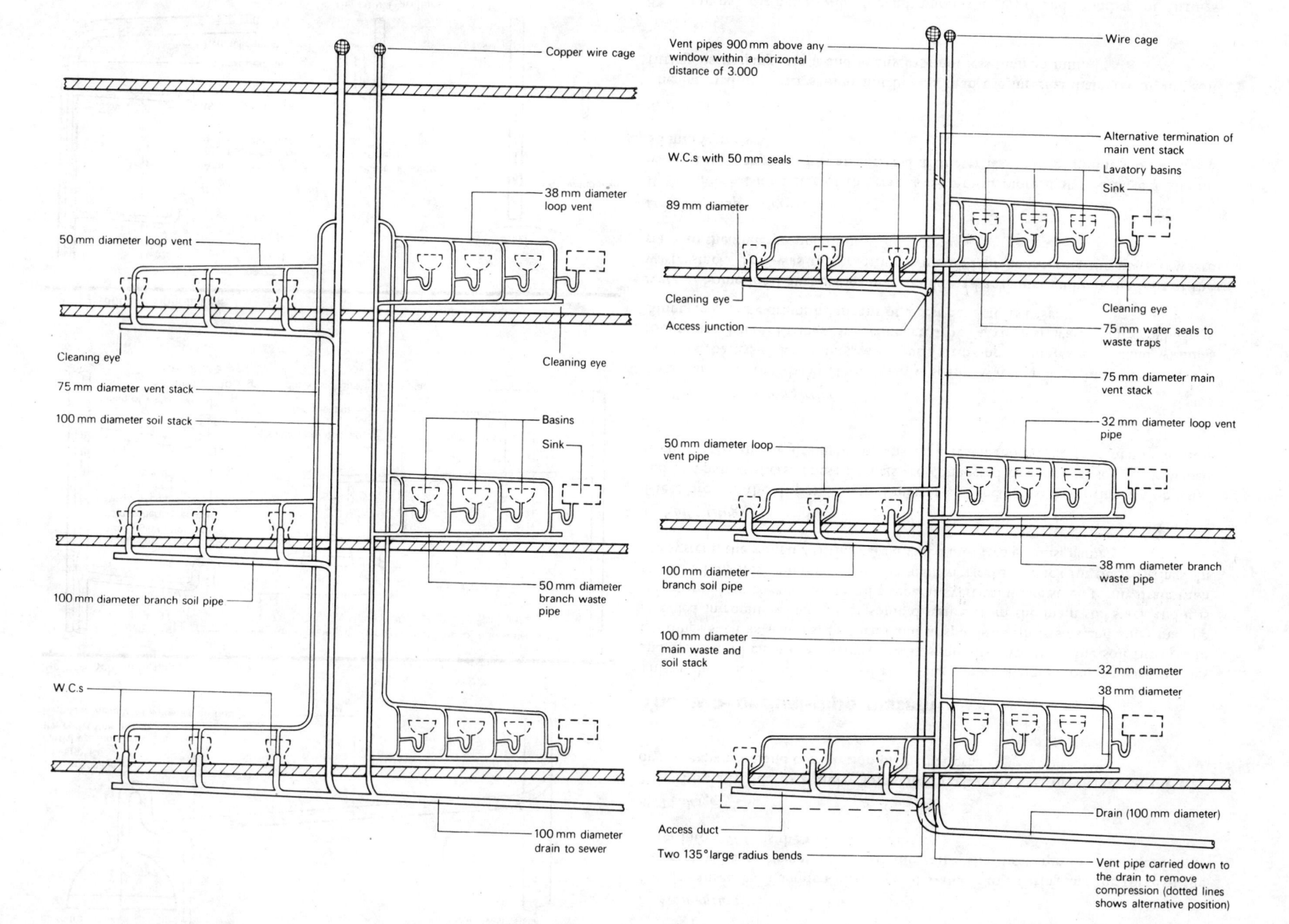

Copper wire cage

38 mm diameter loop vent

50 mm diameter loop vent

Cleaning eye

Cleaning eye

75 mm diameter vent stack

100 mm diameter soil stack

Basins

Sink

100 mm diameter branch soil pipe

50 mm diameter branch waste pipe

W.C.s

100 mm diameter drain to sewer

Fig. 9.11 Two-pipe system

Vent pipes 900 mm above any window within a horizontal distance of 3.000

Wire cage

W.C.s with 50 mm seals

Alternative termination of main vent stack

Lavatory basins

Sink

89 mm diameter

Cleaning eye

Access junction

Cleaning eye

75 mm water seals to waste traps

75 mm diameter main vent stack

50 mm diameter loop vent pipe

32 mm diameter loop vent pipe

100 mm diameter branch soil pipe

38 mm diameter branch waste pipe

100 mm diameter main waste and soil stack

32 mm diameter

38 mm diameter

Access duct

Drain (100 mm diameter)

Two 135° large radius bends

Vent pipe carried down to the drain to remove compression (dotted lines shows alternative position)

Fig. 9.12 Fully vented one-pipe system

Figure 9.11 shows a two-pipe system for a three-storey building having three W.C.s, three wash basins and one sink on each floor. It will be noted that anti-siphonage or vent pipes are provided to all the traps, which is a high-class method of installation. A cheaper method is the fitting of resealing traps to the waste fittings, but some local authorities object to this method, due to the possibility of these traps becoming inoperative by fouling with waste matter.

The one-pipe system

This system was introduced into this country from the USA in the 1930s, but due to the regulations existing at that time, which did not permit the connection of a waste pipe to a soil pipe, it was some time before this country amended the regulations and approved the use of the system. In the system, all soil and waste fittings discharge into common waste and soil stack and all the trap ventilating or anti-siphon pipes connect into one ventilating stack. This ventilating stack is extended down to the horizontal drainage system, to relieve any compressed air which might exist at the base of the waste and soil stack. Alternatively, the relief vent may be connected to the waste and soil stack, near the lowest branch connection. Where the planning of the building provides for close grouping of the waste and soil fittings, the elemination of the waste and waste ventilating stacks required for the two-pipe system results in a saving in installation costs and duct space.

The system is used in hospitals, large offices, schools and factories where there are ranges of fittings on one or several floors of the building. Figure 9.12 shows a fully vented one-pipe system for a three-storey office, or similar type of building. It will be noted that each fitting is provided with a trap and a trap ventilating or anti-siphon pipe.

The Building Research Establishment has carried out extensive research on the one-pipe system, which has resulted in the reduction in the amount of ventilating or anti-siphon pipes. The design is based upon the experience that some loss of trap seal is acceptable and that the soil and waste stack may be designed to restrict air pressure fluctuations to ± 375 N/m^2. A negative pressure of this magnitude corresponds to about 25 mm loss of seal from a washdown W.C. This loss of trap seal would be acceptable in practice, but the local authority would have to be consulted in order to ensure that the regulations would not be contravened.

Ranges of W.C.s

The research has shown that branch pipes from a range of W.C.s which are normally 100 mm diameter do not run full bore and thus there is no risk of induced siphonage; therefore branch ventilating or anti-siphon pipes may be omitted. It is possible that up to eight W.C.s may be installed in a range, with a straight branch pipe at an angle of ½° to 5°, without the need for anti-siphon pipes. The angle is not critical, but where there are bends in the branch pipe it may be necessary to fit a ventilating pipe to the fitting furthest from the waste and soil stack.

Ranges of lavatory basins

There is a greater risk of the branch pipe from lavatory basins flowing full than a branch soil pipe from W.C.s, and therefore more risk of induced siphonage of the traps. With normal taps where the basin is filled with water, it may be necessary to provide trap ventilating or anti-siphon pipes, but if spray taps are used which allows hand washing under the spray the branch waste pipe will run only partly filled. Experiments have shown that where spray taps are used, a 32 mm internal diameter branch waste pipe may serve up to eight lavatory basins without the need of trap ventilating or anti-siphon pipes. The use of spray taps, however, results in the formation of sediment in the waste pipe and regular cleaning is usually necessary. Other methods of avoiding the installation of anti-siphon pipes in ranges of basins are: (a) the use of special resealing traps to each basin, (b) the use of a single running trap at the end of the branch waste pipe, (c) the use of P traps for up to four basins discharging into a 50 mm diameter branch waste pipe, installed at an angle of between 1¼ to 1½° (see Fig. 9.13). Figure 9.14 shows the method used to reduce the amount of trap ventilating pipework for up to five basins.

Ranges of urinals

Trap ventilating pipes are not normally necessary from the consideration of flow of water, but they are available in order to obtain a flow of air through the soil pipe, which helps to reduce the build up of sediment inside the pipe.

Diameters of stacks

The internal diameters of the main soil and waste and vent stacks depends upon the number and type of fittings, pattern of use and the height of the building. Internal diameters of 100 and 150 mm are commonly used for the waste and soil stack. Table 9.1 covers 100 mm internal diameter stacks used for up to twelve floors and 150 mm internal diameter stacks used for up to twenty-four floors. It gives the minimum vent stack sizes recommended for use with vertical waste and soil stacks, serving equal ranges of W.C.s, and basins.

Example 9.1. *Find the internal diameters of the soil and waste stack for an eight-storey office, having five W.C.s, and five basins on each floor, assuming public use of fittings.*

Table 9.1 shows that a 100 mm diameter soil and waste stack with a 40 mm internal diameter vent stack would be suitable with a cross vent on each floor, as shown in Fig. 9.15.

Example 9.2. *Find the internal diameter of the soil and waste stack for a four-storey office having four W.C.s, and four basins on each floor, assuming public use of fittings.*

Table 9.1 shows that a 100 mm diameter stack may be used without the need of a vent stack. For other types of buildings and design considerations, the Code of Practice 5572 1978, Sanitary Pipework Above Ground, provides tables for the sizing of vertical stacks based on the discharge unit values for sanitary appliances. Table 9.2 gives the discharge unit table, which is used in conjunction with Table 9.3 to find the diameter of the stack.

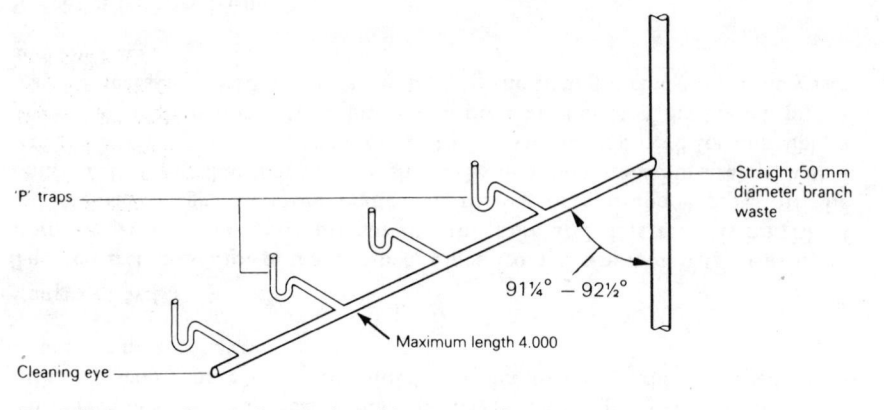

'P' traps

Straight 50 mm diameter branch waste

91¼° – 92½°

Maximum length 4.000

Cleaning eye

Fig. 9.13 Range of up to four basins without vent pipes

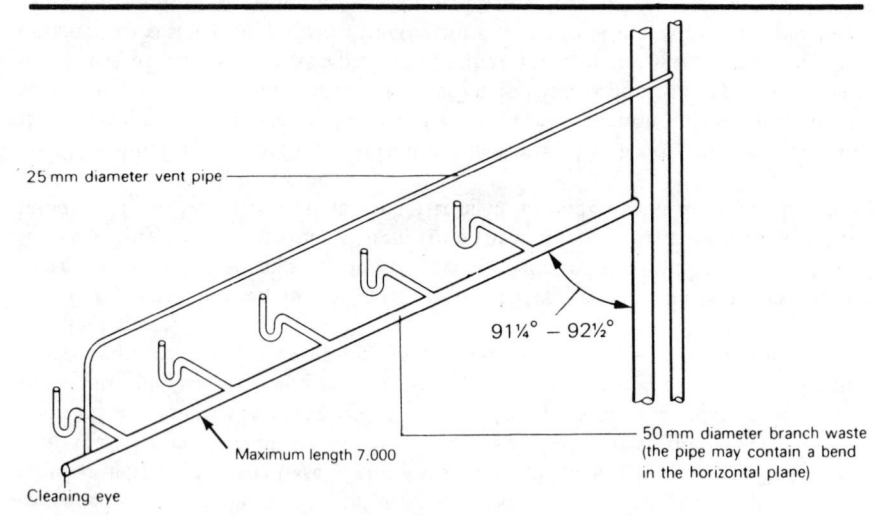

25 mm diameter vent pipe

91¼° – 92½°

50 mm diameter branch waste (the pipe may contain a bend in the horizontal plane)

Maximum length 7.000

Cleaning eye

Fig. 9.14 Range of up to five basins with vent pipe

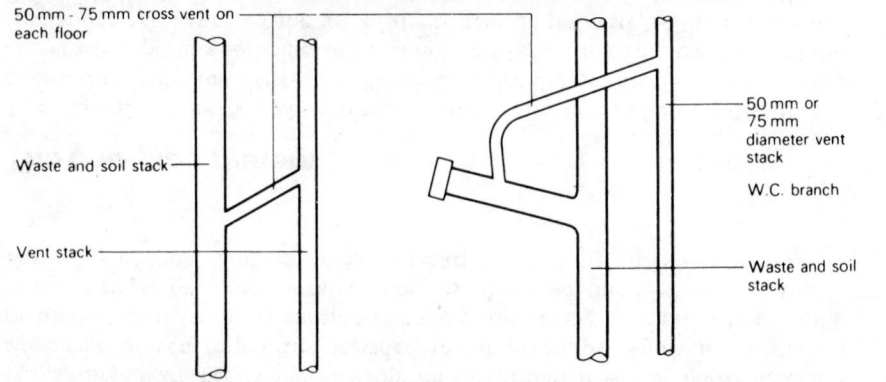

50 mm - 75 mm cross vent on each floor

Waste and soil stack

Vent stack

50 mm or 75 mm diameter vent stack

W.C. branch

Waste and soil stack

Fig. 9.15 Cross vents to prevent air pressure fluctuations in waste and soil stacks

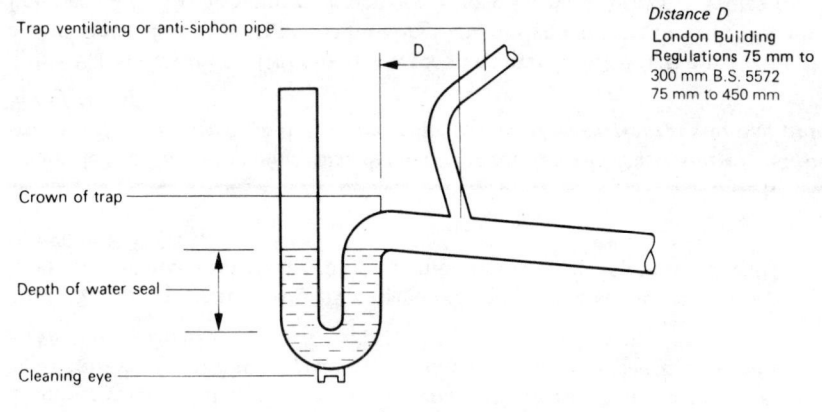

Trap ventilating or anti-siphon pipe

D

Distance D
London Building Regulations 75 mm to 300 mm B.S. 5572 75 mm to 450 mm

Crown of trap

Depth of water seal

Cleaning eye

Fig. 9.16 Anti-siphon pipe connected to 'P' trap

Note: The diameter of the anti-siphon pipe to be two-thirds the diameter of the waste pipe

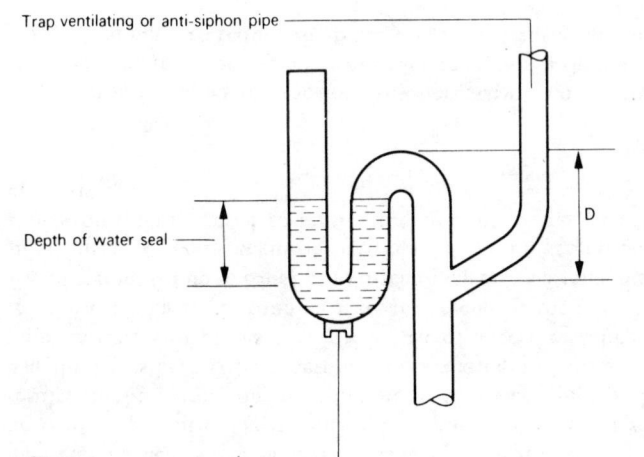

Trap ventilating or anti-siphon pipe

Distance D
London Building Regulations 75 mm to 300 mm B.S. 5572 75 mm to 450 mm

Depth of water seal

D

Cleaning eye

Fig. 9.17 Anti-siphon pipe connected to 'S' trap

Example 9.3. *Determine the diameter of the main waste and soil stack for a five-storey hotel, having 6 W.C.s, 8 wash basins, 3 urinals and 2 sinks on each floor.*

From Table 9.2

30 W.C.s. × 14	= 420·0	Discharge units
40 basins × 3	= 120·0	Discharge units
15 urinals × 0·3	= 4.5	Discharge units
10 sinks × 14	= 140.0	Discharge units
Total	= 684·5	

From Table 9.3 it can be seen that a 100 mm diameter stack will be satisfactory.

Table 9.1 Vent stack sizes (diameter in mm) for office buildings. (From the *Building Research Establishment Digest*, p. 115.)

Diameter of drainage stack	100 mm			150 mm				
Number of floors	4	8	12	8	12	16	20	24
W.C.s and basins								
(a) 10-min. interval								
'Public' use 1 + 1	0	0	30	0	0	0	0	0
2 + 2	0	0	30	0	0	0	0	0
3 + 3	0	30	40	0	0	0	0	0
4 + 4	0	40	40	0	see note			
5 + 5	0	40	see note	0				
(b) 5-min. interval								
'Peak' use 1 + 1	0	0	30	0	0	0	0	0
2 + 2	0	50	50	0	0	0	0	0
3 + 3	0	50		0	0	0	50	65
4 + 4	30	see note		0	Note: for situations outside the range of the table, refer to B.S. 5572.			
5 + 5	30			0				

Main assumptions: 0 means no vent stack needed; Cast iron drainage stack and fittings BS 416; Washdown W.C. to BS 1213 with 9-litre flush; No offset in 'wet' part of drainage stack; Large-radius bend at foot of stack.

Note: The 'public' use (10 min.) table should be sufficient for most purposes and the 'peak' use (5 min.) table is included to cover special situations where concentrated peak use may be expected. The table does not cover offsets in the 'wet' part of the waste and soil stack, nor a series of changes of direction between the lowest point of connection to the stack and the public sewer, which may increase back pressure above that likely in a simpler situation. In cases where above-normal back pressures are likely to arise as a result of downstream conditions, a relief vent connected to the stack near the lowest branch connection is recommended, especially when Table 9.1 shows no vent stack needed. A

50 mm diameter vent pipe for a 100 mm stack and a 75 mm vent pipe for a 150 mm stack are usually sufficient. The use of a large-radius bend at the foot of the stack is required to the root of 150 mm, but two 135° large-radius bends are preferred. With a 100 mm stack, a bend and a drain of 150 mm diameter is recommended.

Table 9.2 Discharge unit values for sanitary appliances, B.S. 5572 *1978*

Type of appliance	Frequency of use (minutes)		Discharge unit value
Spray tap (basin)	Add 0.06 litre per second per tap		—
9 litre water closet	20		7
	10		14
	5		28
Sink	20		6
	10		14
	5		27
Wash basin	20		1
	10		3
	5		6
Bath	75	(domestic)	7
	30	(commercial and congested)	18
	—		—
Shower (per head)	Add 0.1 litre per second per spray		—
Urinal (per stall or bowl)	20	(commercial and congested)	0.3
One group consisting of one W.C. one bath, 1 or 2 basins, sink			14
Washing machine (automatic)	250		4

20 min corresponds to peak domestic use
10 min corresponds to peak commercial use
5 min corresponds to congested use in schools etc

Table 9.3. Maximum number of discharge units to be allowed on vertical stacks (see note) B.S. 5572 1978

Nominal internal diameter of pipe (mm)	Discharge units
50	10
63	60
76	200 (not more than 1 W.C.)
89	350
100	750
125	2500
150	5500

Note: Discharge pipes sized by this method give the minimum size necessary to carry the expected flow load. Separate ventilating pipes may be required. It may be worthwhile to consider oversizing the discharge pipes to reduce the ventilating pipework required.

Trap water seals

1. Traps up to 64 mm internal diameter not provided with a ventilating or anti-siphon pipe should have a 75 mm seal.
2. Traps up to 64 mm internal diameter provided with a ventilating or anti-siphon pipe may have a 38 mm seal (depending upon the local authority approval).
3. Traps from 75 to 100 mm internal diameter should have a 50 mm seal, whether or not they are provided with a ventilating or anti-siphon pipe.

Diameter of traps

Table 9.4 gives the minimum internal diameters of traps for sanitary fittings.

Table 9.4 Minimum internal diameter of traps

Domestic type of fitting	Diameter (mm)	Non-domestic type of fitting	Diameter (mm)
Wash basin	32	Drinking fountain	19
Bidet	32	Bar well	32
Sink	40	Hotel and canteen sink	40
Bath	40	Urinal bowl	40
Shower	40	Urinal stall (1 or 2)	50
Wash-tub	50	Urinal stall (3 or 4)	65
Kitchen waste disposal		Urinal stall (5 or 6)	75
unit	40	Waste disposal unit	50

Connection of anti-siphon pipe

Figures 9.16 and 9.17 show the methods of connecting the anti-siphon pipe to the branch waste pipe. The connection should be away from the crown of the trap and at an angle in the direction of flow of water to prevent the anti-siphon pipe being blocked by grease or soap.

Terminal velocity in stacks

In the past many designers were concerned that the velocity of the flow of water in high waste and soil stacks would be excessive, with the consequent noise, damage to the pipework and unsealing of traps. In some buildings stacks have been offset to reduce the velocity of flow, but this is unnecessary, because the forces of gravity soon balance when the flow of water takes place and a maximum speed of flow called the 'terminal velocity' is quickly reached. The height of the stack necessary for terminal velocity to be achieved depends upon the diameter of the pipe, amount of flow and the smoothness of the internal bore. The height is only likely to be equal to one-storey and therefore offsets in stacks should be avoided, as these may cause a pressure build-up in the stack, add to the cost of installations and may cause blockages.

The single–stack system

In order to reduce the installation costs of soil and waste systems the Building Research Establishment carried out a great deal of experimental work and as a result the single-stack system was evolved, which has dispensed with the need for almost all the ventilating pipework. The single-stack system is a simplification of the one-pipe system, with the trap ventilating or anti-siphon pipes being either completely omitted or including only in special circumstances. In order to prevent the loss of water seals in the traps due to siphonage or back pressure, a high degree of planning and installation work is required.

Figure 9.18 shows a single-stack system for a two-storey block of flats, Fig. 9.19 a single-stack system for a ten-storey block of flats and Fig. 9.20 shows a view of the pipework on each floor.

The main requirements of the system are as follows:

1. Sanitary fittings must be grouped close to the stack, so that the branch waste and soil pipes are as short as possible.
2. All the sanitary fittings must be individually connected to the main stack. This will prevent the loss of trap water seal by induced siphonage.
3. An offset should not occur below the highest branch, in what is known as the 'wet' part of the stack. This is to prevent compression of air in the stack.
4. The foot of the stack should be connected to the drain with a large radius (150 mm root radius) or, preferably, two large radius 130° bends. This is to prevent compression of air at the base of the stack.
5. The vertical distance between the lowest branch connection to the stack and the invert of the drain should be at least 450 mm for two-storey housing and at least 750 mm for above two storeys. Where this cannot be achieved, ground floor sanitary fittings should be connected directly to the drain. This is to reduce the risk of water in the traps on the ground floor being forced out by compressed air.
6. W.C. connections should be swept in the direction of flow, with a radius at the invert of at least 50 mm. This is to prevent the loss of trap water seal due to induced siphonage of traps connected to the stack below the W.C. connection. A swept connection helps to prevent turbulent flow of water, with a resultant reduction in air pressure at the highest part of the stack.
7. The branch pipe from a bath should be connected to the stack so that its centre line meets the centre of the stack, at or above the level at which the

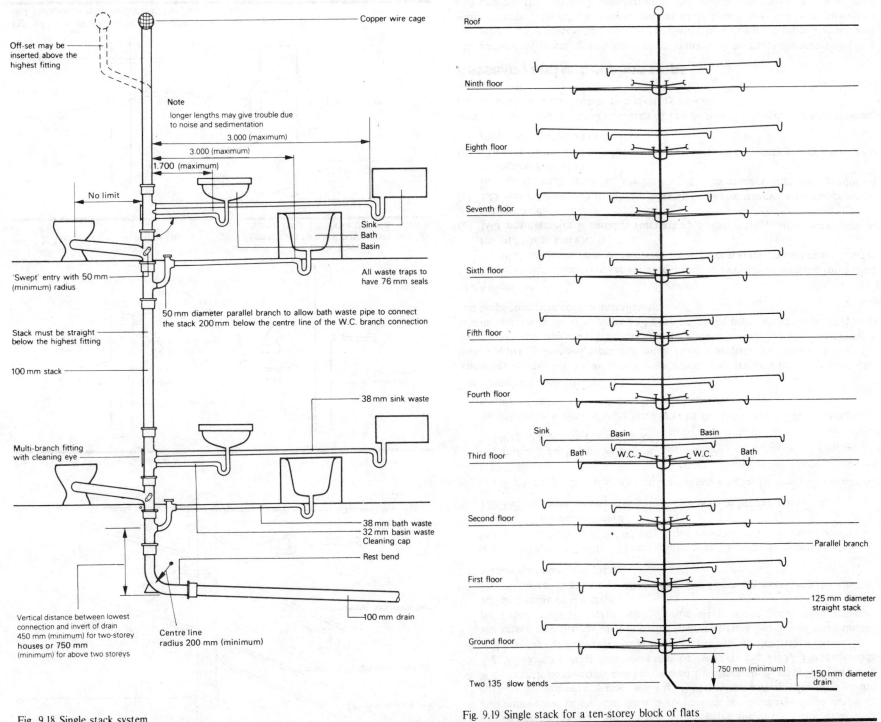

Copper wire cage

Off-set may be inserted above the highest fitting

Note
longer lengths may give trouble due to noise and sedimentation

3.000 (maximum)

3.000 (maximum)

1.700 (maximum)

No limit

Sink
Bath
Basin

All waste traps to have 76 mm seals

'Swept' entry with 50 mm (minimum) radius

50 mm diameter parallel branch to allow bath waste pipe to connect the stack 200 mm below the centre line of the W.C. branch connection

Stack must be straight below the highest fitting

100 mm stack

38 mm sink waste

Multi-branch fitting with cleaning eye

38 mm bath waste
32 mm basin waste
Cleaning cap

Rest bend

Vertical distance between lowest connection and invert of drain 450 mm (minimum) for two-storey houses or 750 mm (minimum) for above two storeys

Centre line radius 200 mm (minimum)

100 mm drain

Fig. 9.18 Single stack system

Roof

Ninth floor

Eighth floor

Seventh floor

Sixth floor

Fifth floor

Fourth floor

Third floor

Sink   Basin   Basin

Bath   W.C.   W.C.   Bath

Second floor

Parallel branch

First floor

125 mm diameter straight stack

Ground floor

750 mm (minimum)

Two 135 slow bends

150 mm diameter drain

Fig. 9.19 Single stack for a ten-storey block of flats

121

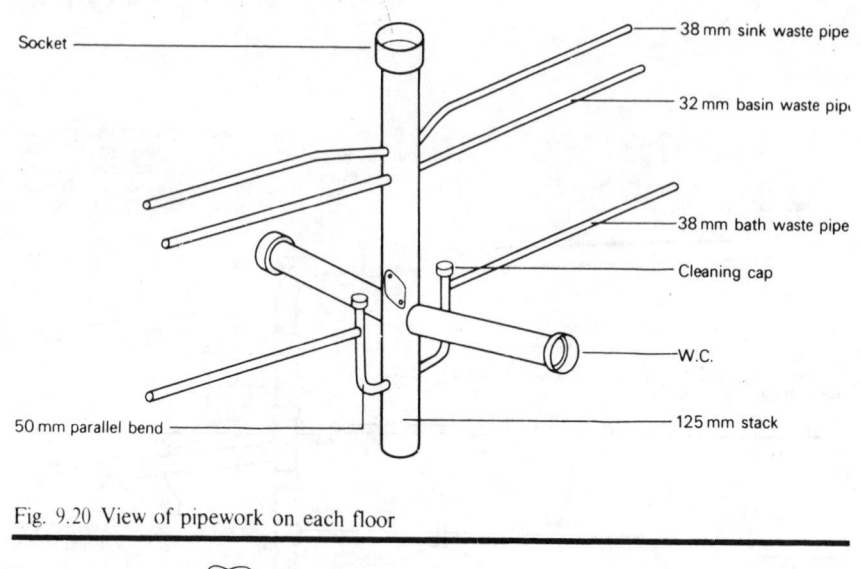

Fig. 9.20 View of pipework on each floor

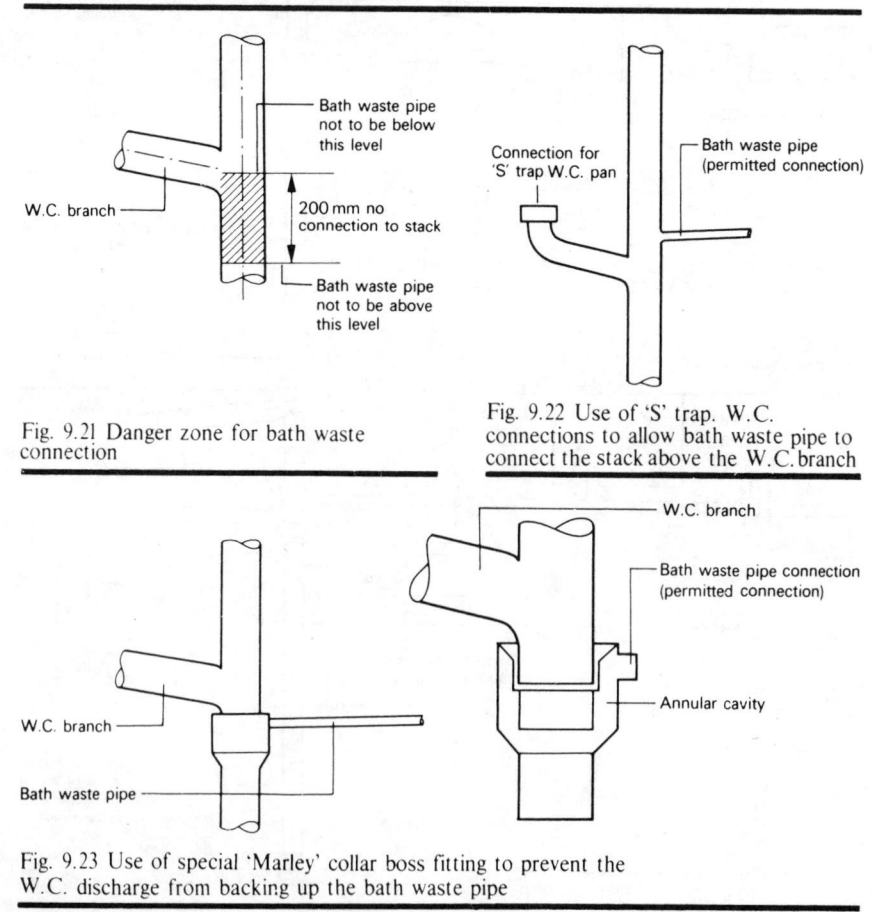

Fig. 9.21 Danger zone for bath waste connection

Fig. 9.22 Use of 'S' trap. W.C. connections to allow bath waste pipe to connect the stack above the W.C. branch

Fig. 9.23 Use of special 'Marley' collar boss fitting to prevent the W.C. discharge from backing up the bath waste pipe

centre line of the W.C. branch meets the centre line of the stack, or at least 200 mm below this point. This is to prevent the discharge from the W.C. backing up the bath waste (see Fig. 9.21). Alternatively, the level of the W.C. branch connection can be lowered by use of an 'S' trap, W.C. pan (see Fig. 9.22) or a special collar boss may be used (see Fig. 9.23). These methods will prevent the discharges from the W.C. backing up the bath waste.

8. Waste sanitary fittings such as baths, lavatory basins, bidets and sinks should be fitted with 'P' traps with 76 mm deep water seals. This prevents self-siphonage of the traps.

9. The Building Regulations 1985 state the following maximum lengths of branch wastes and soil pipes.

(a) *Wash basin:* 1.7 m for 32 mm diameter and 3 m for 40 mm diameter.
(b) *Sink and bath:* 3 m for 40 mm diameter and 4 m for 50 mm diameter.
(c) *W.C.:* No limit to length.

10. The Regulations also state the following slopes of branch pipes.

(a) *Wash basin:* 20−120 mm/m for branch lengths between 600 mm and 1.75 m.
(b) *Sink and bath:* 18−90 mm/m.
(c) *W.C.:* 18 mm/m (minimum).

Note: As the length of pipe increases the slope decreases. This is to prevent the loss of trap water seals by self-siphonage.

Use of multi-branch fittings

These are obtainable to suit most installations and are available in plastic, cast iron, copper, galvanised steel and pitch fibre materials. In blocks of flats the bath, basin and sink may be connected to a multi-branch fitting, thus saving in labour and materials. The slope and position of the pipes are set in these fittings and helps to ensure correct installation.

Stack diameter

(a) 76 or 89 mm diameters are suitable for up to two-storey housing (providing the W.C. outlet is 76 mm diameter for a 76 mm stack and 89 mm diameter for an 89 mm stack).
(b) 100 mm diameter is suitable for flats up to five storeys, with two groups of fittings on each floor.
(c) 125 mm diameter is suitable for flats up to twelve storeys with one group of fittings on each floor, or for flats up to ten storeys with two groups of fittings on each floor.
(d) 150 mm diameter is suitable for flats up to at least twenty storeys, with two groups of fittings on each floor.

Note: One group of fittings consists of one or two W.C.s, with a 9 litre flushing cistern, bath, sink and one or two lavatory basins.

Resealing and anti-siphon traps

The purpose of resealing and anti-siphon traps is to maintain the water seal of a trap without the installation of trap ventilating or anti-siphon pipe. The soil and waste systems is therefore cheaper and neater, but the traps tend to be noisy and also prevent the thorough ventilation of the branch waste pipe, which on long

lengths of pipe may lead to the formation of sediment inside the pipe. They also require more maintenance than ordinary traps.

Figure 9.24 shows the 'Grevak' trap, which works as follows:

(a) Siphonage of the water in the trap takes place and the water level is lowered to point A of the anti-siphon pipe.
(b) Air passes through the anti-siphon pipe and equalises the air pressures on the inlet and outlet sides of the trap, thus breaking the siphonic action.
(c) Water gravitates back from the reserve chamber into the U of the trap and maintains the water seal.

Figure 9.25 shows the 'McAlpine' trap, which works as follows:

(a) Siphonage of water in the trap takes place and the water level is lowered to point A.
(b) Air passes through the pipe thus breaking the siphonic action and also causes water to be retained in the reserve chamber.
(c) Water gravitates back from the reserve chamber into the U of the trap and maintains the water seal.

Figure 9.26 shows the 'Econa' resealing trap, which has a cylindrical reserve chamber. The trap operates on the same principle as the 'McAlpine' trap.

Figure 9.27 shows a bottle trap with an anti-siphon valve, which works as follows:

(a) If a partial vacuum is formed at the outlet side of the trap at A, there is a reduction of air pressure inside chamber B.
(b) The greater air pressure at C lifts the brass valve and flows through to the outlet side of the trap, thus preventing siphonage of water in the trap taking place and the seal is maintained.

Modified one-pipe system

The modified one-pipe system is basically similar to the one-pipe and single-stack systems, but unlike a one-pipe system, the trap ventilating or anti-siphon pipes are omitted. In the system a main vent stack is installed which prevents fluctuations of air pressure inside the main waste and soil stack. A cross vent pipe is connected between the vent stack and waste soil stack, or, alternatively, a loop vent pipe may be connected from the W.C. branch pipe to the vent stack. These cross vent or loop pipes are usually installed on each floor and air may be drawn through them on the upper floors, to prevent a negative pressure from being created inside the waste and soil stack. The vent stack is terminated below the lowest waste or soil branch pipe and the vent stack at this point removes any compressed air at the lowest part of the waste and soil stack.

Figure 9.28 shows a typical air pressure distribution diagram, with water flowing down the stack.

If the negative or positive air pressures in the stack are above or below 373 N/m^2 a modified one-pipe system would be necessary, or a larger diameter waste and soil stack installed. A 150 mm diameter waste and soil stack is satisfactory without venting for up to twenty storeys, with the equivalent loading of two W.C.s, two sinks, two basins and two baths on each floor, or up to twenty-five storeys with one W.C. bath, basins and sink on each floor. Since a 150 mm diameter stack is the maximum size obtainable, a building above twenty or twenty-five storeys having the same number of fittings on each floor will require a modified one-pipe system.

Figure 9.29 shows a modified one-pipe vented stack system for a twelve-storey building. Table 9.5 gives the recommendations of the Building Research Establishment for the installation of vented stacks, or modified one-pipe systems.

Table 9.5   Minimum stack sizes and vents required for various loading conditions

Type of building	Stack diameter (mm)	Requirements
Flats	Stack serving one group on each floor	
6 to 10 storeys	100	50 mm vent stack with one cross connection on alternate floors
11 to 15 storeys	100	50 mm vent stack with one cross connection on each or alternate floors
16 to 20 storeys	100	64 mm vent stack with one cross connection on each or alternate floors
12 to 15 storeys	125	50 mm vent stack with one cross connection on alternate floors
Maisonettes	Stack serving one group on alternate floors	
11 to 15 storeys	100	50 mm vent stack with one cross connection on alternate bathroom floors
16 to 20 storeys	100	64 or 50 mm vent stack with one cross connection on alternate bathroom floors

Materials used for soil and waste systems

Lead

This is the traditional material for soil and waste systems and was sometimes used for the entire installation. Lead is now used for short branch pipes to the main case iron stacks and has the advantage of adaptability, especially in restricted duct spaces. It has a smooth internal bore and is resistant to acid attack, but it can be attacked by Portland cement, lime, plaster, brickwork and magnesite and therefore where lead pipes pass through floors or walls they should be protected by Denso tape, or waterproof paper wrapping, or coated with bitumen. Lead is heavy and easily damaged and is liable to thermal movement. The pipes require frequent or event continuous support and allowance must be made for expansion and contraction.

Jointing: pipes are jointed by either lead welding or soldered joints, as shown in Figs 9.30 and 9.31. Figure 9.32 shows the method of jointing lead to cast iron and clay pipes.

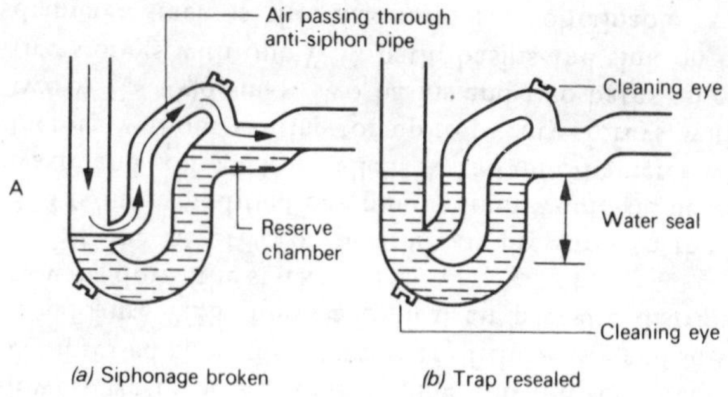

Air passing through anti-siphon pipe

Cleaning eye

Reserve chamber

Water seal

A

Cleaning eye

(a) Siphonage broken

(b) Trap resealed

Fig. 9.24 The 'Grevak' resealing trap

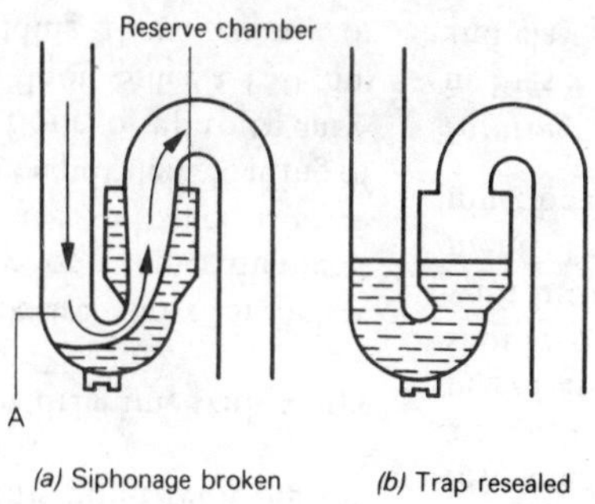

Reserve chamber

A

(a) Siphonage broken

(b) Trap resealed

Fig. 9.25 The 'McAlpine' resealing trap

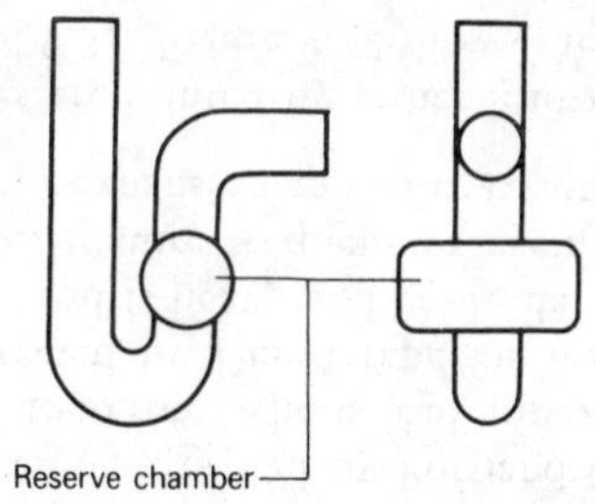

Reserve chamber

Fig. 9.26 'Econa' resealing trap

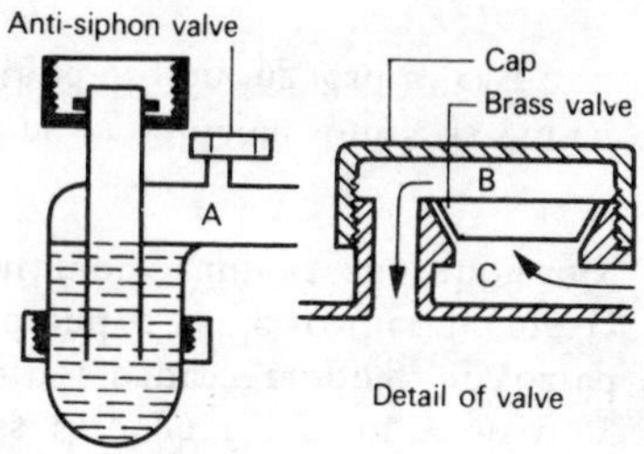

Anti-siphon valve

Cap

Brass valve

A

B

C

Detail of valve

Fig. 9.27 Bottle trap with anti-siphon valve

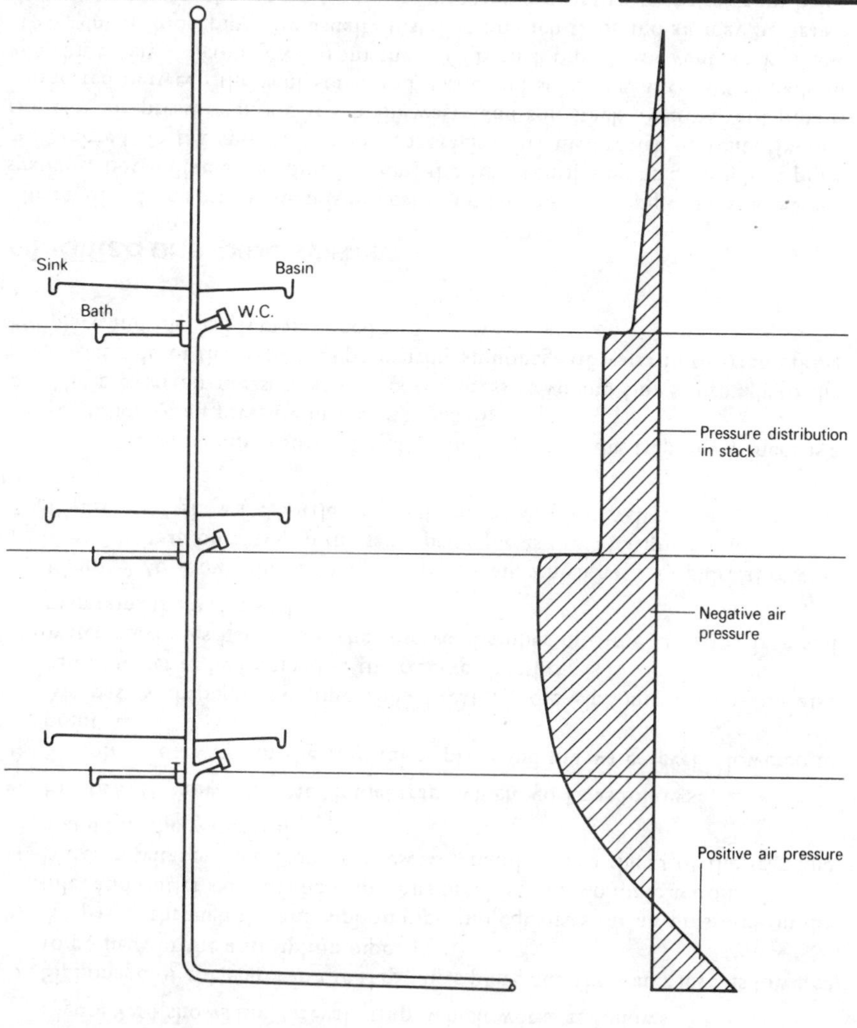

Sink

Basin

Bath

W.C.

Pressure distribution in stack

Negative air pressure

Positive air pressure

Fig. 9.28 Air pressure distribution with water flowing down the stack

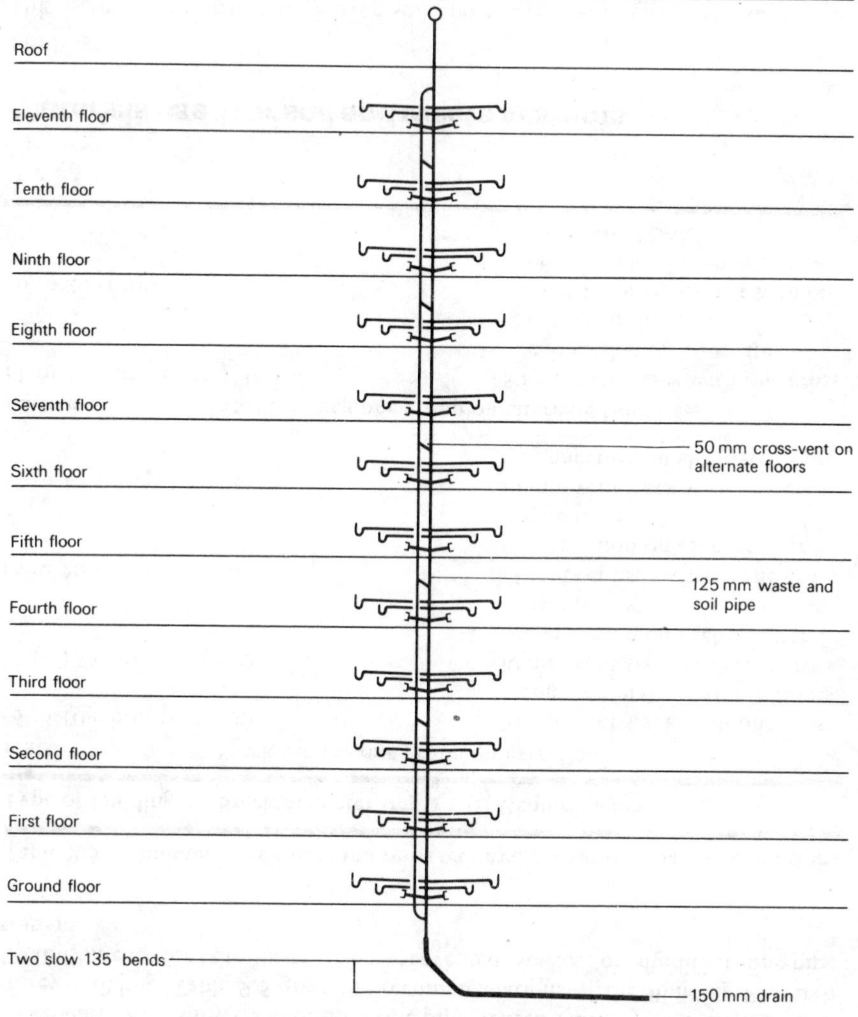

Roof

Eleventh floor

Tenth floor

Ninth floor

Eighth floor

Seventh floor

Sixth floor

50 mm cross-vent on alternate floors

Fifth floor

Fourth floor

125 mm waste and soil pipe

Third floor

Second floor

First floor

Ground floor

Two slow 135 bends

150 mm drain

Fig. 9.29 Modified one-pipe vented stack system for a twelve-storey building

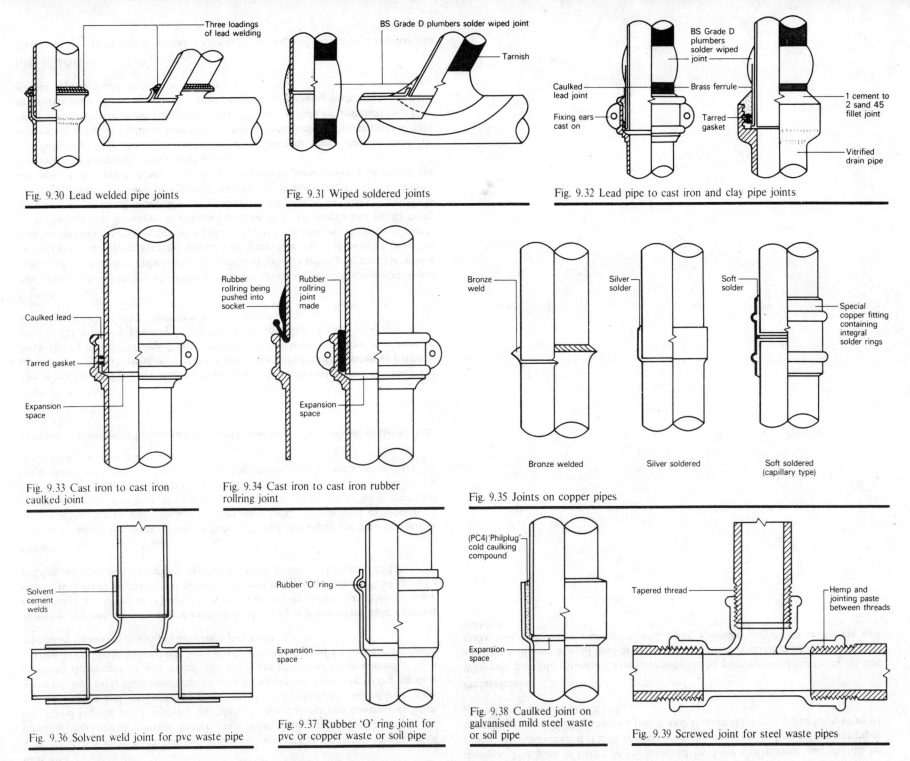

Fig. 9.30 Lead welded pipe joints

Fig. 9.31 Wiped soldered joints

Fig. 9.32 Lead pipe to cast iron and clay pipe joints

Fig. 9.33 Cast iron to cast iron caulked joint

Fig. 9.34 Cast iron to cast iron rubber rollring joint

Fig. 9.35 Joints on copper pipes

Fig. 9.36 Solvent weld joint for pvc waste pipe

Fig. 9.37 Rubber 'O' ring joint for pvc or copper waste or soil pipe

Fig. 9.38 Caulked joint on galvanised mild steel waste or soil pipe

Fig. 9.39 Screwed joint for steel waste pipes

Cast iron

The metal is widely used for soil and waste systems and has the advantage of resisting mechanical damage better than most other materials. The pipes are heavy and require good support, but they do not expand and contract as much as lead, copper and unplasticised polyvinyl chloride pipes. If caulked lead joints are used, however, they may require recaulking periodically, due to the pipeline receiving discharges of hot water. The pipes are protected from corrosion by a coating both inside and outside with pitch. A wide range of fittings are available, especially various types of multi-branch junctions.

Jointing: pipes are jointed by caulked lead, or by a proprietary cold caulking compound as shown in Fig. 9.33. A rollring flexible joint may also be used, which is quicker to make and permits thermal movement. Figure 9.34 shows a rollring joint on cast iron pipe, supplied by Burn Brothers (London) Ltd.

Copper

This is an adaptable metal of medium weight and the pipes can be obtained in long lengths, which reduces the number of joints. The thermal expansion of copper is higher than that of cast iron and expansion joints may be required on long runs. The pipe has a smooth internal bore, is strong, rigid and resists attack from most building materials. Where the pipes can be seen, they may be chromium plated or polished.

Jointing: pipes may be jointed by bronze welding, silver or soft soldering (see Fig. 9.35).

Pitch fibre

The pipes are used for the main stacks and are light and easy to handle. They are subject to larger thermal movement than copper, but not so large as plastics. Pitch fibre pipes are strong, have a smooth internal bore and there is no need to protect the pipes by painting. They should be installed in a fire-protected duct. Taper or rubber 'O'-ring joints may be used for jointing.

Plastics

Pipes made of a number of thermoplastic materials are used for soil and waste pipes. These include unplasticised polyvinyl chloride (upvc) polypropylene and acylonitrile butadiene styrene (abs). The pipes are light in weight, easy to handle, smooth internal bore and highly resistant to corrosion. Their coefficients of expansion are, however, much higher than those of metals and much more provision for thermal movement is required. Unplasticised polyvinyl chloride is the most commonly used plastic material for larger diameter stacks, but due to the risk of distortion it should not be used where large volumes of water are discharged at temperatures exceeding 60 °C.

Polypropylene and acylonitrile butadiene styrene pipes may be used for conveying water at higher temperatures, but manufacturer's instructions should be consulted where these conditions exist. Plastics should be installed in a fire-protected duct. Jointing may be accomplished by welding techniques, or by a rubber 'O'-ring (see Figs 9.36 and 9.37).

Galvanised steel

This is lighter than cast iron and is extremely strong and resistant to mechanical damage. The pipe is liable to be attacked by lime and plaster and should be coated with bitumen if it is in contact with these materials. Large diameter pipes have spigot and socket caulked joints and smaller diameter pipes have screwed joints (see Figs 9.38 and 9.39).

Prefabrication

Wherever possible pipework assemblies should be prefabricated for fixing on site and this will save a good deal of site work and time. A number of manufacturers make soil and waste assemblies and offer a design service to architects and builders.

Chapter 10

Sanitary appliances

Types of sanitary appliances

Many types of sanitary appliances are required in buildings to fulfil a variety of specialised functions and are supplied with water either directly from the main, or from hot or cold storage vessels. It is essential that the supply water is not contaminated by foul water and for this reason in most cases the taps are designed to discharge above the flooding level of the appliance, to prevent the risk of back siphonage of the foul water into the supply pipe. A sanitary appliance should be designed so that its fouling area is reduced to the minimum and should have durable, easily cleaned and non-absorbent surfaces.

Materials

Ceramics

The strength and degree of impermeability of the materials depends upon the composition of the clay mixture and the temperature at which they are fired.

Glazed earthenware: this produces appliances of good colour, lends itself well to the formation of complicated shapes and is relatively cheap. It is used mainly for sinks and W.C. pans.

Glazed fireclay: this produces a tough appliance which is resistant to knocks and hard wear. Fireclay appliances such as urinals, sinks and W.C. pans are often specified for use in schools and factories.

Glazed stoneware: this produces a tough appliance which is resistant to knocks and hard wear, but, unlike earthenware and fireclay, the material is non-absorbent even when it is unglazed. It is mainly used for channels, sinks and urinal stalls.

Vitreous china: this lends itself to fine detail and good finish, but is not as strong as fireclay and is therefore unsuitable for buildings where hard wear is expected, such as schools and factories. It is, however, used extensively in houses and hotels. The material does not absorb water even when the glaze is broken and can be used for the manufacture of almost all types of appliances, in which various colours may be obtained.

Pressed metal

Mild steel, stainless steel and monel metal sheets are moulded in a press to form one-piece units. Mild steel should be galvanised or enamelled to protect the metal from corrosion and it is used for troughs and sinks where cheapness is of paramount importance. Stainless steel is used for sinks, urinals, wash basins, W.C. pans and draining units. It is rather expensive but has a good appearance and is highly resistant to hard wear. Monel metal has many of the properties of stainless steel, but does not have quite as good an appearance. It is used for sinks and is generally cheaper than stainless steel.

Note: Stainless steel and monel metal bar sinks are more resilient and therefore breakages of glass does not occur as frequently as in the case of ceramic sinks.

Acrylic plastic (perspex)

This is produced in many colours, is light in weight and relatively cheap. It takes a hard gloss finish and has an excellent appearance. Hot water, however, tends to soften the material and baths made from it must be supported with metal cradles.

Glass-reinforced polyester

This material is more expensive than acrylic plastic, but is much stronger. A good gel coat finish is essential to protect the reinforcing fibres and various colours may be obtained.

Note: Baths made from acrylic plastic or glass-reinforced polyester have a lower thermal conductivity than baths made from cast iron and therefore tend to feel warmer to the touch.

Cast iron

This is used mainly for large appliances such as baths, which would be too heavy if made in ceramics. The material is strong but heavy and may have either a white or a coloured vitreous enamel finish fired on. It is a very cheap material which is also used extensively for sink and wash basin brackets.

Terrazzo

The manufacture of small sanitary appliances from the material is very difficult and uneconomical, but it is often used for large appliances formed *in situ*. In special circumstances washing fountains, sunken baths and shower trays may be formed on site to the architect's design.

Water closets

The most widely used pattern is the 'wash-down', in which the contents of the pan are removed by a gravity water flush. These W.C.s, are cheap, simple and efficient and are rarely blocked by misuse. The pan shape has been developed from the earlier long and short hopper types, to provide the minimum of fouling area. They are designed to maintain a 50 mm minimum water seal.

Figure 10.1 shows a section through a wash-down W.C. pan. The outlet may be obtained left or right hand and also P or S as shown; it may be flushed from a high- or low-level flushing cistern depending upon the circumstances. The high-level cistern provides a more effective flush, but is not as neat and as modern as the low-level cistern. Siphonic water closets are more silent and positive in action than the wash-down type, but are more prone to blockages.

Figures 10.2 and 10.3 show sections of two types of single trap siphonic W.C.s, which operate as follows:

1. The cistern is flushed and water passes through the pan to the long leg of the siphon, which is shaped to restrict the flow of water.
2. The long leg of the siphon is momentarily filled with water and a siphonic action set up, which empties the water and any contents from the pan.
3. Water contained in the after flush chamber reseals the trap.

Figure 10.4 shows the method used to joint flush and outlet pipes to both wash-down and siphonic W.C.s.

Figure 10.5 shows a two-trap type siphonic W.C. which operates as follows:

1. The cistern is flushed and water passes through the pressure reducing fitment A, which draws air through pipe G and reduces the air pressure in chamber B.
2. A siphonic action is set up, which empties the water and any contents from the pan through the sealed traps C and D. At the same time, the sides of the pan are thoroughly washed by streams of water from the perforated rim E.
3. Water contained in the after-flush chamber F reseals the trap.

Urinals

Figures 10.6 and 10.7 show the installations of ceramic slab and stall-type urinals respectively. The slab type is cheaper than the stall type, but it does not provide the same degree of privacy. Vertical partitions which increase privacy, however, may be obtained with some types of slab urinals.

Figure 10.8 shows the installation of ceramic bowl-type urinals, which have less fouling area than the slab and stall urinals. Pipework is also more accessible and there is no problem with traps below the floor level.

Figure 10.9 shows alternate arrangement for bowl type urinals.

Note: The anti-siphon trap method is more expensive than the inverted trap method but is more hygienic.

Slop hopper

This is a very useful appliance for use in hospitals and hotels for the efficient disposal of slops. The hinged brass grating allows for the filling of a bucket from the hot and cold taps. Figure 10.10 shows the installation of a slop hopper and Fig. 10.11 shows a combined washup sink and slop hopper, also for use in hospitals.

Flushing cisterns

The capacity of the cistern is usually 9 litre but to conserve water, a 7 litre capacity cistern will be phased in in 1991. Figure 10.12 shows the piston type flushing cistern, which operates as follows:

1. When the lever is depressed sharply, the piston is lifted, which displaces water over the siphon.
2. Water discharging down the flush pipe takes some air with it and creates a partial vacuum in the siphon.
3. The greater air pressure acting upon the water in the cistern forces water through the siphon until air is admitted under the piston, which breaks the siphonic action. A dual flush siphon, which is shown as part of Fig. 10.12, reduces the water consumption.

Operation

1. When the flush lever is depressed and immediately released, air is drawn through the air pipe at A, which breaks the siphonic action at this level and only 4.5 litre or 3.5 litre of water is used.
2. When the flush lever is depressed and then held down, the piston closes the air pipe and a 9 litre or 7 litre flush takes place.

Figure 10.13 shows a bell-type flushing cistern which operates as follows:

1. The lever is depressed, which lifts the bell inside the cistern.
2. The lever is released, which allows the bell to fall, thus displacing water under the bell down the stand pipe.
3. Water flowing down the stand pipe takes some of the air with it and creates a partial vacuum in the pipe, which starts the siphonic action and the cistern is emptied.

Automatic flushing cistern

This is used for the automatic flushing of urinals. It is usual to allow a capacity of 4.5 or 3.5 litre of water per stall, or 600 mm of slab and for an interval of 20 minutes between each flush. Some water authorities require the water supply to automatic flushing cisterns to be metered. Figure 10.14 shows a section through an automatic flushing cistern, which operates as follows:

1. Water rises evenly inside and outside the bell until it reaches the air hole.
2. Air inside the bell is thus trapped and becomes compressed as the water level rises outside the bell.
3. When the water level reaches a certain height above the dome, the compressed air is sufficient to force water out of the U-tube and reduce the air pressure inside the dome.
4. The reduced air pressure inside the dome immediately allows water to flow through the siphon and a siphonic action is set up, which empties the cistern.
5. When the flush is finished, water from the reserve chamber is siphoned through the siphon tube, which refills the lower well and U-tube.

Flushing troughs

This may be used as an alternative to separate cisterns, for the flushing of a range of W.C.s. It has the advantage over separate W.C.s of reducing the supply pipework, connections and valves, but if a repair is required to one of the siphons or

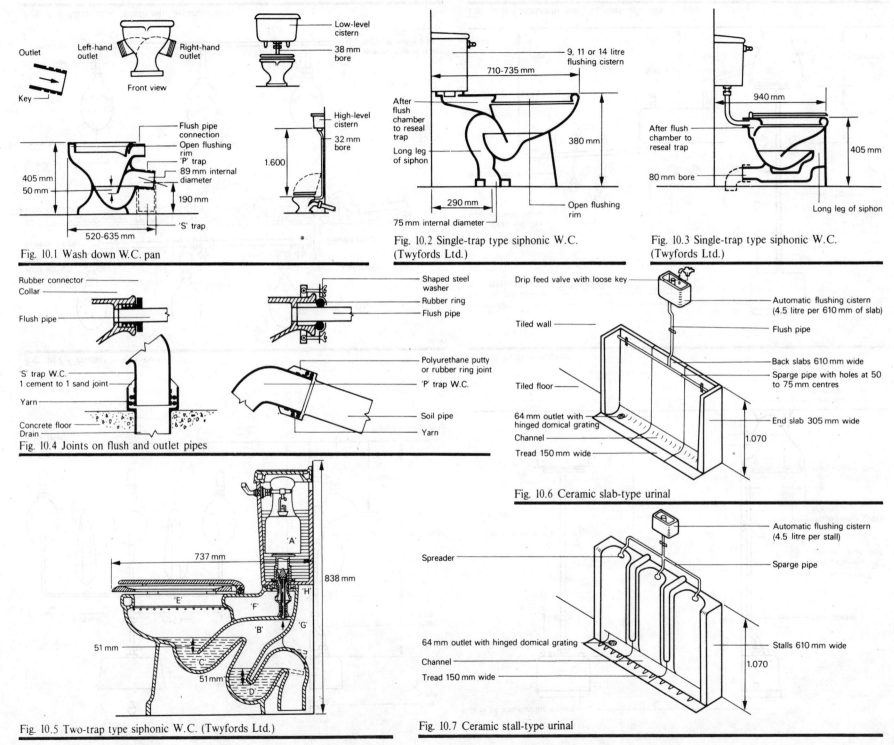

Outlet

Key

Left-hand outlet

Right-hand outlet

Front view

Low-level cistern

38 mm bore

High-level cistern

32 mm bore

1.600

Flush pipe connection

Open flushing rim

'P' trap

89 mm internal diameter

405 mm

50 mm

190 mm

'S' trap

520-635 mm

Fig. 10.1 Wash down W.C. pan

9, 11 or 14 litre flushing cistern

710-735 mm

After flush chamber to reseal trap

Long leg of siphon

380 mm

290 mm

Open flushing rim

75 mm internal diameter

Fig. 10.2 Single-trap type siphonic W.C. (Twyfords Ltd.)

940 mm

After flush chamber to reseal trap

80 mm bore

405 mm

Long leg of siphon

Fig. 10.3 Single-trap type siphonic W.C. (Twyfords Ltd.)

Rubber connector

Collar

Flush pipe

'S' trap W.C.
1 cement to 1 sand joint

Yarn

Concrete floor

Drain

Shaped steel washer

Rubber ring

Flush pipe

Polyurethane putty or rubber ring joint

'P' trap W.C.

Soil pipe

Yarn

Fig. 10.4 Joints on flush and outlet pipes

Drip feed valve with loose key

Tiled wall

Tiled floor

64 mm outlet with hinged domical grating

Channel

Tread 150 mm wide

Automatic flushing cistern (4.5 litre per 610 mm of slab)

Flush pipe

Back slabs 610 mm wide

Sparge pipe with holes at 50 to 75 mm centres

End slab 305 mm wide

1.070

Fig. 10.6 Ceramic slab-type urinal

737 mm

'A'

838 mm

'E'

'H'

'F'

'B'

'G'

51 mm

'C'

51 mm

'D'

Fig. 10.5 Two-trap type siphonic W.C. (Twyfords Ltd.)

Automatic flushing cistern (4.5 litre per stall)

Spreader

Sparge pipe

64 mm outlet with hinged domical grating

Channel

Tread 150 mm wide

Stalls 610 mm wide

1.070

Fig. 10.7 Ceramic stall-type urinal

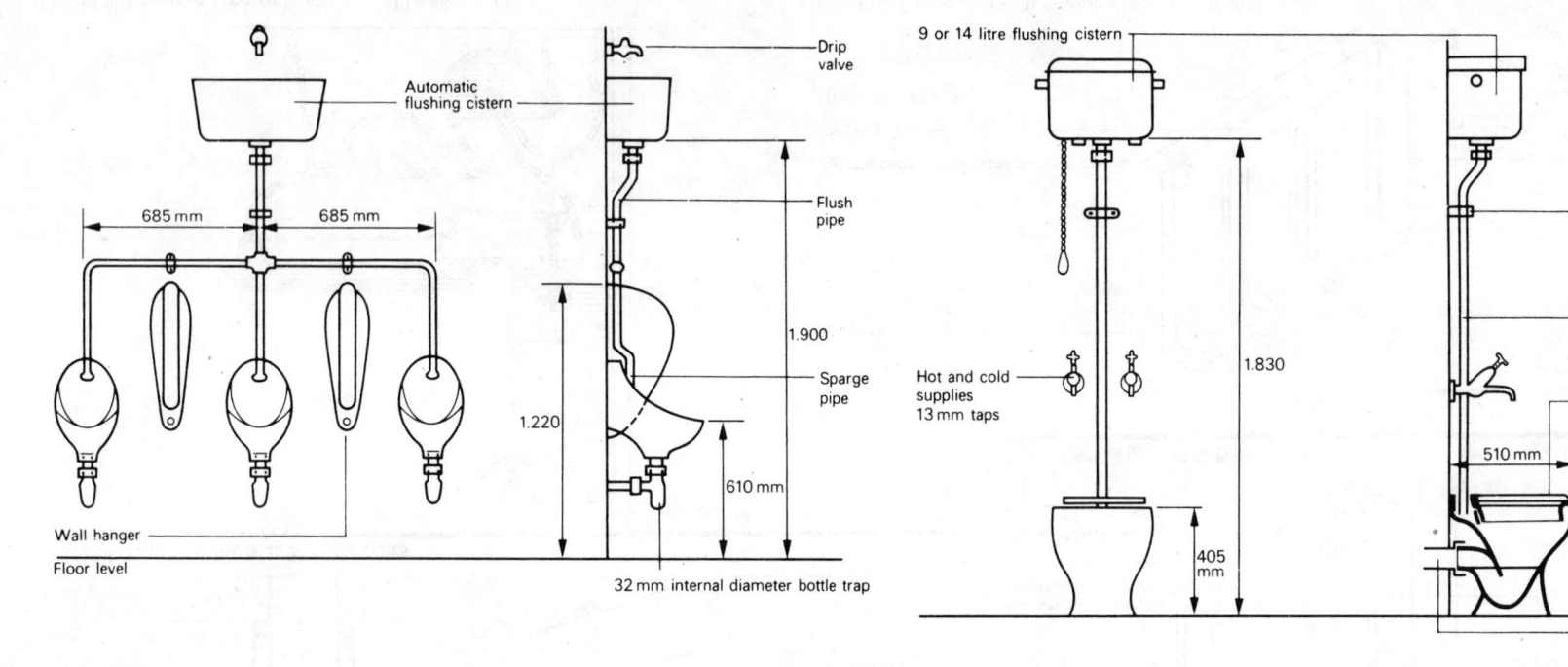

Fig. 10.8 Ceramic bowl-type urinals

Fig. 10.10 Slop hopper

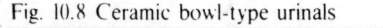

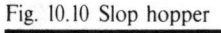

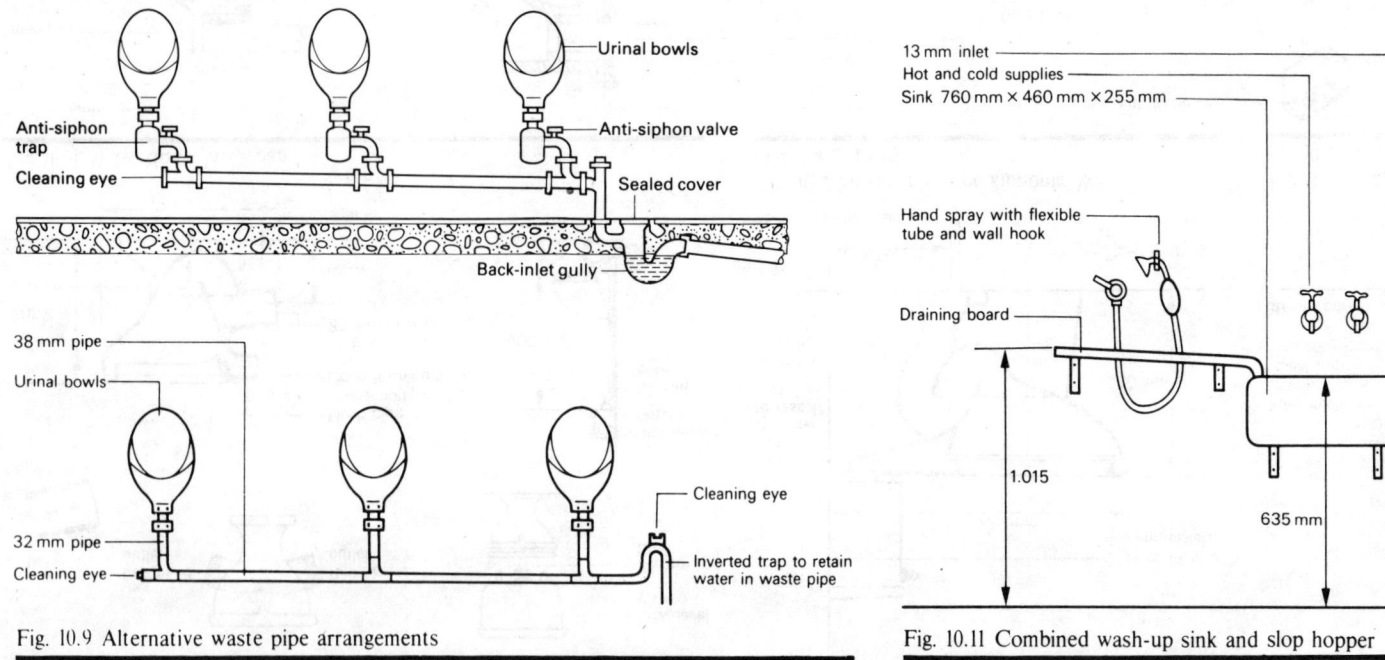

Fig. 10.9 Alternative waste pipe arrangements

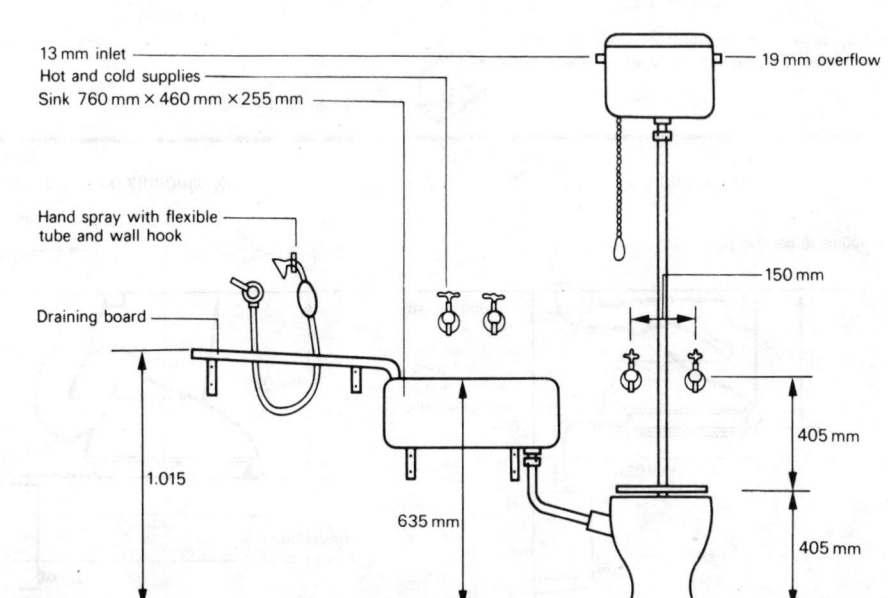

Fig. 10.11 Combined wash-up sink and slop hopper

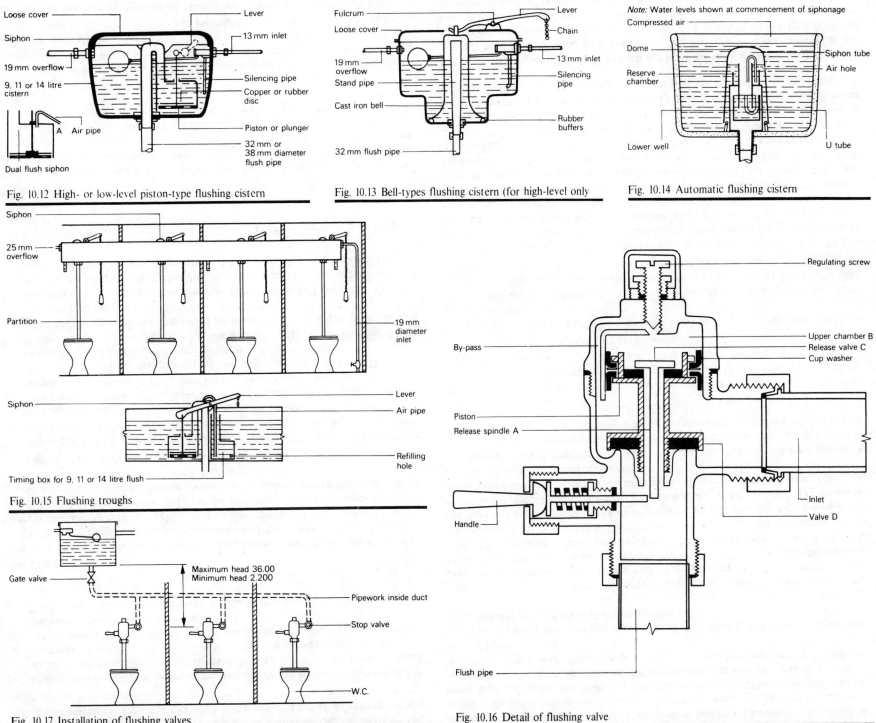

Loose cover

Siphon

19 mm overflow

9, 11 or 14 litre cistern

Lever

13 mm inlet

Silencing pipe

Copper or rubber disc

Piston or plunger

32 mm or 38 mm diameter flush pipe

A   Air pipe

Dual flush siphon

Fig. 10.12 High- or low-level piston-type flushing cistern

Fulcrum

Loose cover

19 mm overflow

Stand pipe

Cast iron bell

32 mm flush pipe

Lever

Chain

13 mm inlet

Silencing pipe

Rubber buffers

Fig. 10.13 Bell-types flushing cistern (for high-level only

Note: Water levels shown at commencement of siphonage

Compressed air

Dome

Reserve chamber

Lower well

Siphon tube

Air hole

U tube

Fig. 10.14 Automatic flushing cistern

Siphon

25 mm overflow

Partition

19 mm diameter inlet

Siphon

Lever

Air pipe

Refilling hole

Timing box for 9, 11 or 14 litre flush

Fig. 10.15 Flushing troughs

Regulating screw

By-pass

Piston

Release spindle A

Handle

Upper chamber B

Release valve C

Cup washer

Inlet

Valve D

Flush pipe

Fig. 10.16 Detail of flushing valve

Gate valve

Maximum head 36.00
Minimum head 2.200

Pipework inside duct

Stop valve

W.C.

Fig. 10.17 Installation of flushing valves

the trough, the range of W.C.s cannot be used. It is used in schools, factories and public conveniences. Figure 10.15 shows the installation of a flushing trough, which operates as follows:

1. The siphon is operated in the same manner as the individual cistern and as water flows down the long leg of the siphon air is withdrawn from the small diameter air pipe, which allows water in the timing box to be siphoned out.
2. Air is admitted through the top of the timing box and the main siphonic action is broken at a predetermined time.
3. Water flows through the refilling hole and the timing box is refilled with water.

Flushing valve

This is an alternative method to the flushing cistern or trough, for the flushing of W.C.s, and is very popular on the Continent and USA. The water authorities in Great Britain, however, are reluctant to allow the use of the valve, due to the possibility of waste of water and also the danger of back siphonage. The valve should be supplied from a storage cistern and no other sanitary fitting should be connected to the supply pipe to the valve.

Figure 10.16 shows a section of a flushing valve, which operates as follows:

1. When the handle is operated, the release spindle A and release valve C are tilted and water escapes from the upper chamber B through to the flush pipe, quicker than water can enter through the bypass.
2. The pressure in the upper chamber B is lowered and the greater upward pressure of water under the piston lifts valve D from its seating.
3. When the piston reaches the top of the upper chamber B, valve D is fully open and water passes down the flush pipe.
4. Water also passes through the bypass and fills the upper chamber B and equalises the upward and downward pressures on the piston, thus permitting the piston to fall gradually under its own weight.
5. The amount of water passing down the flush pipe is determined by the time it takes to fill the upper chamber B with water through the bypass. The regulating screw is used to determine the amount of water discharged by the valve.

Figure 10.17 shows the method of installing the flushing valve for a range of three W.C.s.

Waste appliances

Wash basins: many wash basin designs are available, ranging from surgeon's basins to small hand rinse basins. They can be obtained to fit into a corner of the room and may be supported on brackets, a pedestal or by a 'built in' corbel. Figure 10.18 shows a vertical section and a plan of a typical wash basin supported on a pedestal, which allows the supply pipes to be concealed from view.

Sinks: Figure 10.19 shows the Belfast, London and combination type glazed fireclay sinks and Fig. 10.20 shows a view of a stainless steel sink having a cupboard with sliding doors below. Figure 10.21 shows a cleaner's bucket sink which is usually installed in a cleaner's cubicle. Figure 10.22 shows an ablution fountain, which may be used as an alternative to a range of eight wash basins in

schools and factories. the central chromium plated or stainless steel pillar is fitted with a thermostatically controlled umbrella spray, or, in some fountains, spray taps are fitted, fed from hot and cold supply pipes. The minimum head of water above the spray is 3 m and the water supply is usually controlled by foot pedals. A liquid soap dispenser or a soap tray may be fitted at the top of the central pillar.

The fountain is a very hygienic method of washing, but is only economical where several persons wash at the same time.

Baths: although a shower is a more efficient and hygienic means of washing than the bath, most people find a bath more relaxing and it is therefore more popular. Figure 10.23 shows a typical installation of a bath and also various siting arrangements. The water supply may be by pillar taps as shown, or by a special fitting incorporating a diverter and a shower. For domestic installations, the taps and supply pipes are 19 mm internal diameter, but for institutions these are sometimes enlarged to 25 mm, to increase the speed of filling. The trap and waste pipes are 38 mm internal diameter for domestic installations and 50 mm internal diameter for institutions.

Figure 10.24 shows the 'sitz' bath, which is short, deep and incorporates a seat. It may be used where space is limited, but is not conducive to relaxation. It is, however, suitable for old people since the user may maintain a normal sitting position.

Showers: these are more hygienic, quicker to use and require less water than the bath. For an efficient shower, a minimum head of water of 1 m above the spray is required. A head of water of 1.5 m is however, preferable.

Two types of spray heads are obtainable.

1. A traditional rose type, which delivers water through a perforated disc fitted at high level.
2. An adjustable umbrella spray, which is usually fitted at chest level. Some people, especially women, find the umbrella spray preferable to the disc, because the shower may be used without wetting the hair.

Hot and cold water to the shower should be made through 13 or 19 mm internal diameter pipes and a thermostatically controlled mixing valve is recommended to prevent scalding. The shower tray may be of glazed fireclay, or acrylic plastic as shown in Fig. 10.25. Figure 10.26 shows the installation of the shower, including a view of the thermostat-controlled mixing unit. The modern type of mixing unit does not require non-return or outlet valves. A flexible chromium plated outlet pipe is shown, with two wall brackets for the spray, which allows the spray unit to be used at adult height, or at a lower position for children. The spray unit may also be held in the hand for shampooing the hair. Figure 10.27 shows the hot and cold pipework required for a shower, including the minimum head of water above the spray. Where this minimum head of water is not available a small low-voltage pump may be inserted in the outlet pipe.

Bidet: (pronounced 'beeday': French for little horse) is used for perineal washing and may also be used as a footbath. The hot and cold water supplies are 13 mm internal diameter and hot water may be supplied to the rim. A mixed supply of hot and cold water is also connected to an ascending spray, either through individual control valves or by a thermostatic valve, which eliminates

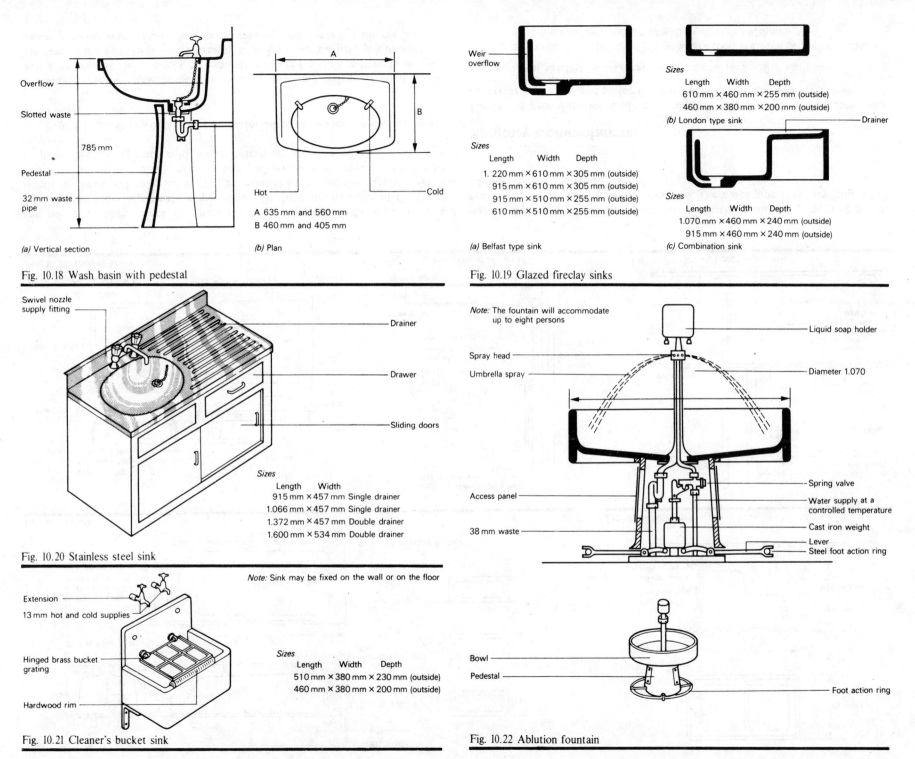

Overflow

Slotted waste

785 mm

Pedestal

32 mm waste pipe

(a) Vertical section

A

B

Hot — Cold

A 635 mm and 560 mm
B 460 mm and 405 mm

(b) Plan

Fig. 10.18 Wash basin with pedestal

Weir overflow

Sizes

Length	Width	Depth
1. 220 mm	× 610 mm	× 305 mm (outside)
915 mm	× 610 mm	× 305 mm (outside)
915 mm	× 510 mm	× 255 mm (outside)
610 mm	× 510 mm	× 255 mm (outside)

(a) Belfast type sink

Sizes

Length	Width	Depth
610 mm	× 460 mm	× 255 mm (outside)
460 mm	× 380 mm	× 200 mm (outside)

(b) London type sink

Drainer

Sizes

Length	Width	Depth
1.070 mm	× 460 mm	× 240 mm (outside)
915 mm	× 460 mm	× 240 mm (outside)

(c) Combination sink

Fig. 10.19 Glazed fireclay sinks

Swivel nozzle supply fitting

Drainer

Drawer

Sliding doors

Sizes

Length	Width
915 mm	× 457 mm Single drainer
1.066 mm	× 457 mm Single drainer
1.372 mm	× 457 mm Double drainer
1.600 mm	× 534 mm Double drainer

Fig. 10.20 Stainless steel sink

Note: Sink may be fixed on the wall or on the floor

Extension

13 mm hot and cold supplies

Hinged brass bucket grating

Hardwood rim

Sizes

Length	Width	Depth
510 mm	× 380 mm	× 230 mm (outside)
460 mm	× 380 mm	× 200 mm (outside)

Fig. 10.21 Cleaner's bucket sink

Note: The fountain will accommodate up to eight persons

Liquid soap holder

Spray head

Umbrella spray

Diameter 1.070

Access panel

Spring valve

Water supply at a controlled temperature

38 mm waste

Cast iron weight

Lever

Steel foot action ring

Bowl

Pedestal

Foot action ring

Fig. 10.22 Ablution fountain

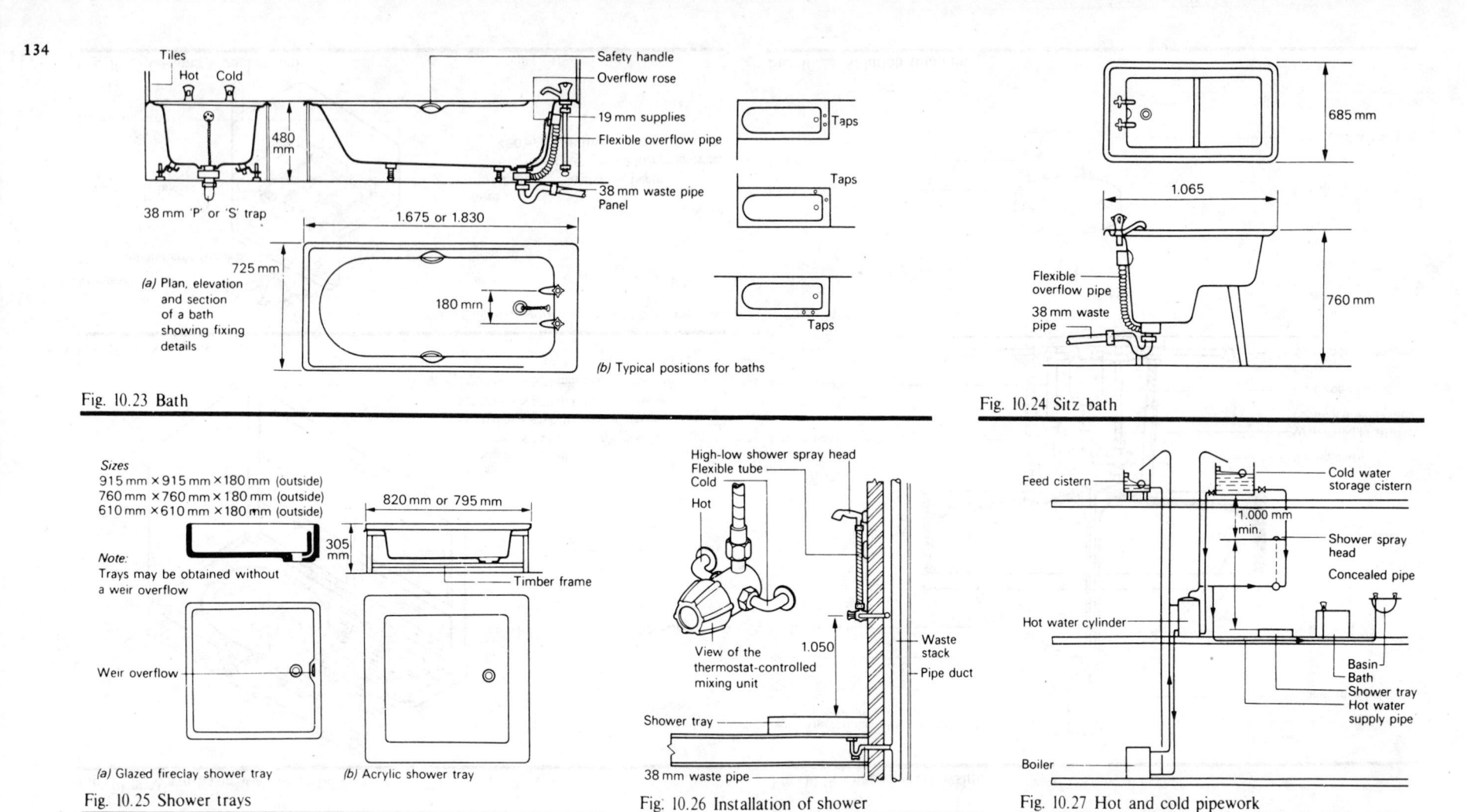

Fig. 10.23 Bath

Fig. 10.24 Sitz bath

Fig. 10.25 Shower trays

Fig. 10.26 Installation of shower

Fig. 10.27 Hot and cold pipework

any risk of scalding. Since the spray may be submerged if the waste became blocked, the bidet represents a special risk to the contamination of the water supply by back siphonage; this may be reduced by having the supply pipes at least 2 m above the bidet. Some water authorities may require bidets to be supplied from a separate cold water cistern, or separate hot and cold water supplies.

Figure 10.28 shows details of a bidet, including the water supplies and waste connections.

Drinking fountain: this is used in most business and industrial buildings, as well as in schools and colleges. Several patterns are available in fireclay or stainless steel. Figure 10.29 shows a fireclay drinking fountain having a hooded jet, which prevents water from the mouth coming into contact with the jet. The

water supply should always be directly from the main and a spray-operated, non-concusive self-closing valve is essential. The waste pipe sizes vary from 19 to 32 mm internal diameter.

Sanitary conveniences

The list of provisions required to comply with the Building Regulations 1985 with regards to entry and performance of sanitary conveniences are as follows:

1. Sufficient sanitary conveniences shall be provided which shall be:

   (*a*) in rooms separated from places where food is stored or prepared; and
   (*b*) designed and installed so as to allow effective cleaning.

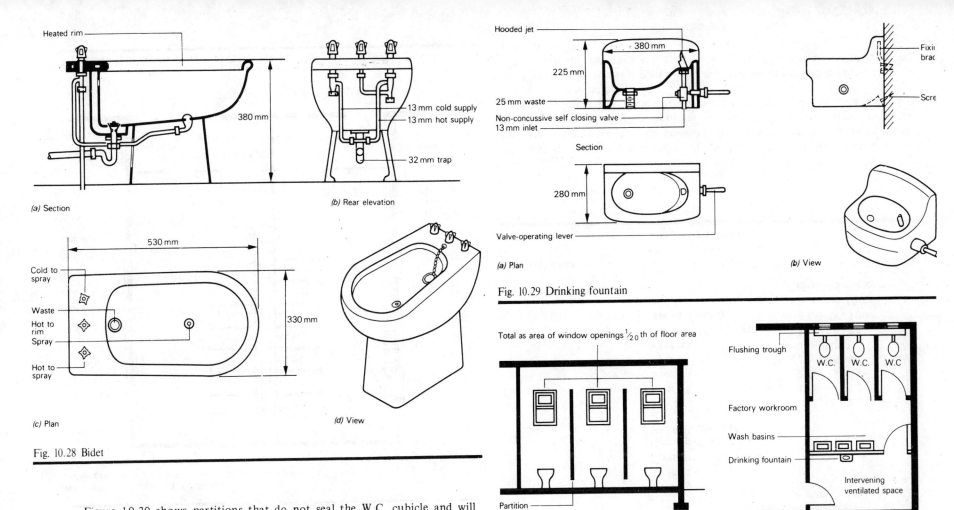

(a) Section

(b) Rear elevation

13 mm cold supply
13 mm hot supply
32 mm trap

Heated rim
380 mm

Cold to spray
Waste
Hot to rim
Spray
Hot to spray

530 mm

330 mm

(c) Plan

(d) View

Fig. 10.28 Bidet

Hooded jet
380 mm
225 mm
25 mm waste
Non-concussive self closing valve
13 mm inlet

Section

280 mm

Valve-operating lever

(a) Plan

Fixing bracket
Screw

(b) View

Fig. 10.29 Drinking fountain

Total as area of window openings ¹/₂₀ th of floor area

Partition

Partitions not sealing the cubicles and allowing
free circulation of air throughout the whole room

Fig. 10.30 Ventilation

Flushing trough
W.C.   W.C.   W.C
Factory workroom
Wash basins
Drinking fountain
Intervening
ventilated space

Sanitary accommodation entered from a kitchen
or workroom

Fig. 10.31 Entry

Figure 10.30 shows partitions that do not seal the W.C. cubicle and will allow effective cleaning of the floor and also permit good ventilation. Figure 10.31 shows the method of entering a snaitary convenience from a room where food is stored or prepared. The ventilated space should effectively separate the convenience from the other room. Figure 10.32 shows the entry to a bathroom containing a WC, direct from a bedroom or dressing room. It is good practice, however, to separate the bathroom from the bedroom or dressing room by means of an intervening ventilated space, especially if the only W.C. in the premises is in the bathroom. A landing or corridor will act as an intervening ventilated space — see Fig. 10.33

2. *Acceptable level of performance.* In order to reduce the risk to health of persons in buildings the Regulations require closets to be provided which are:

(a) in sufficient number and of the appropriate type for the sex and age of the persons using the building; and
(b) sited, designed and installed so as not to be prejudicial to health.

3. *Ventilation.* If mechanical ventilation of a sanitary convenience is required the Building Regulations 1985 require a minimum air change of three per hour. The ventilation may be intermittent but should run for 15 minutes after the convenience has been vacated.

If natural ventilation is required this can be by means of a window, skylight or similar means of ventilation which open directly into the external air.

Planning of sanitary apartments

Legislation such as the Standards for School Premises, Public Health Act, Factory Act and the Offices, Shops and Railway Premises Act require that

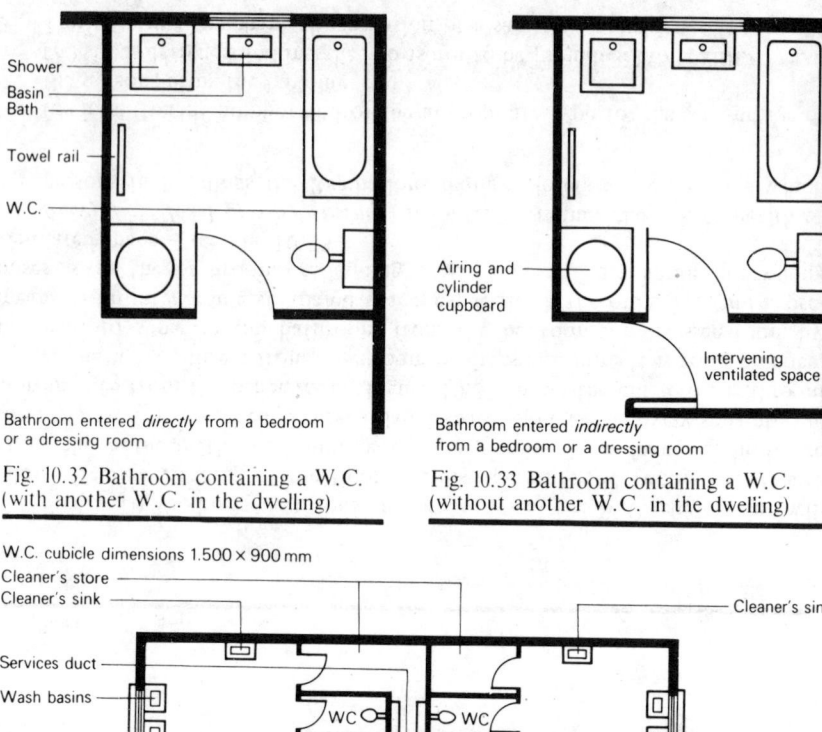

Bathroom entered *directly* from a bedroom or a dressing room

Fig. 10.32 Bathroom containing a W.C. (with another W.C. in the dwelling)

Bathroom entered *indirectly* from a bedroom or a dressing room

Fig. 10.33 Bathroom containing a W.C. (without another W.C. in the dwelling)

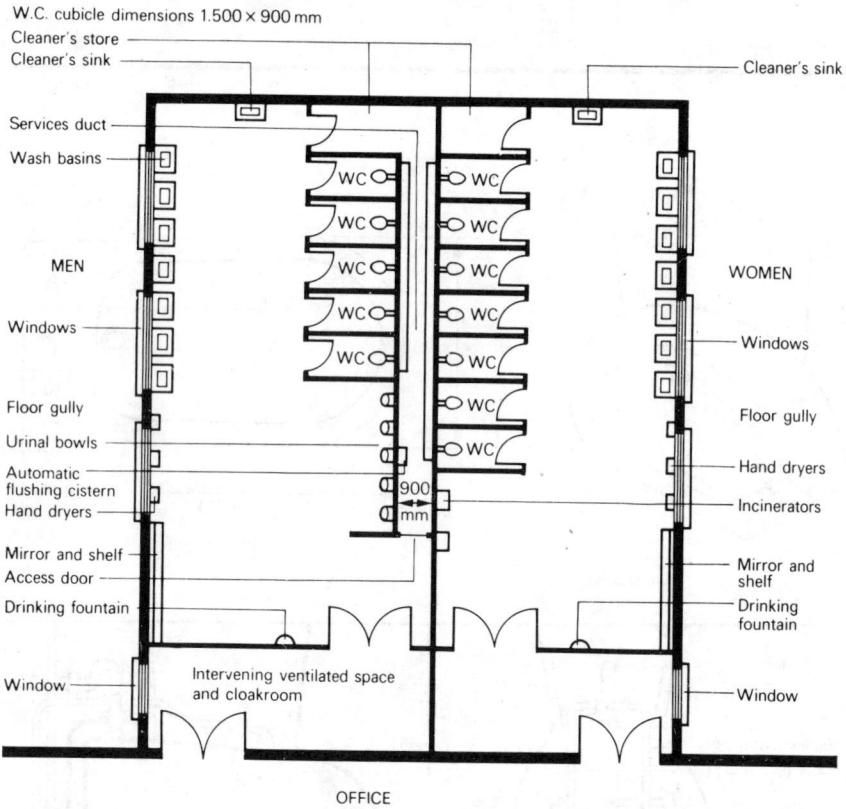

W.C. cubicle dimensions 1.500 × 900 mm

Fig. 10.34 Plan of sanitary accommodation for an office for 120 women and 130 men

Table 10.1 Office Buildings, British Standards Code of Practice CP 3, Engineering and Utility Services

Fitments	For accommodation other than for principals, etc.	
	Male personnel	Female personnel
W.C.s	1 for 1—15 2 for 16—35 3 for 36—65 4 for 66—100 For over 100 add at the rate of 3 per cent	1 for 1—12 2 for 13—25 3 for 26—40 4 for 41—57 5 for 58—77 6 for 78—100 For over 100 add at the rate of 5 per cent
Urinals	Nil up to 6 1 for 7—20 2 for 21—45 3 for 46—70 4 for 71—100 From 101—200 add at the rate of 3 per cent. For over 200 add at the rate of 2½ per cent	
Lavatory basins	1 for 1—15 2 for 16—35 3 for 36—65 4 for 66—100 For over 100 add at the rate of 3 per cent	1 for 1—12 2 for 13—25 3 for 26—40 4 for 41—57 5 for 58—77 6 for 78—100 For over 100 add at the rate of 5 per cent
Cleaners' sinks	At least 1 per floor, preferably in or adjacent to a sanitary apartment	

sufficient sanitary accommodation should be provided. In the absence of such regulations, reference may be made to the Code of Practice, Engineering and Utility Services, which gives recommended sanitary accommodation for various types of buildings and Table 10.1 gives the recommended sanitary accommodation for office buildings.

Figure 10.34 shows a plan of the sanitary accommodation for an office for 120 women and 130 men and Fig. 10.35 a plan of back-to-back bathrooms and kitchens for flats.

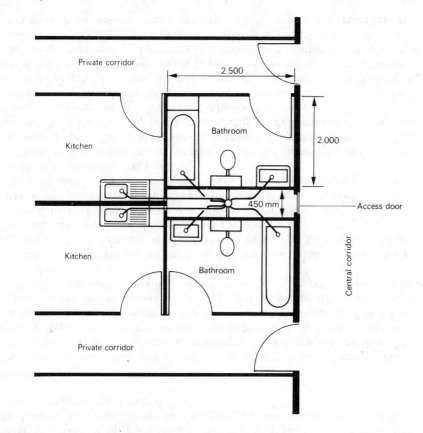

Private corridor

2.500

Bathroom

Kitchen

2.000

Kitchen

450 mm

Access door

Bathroom

Central corridor

Private corridor

Fig. 10.35 Plan of back to back bathrooms and kitchens for flats

Chapter 11

Refuse disposal

Types of refuse

1. *Domestic:* includes ashes, food, paper, bottles, tins, rags and cardboard.
2. *Commercial:* includes some of the domestic rubbish, but it mostly consists of waste paper from office stationery, catering waste and cartons.
3. *Industrial:* includes various waste materials produced by industries, some of which may be toxic, offensive and dangerous. Some of the waste may be salvageable and valuable; for example, metal waste may be melted down and re-used.

Storage

The storing of refuse until collection in disposable plastic or paper bags is becoming increasingly more popular and is generally more hygienic than storing refuse in a metal dustbin. The re-use of bags is not considered acceptable from the hygienic point of view and the bags are therefore disposed of with the refuse. A bag filled with refuse weighs much less than a full metal dustbin, and since they have not to be returned to the premises there is a saving in collection time and labour. Bags can be used as liners for metal bins and this is probably the most hygienic method of storing refuse. Free-standing metal frames are also available for holding refuse-bags and, if required, the frame may have a metal mesh surround to protect the bags from attack by rodents (see Fig. 11.1).

Metal bins can be made quieter by using rubber lids and base rings. Since refuse is becoming lighter in weight and more bulky, a large dustbin should be

138

recommended. A large British Standard bin is 500 mm in diameter and 730 mm high and will hold about 0.09 m³ of refuse.

Table 11.1 gives a comparison of mass of full 0.09 m³ refuse containers.

Table 11.1

Type of container	Mass (kg)
Metal dustbin with lid	28 to 31
Plastic or paper bag inside a metal frame	20
Plastic or paper bag	15

Refuse chutes

In high-rise flats it is not practical or hygienic to carry dustbins or bags down to the ground floor for subsequent collection. A method of overcoming this problem is to provide a refuse chute carried vertically through the building, with an inlet hopper on each floor. The hoppers must be designed to close the chute when they are opened to receive refuse, or otherwise people on the lower floors might be covered with refuse from above when they put their own refuse into the hopper. This type of hopper also prevents dust, smoke and smells from passing through to the floors.

Planning

There may be more than one chute in a building and these discharge into a refuse container, or an incinerator, inside specially constructed chambers. The Building Regulations require a refuse chute to have a circular cross section, with a minimum internal diameter of 375 mm. The British Standard 1703 recommends that the internal diameter of a refuse chute should be at least 457 mm, in blocks of over twelve storeys high, or where four or more refuse containers are provided in a storage chamber.

Refuse chutes should be sited on well-ventilated landings, balconies or adjacent to the kitchens and storage spaces. It should not be sited in a kitchen. For sound insulation, any wall separating a refuse chute from a habitable room must weight 1318 kg/m² and this is satisfied with a 675 mm brick wall, or about 210 mm of concrete. Walls separating chutes from other rooms need only weigh 220 kg/m² and a 113 mm brick wall is adequate. Bends or offsets in chutes should be avoided whenever possible.

Washing down facilities may be provided for the chute, by means of a dry riser with jet heads fitted inside the chute at each floor level. Some authorities, however, hold that bacteria breed more readily in the presence of water and therefore washing down of chutes should be avoided. Users should be advised to wrap the refuse in order to prevent soiling of the chute. A chute can be arranged to discharge into two bins by bifurcating the end with a cut-off damper, operated by a caretaker when one is filled. One chute can also be made to serve four to eight bins or bags, by means of a roundabout in which the bins or bags move either manually, or automatically around a central spindle. A special machine may also be used, which automatically compresses the refuse into the bags.

Ventilation

The chute should be ventilated by means of a pipe, or duct, having a cross sectional area of not less than 17 500 mm². The ventilator should be of non-combustible material and carried up high enough to avoid foul air causing a nuisance.

Materials

Refuse chutes should be of non-combustible and acid-resistant materials. Glazed stoneware, spun concrete or asbestos cement pipes may be used. Hoppers should also be made of non-combustible materials not subject to corrosion or abrasion. Hoppers are manufactured from cast iron, wrought and cast aluminium and steel. Steel hoppers are galvanised and cast iron hoppers painted.

Chambers

Refuse storage chambers must be surrounded by floors and walls, with at least 1 hour fire rating. Surface finishes must be non-combustible, moisture-proof and easy to clean. A ½ hour fire rating lockable, dust and fume-proof door must be provided and the floor must be laid to falls to an external trapped gully. A tap should be fitted outside, so that the chambers can be hosed down. To reduce noise, the chambers should be structurally isolated from the rest of the building, by means of double walls or separate floor slabs.

Figure 11.2 shows a refuse chute installation, with a refuse container at the base and Fig. 11.3 a refuse chute installation having an incinerator at the base.

Garchey system

The system was invented by a Frenchman, M. Louis Garchey, and has been improved by a British company. The original French system required underground suction pipework and a central incinerator plant, which increased the capital and running costs of the system. The new improved British system uses a refuse-collection vehicle, which replaces the underground suction pipework and central incinerator plant. The vehicle is much cheaper and can receive 7 days' refuse from over 200 flats, virtually serving an entire housing estate in one day or night shift of about 2½ hours. Whilst the system is particularly suitable for high density multi-storey flats, especially if it is installed whilst building work is in progress, it can also be installed in existing buildings, including single-storey dwellings.

In the system a special sink has a grid and waste plug, which fits over the outlet and this enables the sink to be used for normal purposes. To deposit refuse, the housewife lifts off the sink grid and places the refuse in a waste tube. In order to remove the refuse, the tube is lifted by the grid and the refuse, together with retained waste water, are carried through a trap into a vertical stack pipe. A concrete collection chamber below ground receives the refuse and waste water and the chamber is emptied by the refuse vehicle at regular intervals, usually weekly, and taken directly to a tip. Any type of refuse can be handled by the system, including bottles and tins, providing that these will fit inside the tube. Larger items, however, must be taken to a communal container and collected in the normal manner. Figure 11.4 shows a section through the special sink unit and Fig. 11.5 a diagrammatical layout of the system.

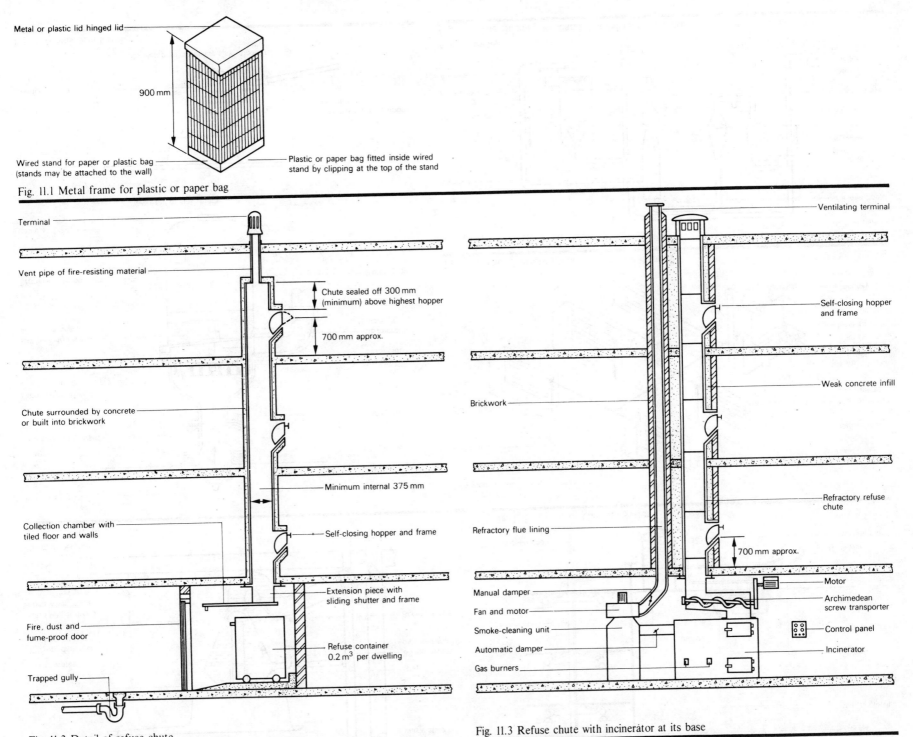

Metal or plastic lid hinged lid

900 mm

Wired stand for paper or plastic bag (stands may be attached to the wall)

Plastic or paper bag fitted inside wired stand by clipping at the top of the stand

Fig. 11.1 Metal frame for plastic or paper bag

Terminal

Vent pipe of fire-resisting material

Chute sealed off 300 mm (minimum) above highest hopper

700 mm approx.

Chute surrounded by concrete or built into brickwork

Minimum internal 375 mm

Collection chamber with tiled floor and walls

Self-closing hopper and frame

Extension piece with sliding shutter and frame

Fire, dust and fume-proof door

Refuse container 0.2 m³ per dwelling

Trapped gully

Fig. 11.2 Detail of refuse chute

Ventilating terminal

Self-closing hopper and frame

Weak concrete infill

Brickwork

Refractory refuse chute

Refractory flue lining

700 mm approx.

Manual damper

Motor

Fan and motor

Archimedean screw transporter

Smoke-cleaning unit

Control panel

Automatic damper

Incinerator

Gas burners

Fig. 11.3 Refuse chute with incinerator at its base

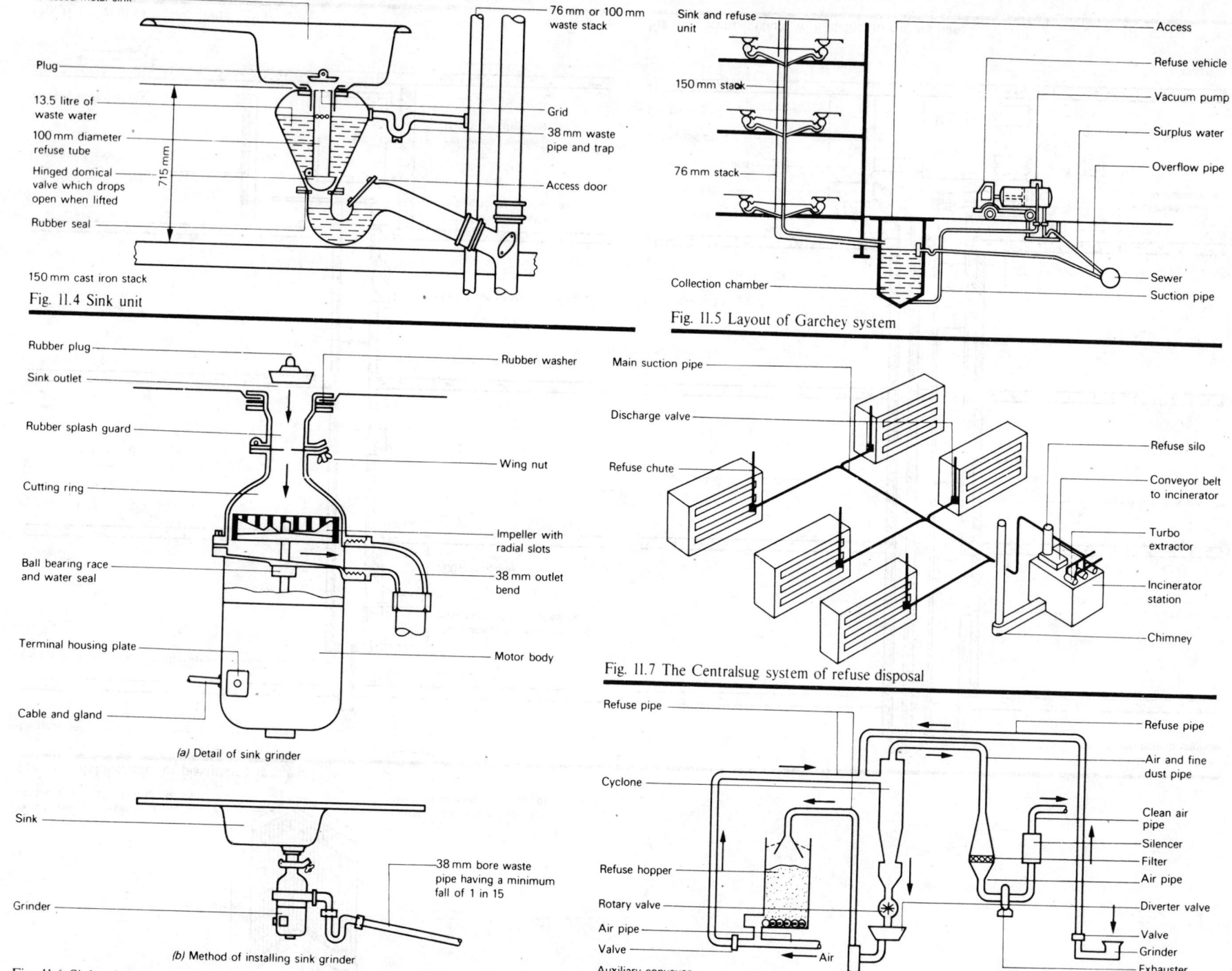

Fig. 11.4 Sink unit

Fig. 11.5 Layout of Garchey system

(a) Detail of sink grinder

(b) Method of installing sink grinder

Fig. 11.6 Sink grinder

Fig. 11.7 The Centralsug system of refuse disposal

Fig. 11.8 The Building Research Establishment experimental refuse disposal system

The collection vehicle

After 5 years of experiments, a special road vehicle was successfully developed by the British company. The vehicle is sold by the company to Local Authorities; it includes a patented pusher plate, which compresses the refuse inside a cylindrical tank.

Advantages

The system has proved satisfactory and popular with the user and, unlike the refuse chute, the refuse can be removed without taking it to a landing or balcony. The refuse is taken directly from the kitchen sink, to the tip, through an enclosed water-sealed system and it is therefore more hygienic than the chute and there is less risk of fire spread. The refuse collection is easier, quicker, quieter and the damp refuse does not cause spread of dust at the tip.

Sink grinders

Several types of sink grinders are available, which are used mainly for the disposal of food waste, but cannot be used for metals, plastics, ceramics or string. The grinder, which is fitted under the sink, reduces the food waste to fine particles, which can be carried away by the waste water to the foul water drain. Figure 11.6 shows a detail of a sink grinder and the method of installation.

Electrical

Some grinders have devices for freeing stoppages by automatically reversing the motor and they also have a thermal overload device, to prevent the motor burning out if it jams. The electrical current used for the motor is about 3 amperes, but in the event of overloading, due to improper use, the current will rise to 13 amperes before the thermal overload operates. Domestic grinder motors are available in the following voltage range: 110, 200, 220, 230 and 250 alternating current.

Large catering and industrial grinders may require a 415 volt three-phase alternating current supply. Controls consist of an on-off switch, reversing switch, reset button and sometimes an indicator light. The controller and the disposer should be earthed independently, in accordance with the Institution of Electrical Engineers Regulations, and there must be no danger of water or dampness affecting the electrical wiring.

Safety

It should be impossible for the users to insert his or her fingers into the shedding compartment and blockages should not be freed by hand.

Noise

Noise and vibration should not be a problem, provided that grinders are fitted properly. Most grinders are supplied with sound-deadening jackets, and it is important that the disposer should hang freely underneath the sink. Resilient neoprene disposer mountings on the underside of the sink reduce vibration.

Waste pipe

A 38 mm minimum internal diameter waste pipe and an ordinary trap should be used, which should discharge into a soil stack, or back inlet gully. The waste pipe should have a minimum fall of 1 in 15 and a bottle trap should not be used. The drain may either discharge into a public sewer or a septic tank.

Operation

Cold water is run into the sink, the disposer is switched on and waste food or paper is fed into the disposer. The waste falls on to an impeller, which revolves at a speed of 24 revolutions per second and the waste is thrown against a cutting ring by a centrifugal force. The waste is shredded into fine particles and is carried by the water into the drain. The impeller is made of high carbon chrome alloy and the cutter ring of carbon chrome vanadium steel.

Pneumatic transport of refuse

Centralsug system: this system originated in Sweden for the transport of crude refuse. The refuse is discharged into a normal refuse chute, but is allowed to accumulate in about 3 m of the chute below the lowest hopper. At predetermined intervals, usually twice per day, the refuse is released into a 525 mm bore main horizontal suction pipe. The refuse is drawn through the pipe at about 25 m/s to a silo and released as required into an incinerator.

BRE system: The Building Research Establishment has carried out extensive experiments on a pneumatic transport system which would pulverise the refuse first, in order to use smaller transporting pipes. In the system, the refuse is disintegrated by a special grinder inside the dwelling, or located near a group of dwellings. The refuse is cut by the grinder into pieces of about 10 mm across and is then blown a short distance down a 75 mm bore pipe, where it lies. At predetermined times a valve is opened and the refuse is drawn into a 225 mm bore horizontal pipe and transported by an air stream about 1 km to a collection point. The refuse is then transferred to a positive-pressure pneumatic system and blown to a treatment plant.

Figures 11.7 and 11.8 show a diagrammatical layout of the Centralsug and the Building Research Station Establishment systems.

Chapter 12

Sewage disposal

Whenever sewage disposal is not provided by a municipal sewerage system, it is necessary to install a small sewage purification plant. The problem is one which seldom arises in towns and urban districts, as most of these places are provided with a sewerage system and it is therefore advantageous to the building owner to connect his property to the system. The local authority may compel an owner of a property to connect his foul water drain to the public sewer, where such a sewer is within 30 m of the site boundary in which the property is built. In rural areas, however, considerations of expense often prevent the local authority from providing sewers and sewage purification plants and it becomes necessary to install a small sewage purification plant for one or more isolated buildings.

Methods of disposal

The disposal of sewage from isolated buildings may be classified under the following headings:

1. *Irrigation:* in which the sewage is first liquified before being discharged to an irrigation field.
2. *Discharge to a river or stream:* in which the sewage is first purified before being discharged to the water course.
3. *Conservancy:* in which the sewage is discharged into a cesspool and periodically removed, or is treated in a chemical or earth closet.
4. *Dilution:* in which the sewage is discharged into the sea, or an estuary of a tidal river.

Irrigation systems

Septic or settlement tanks are used where the crude sewage is liquified and the effluent conveyed by a channel or a pipe to an irrigation field. The liquid is allowed to percolate into the soil, by means of surface or sub-surface irrigation. Surface irrigation is achieved by discharging the liquid into channels formed in the ground (see Fig. 12.1) or by flooding a prepared area of ground, contained within earth banks to confine the liquid. Sub-surface irrigation is carried out by a system of sub-soil drainage pipes, laid in either a grid or a herring-bone formation (see Fig. 12.2) using 100 mm bore open-jointed pipes. Trenches 600 mm wide and 600 mm deep should be dug and the bottom filled with 150 mm of heavy polythene shingle, hard-burnt clinker or hard-broken stone. The pipes should be laid on this bed to a fall of about 1 in 200 and the open joints covered with strips of tarred felt, or pieces of broken pipes to prevent entry of silt. The same material used for the pipe bed should be carefully hand packed around the pipes and the trench filled with the material to within 150 mm of the ground level covered with soil and turfed over (see Fig. 12.3).

Site investigation

Before using irrigation for the disposal of the final effluent, it is necessary to make a careful investigation into the porosity of the soil. A dry day should be chosen during a period when the weather is not unusually wet or dry. A hole is dug 600 mm x 600 mm x 300 mm deep in the area where it is intended to put the irrigation pipes and the hole filled with water in the afternoon and left to the following morning. Water is then poured gently into the empty hole until it is 250 mm deep and careful observations made of the time it takes for the water to drain away. It will be noticed that the first few millimetres will disappear more rapidly than the remainder and it is therefore necessary to find the average time it takes for the water level to fall 1 mm. This average time will be found by dividing the total time by 250. The length of a sub-surface irrigation pipe may be found from Table 12.1.

Table 12.1 Length of a sub-surface irrigation pipe

Average time for the water to fall 1 mm (seconds)	Length of irrigation pipe per dwelling (metres)
5	10.00
7	13.75
10	16.00
12	18.30
24	24.40
36	30.00
44	41.00
72	55.00

Septic or settlement tank

A septic tank is a watertight chamber in which the sewage is acted upon by anaerobic bacteria, which breaks up the crude sewage and converts it into gases

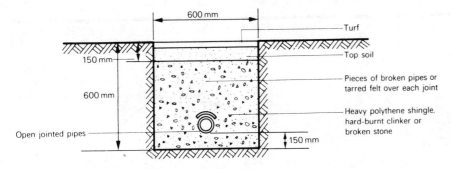

Fig. 12.3 Method of laying sub-surface drains

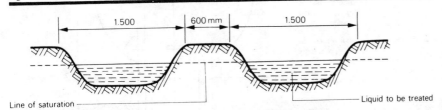

Fig. 12.1 Surface irrigation

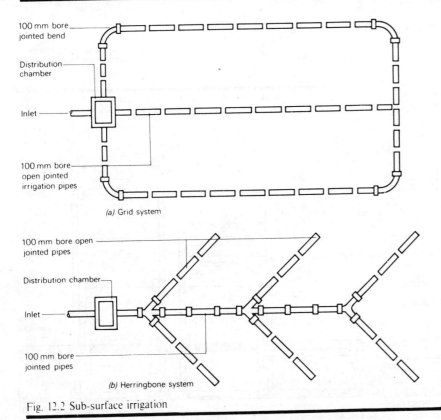

(a) Grid system

(b) Herringbone system

Fig. 12.2 Sub-surface irrigation

and liquids. The anaerobic bacteria or anaerobes work in the absence of oxygen and therefore the septic tank should be covered. Sludge falls to the bottom of the tank and scum floats on top of the liquid. The scum acts as an air excluder and the liquid will flow out of the tank as new sewage enters. The inlet and outlet pipes of the tank should be submerged to ensure that the sewage cannot flow across the top of the tank without being broken down by the bacteria and also to prevent the scum which forms on top of the liquid from being broken.

Size

The sewerage should be retained in the tank from between 16s and 48 hours and therefore thes tank should be sized accordingly. The Code of Practice BS 6297 Small Sewage Treatment Works 1983, gives the following formula for the total capacity of a septic tank, where desludging is carried out at not more than 12-monthly intervals.

$$C = 180\,P + 2000$$

where

C = Capacity of tank in litres with a minimum value of 2720 litres
P = Number of people served with a minimum of four

When the number of people served is four, the capacity of the tank is:

$$C = 180 \times 4 + 2000$$
$$C = 2720 \text{ litres}$$

This is in line with the Building Regulations 1985, where the minimum capacity of a septic tank is 2700 litres.

The CP 6297 recommends that a septic tank to serve a population of up to 100 persons should be divided into two compartments. The first compartment should have a length of twice its breadth and should provide about two-thirds of the total capacity. The second compartment should be square and sized to hold about one-third of the total capacity. The compartments can be two separate units, or a single unit separated by a dividing wall. Instead of brickwork or *in situ* concrete, two circular precast concrete pipe sections may be used to make the two chamber tanks (see Fig. 12.4).

The depth of liquid in the tanks should not be less than 1.5 m. The anaerobic action which takes place in a septic tank is a partial breakdown of the unstable compounds only and there must therefore be a secondary process to follow. This secondary process relies on the action of aerobic bacteria, which purify the effluent from the septic tank to a standard sufficient for discharge to a river or stream. Aerobic bacteria thrive in the presence of oxygen and therefore the liquid should be exposed to as much air as possible. The process takes place by allowing the liquid to percolate through the soil, or bypassing the liquid through a biological filter.

Figure 12.5 shows a detail of a brick septic tank, where the effluent flows to a sub-surface irrigation system for final treatment, and Fig. 12.6 shows a detail of a glass fibre settlement or septic tank, which operates as follows:

1. Sewage enters the tank and solids are deposited in chamber 3, which provides a 12-month sludge storage and digestion capacity.
2. From chamber 3, the settled effluent passes upward through three slots into

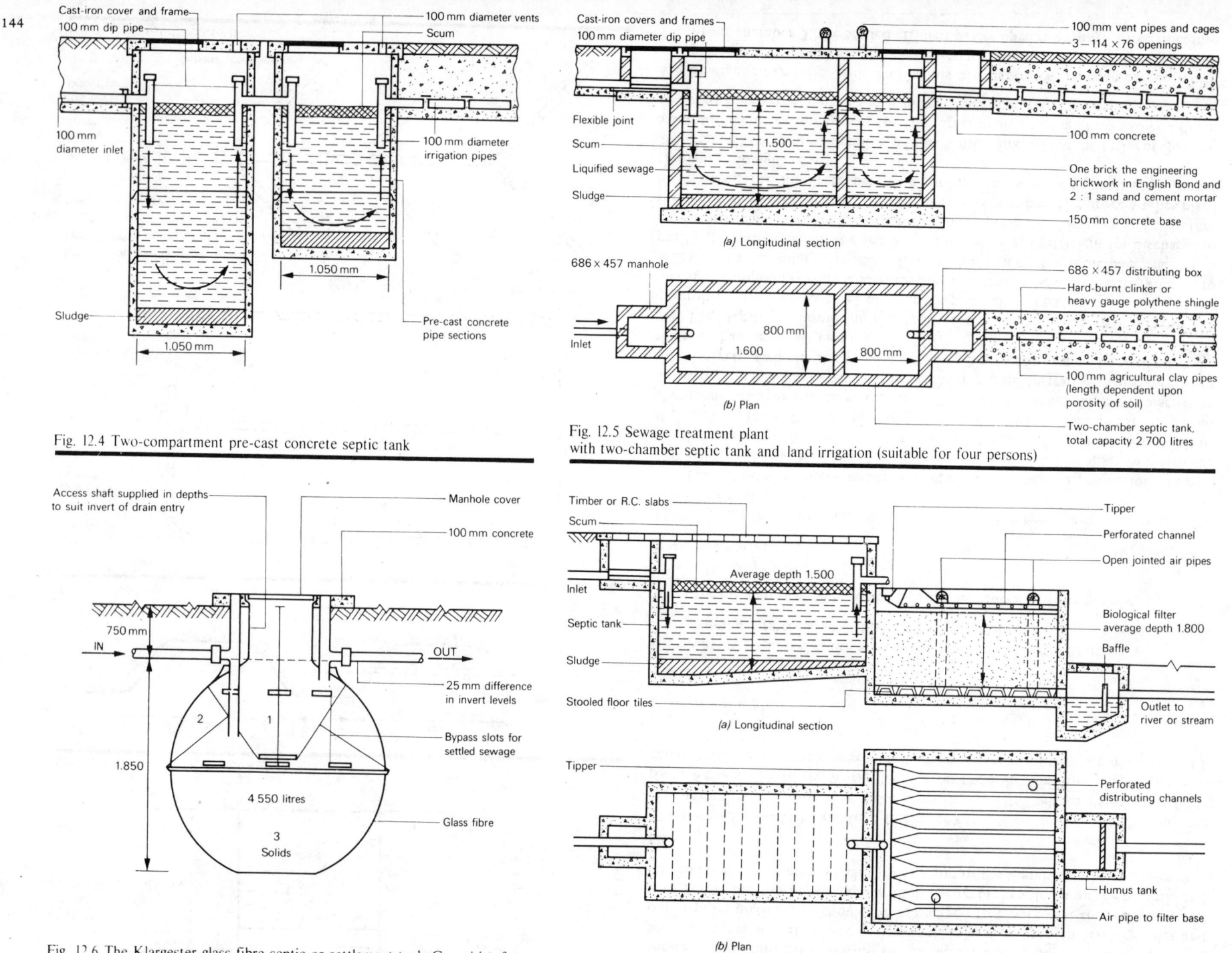

Fig. 12.4 Cast-iron cover and frame — 100 mm dip pipe — 100 mm diameter vents — Scum — 100 mm diameter inlet — 100 mm diameter irrigation pipes — Sludge — Pre-cast concrete pipe sections — 1.050 mm — 1.050 mm

Fig. 12.4 Two-compartment pre-cast concrete septic tank

Fig. 12.5 Cast-iron covers and frames — 100 mm diameter dip pipe — Flexible joint — Scum — Liquified sewage — Sludge — 100 mm vent pipes and cages — 3 – 114 × 76 openings — 100 mm concrete — One brick the engineering brickwork in English Bond and 2 : 1 sand and cement mortar — 150 mm concrete base — 1.500

(a) Longitudinal section

686 × 457 manhole — Inlet — 800 mm — 1.600 — 800 mm — 686 × 457 distributing box — Hard-burnt clinker or heavy gauge polythene shingle — 100 mm agricultural clay pipes (length dependent upon porosity of soil) — Two-chamber septic tank, total capacity 2 700 litres

(b) Plan

Fig. 12.5 Sewage treatment plant with two-chamber septic tank and land irrigation (suitable for four persons)

Fig. 12.6 Access shaft supplied in depths to suit invert of drain entry — Manhole cover — 100 mm concrete — 750 mm — IN — OUT — 25 mm difference in invert levels — Bypass slots for settled sewage — Glass fibre — 1.850 — 2 — 1 — 4 550 litres — 3 Solids

Fig. 12.6 The Klargester glass fibre septic or settlement tank. Capacities from 2 750 to 18 200 litres (Klargester Ltd, College Road, Aston Clinton, Bucks.)

Fig. 12.7 Timber or R.C. slabs — Scum — Inlet — Septic tank — Sludge — Stooled floor tiles — Average depth 1.500 — Tipper — Perforated channel — Open jointed air pipes — Biological filter average depth 1.800 — Baffle — Outlet to river or stream

(a) Longitudinal section

Tipper — Perforated distributing channels — Humus tank — Air pipe to filter base

(b) Plan

Fig. 12.7 Sewage treatment plant with septic tank and biological filter

chamber 2. In this chamber, sedimentation of finer suspended matter occurs and three peripheral slots allow any sediment to return to chamber 3.

3. From chamber 2, highly clarified effluent passes through three slots into chamber 1, from which it is discharged through the outlet square junction, to land drains or a biological filter.

4. Sludge in chamber 3 is removed by a suction emptier after removing the floor of chamber 1 with a handle.

Sludge removal

A service road for the local authority desludging vehicle should be provided for the septic tank. The sludge is usually removed at 6-monthly intervals and about 20 per cent of the sludge should be left in the tank for seeding purposes at each desludging.

Discharge to a river or stream

Before the liquid is allowed to be discharged into a river or stream it must be treated to a standard laid down by the Royal Commission on Sewage Disposal. The standard requires that the liquid shall not contain more than three parts per 100 000 of suspended solid, nor absorb in 5 days at 18 °C, more than two parts per 100 000 of dissolved oxygen. The latter standard is referred to as the biochemical oxygen demand or BOD.

Note: Since dissolved impurities present in sewage are removed by a process of oxidisation, it follows that the greater the degree of oxidisation achieved the purer will be the effluent and less oxygen will be absorbed when it is discharged to the river or stream. The usual method of achieving this standard is by treating the sewage, firstly in a septic tank and secondly by passing the resultant liquid through a biological filter.

Biological filter

This is usually built with a concrete floor and concrete or brick walls. The walls need not be watertight, because the liquid is distributed evenly over the whole surface of the bed and flows through the filter media to the outlet pipe. Open-jointed brickwork may be used which increases the aeration of the liquid and the walls may therefore also be constructed of random rubble. The filter may be rectangular or square on plan, depending upon the arrangement made for the liquid distribution over the bed. The floor should have a good fall towards the outlet and covered with special floor tiles, which allow easy passage of liquid and through aeration to the bed. A wire mesh should be fitted over the filter, to prevent leaves from covering the top of the bed.

Figure 12.7 shows a sewage treatment plant with septic tank and biological filter.

Filtering media

Metallurgical coke, hard-burnt clinker or hand-broken stone may be used, as these have a rough uneven surface and allow more surface contact between the liquid and the air. The media should be roughly spherical in shape, free from fires or dust and uniformly graded. the bottom 300 mm should be of 100–150 mm grade, to allow good ventilation and drainage and the rest of the bed should consist of 40–50 mm grade media.

Humus tanks

Before the effluent from the biological filter is discharged into the water course, the humus, which is a waste product of the bacterial action, should be removed. Although humus is an inoffensive, earth-like material, it would, if discharged into a river or stream, cause banks to be formed which may cause flooding, particularly with a small stream. The humus should be removed from the tank at weekly intervals, but if this is not possible in a small plant it is better to omit the tank and rely upon irrigating the effluent over the surface of a grass patch or scrub, about 3 m^2 per person, depending upon the porisity of the soil.

Size

The following formula for the capacity of the humus tank for retention times, varying from 24 to 8 hours for a population range 10 to 50 and from 8 to 5 hours for a population range 50 to 300 may be used.

$$C = 30P + 1500$$

where

C = capicity of tank in litres
P = the design population.

The Code of Practice BS 6297 gives the following formula for the volume of the filter medium

$$V = 1.5P^{0.83}$$

where

V = volume of medium in m^3
P = the design population.

For very small plants serving up to ten persons, the tank capacity may be about one-third the capacity of the septic tank. The humus is an excellent garden fertiliser and is better for this purpose than dried sludge from the septic tank.

Siting of treatment plant

The site for a sewage-treatment plant must be chosen with care, giving consideration to the proximity of wells for drinking water, inhabited buildings and access for the local authority desludging vehicle. The flow of sewage through the plant, should be by gravity where this is possible and levels should be checked to ensure that flooding from water courses, or ground water, is impossible. Trees should not be near the plant, as the leaves and roots may cause problems of plant maintenance. Fencing should be provided around the plant, to prohibit the entry of unauthorised persons or livestock.

Table 12.2 gives the minimum distances of treatment plant from inhabited buildings and source of drinking water.

Figure 12.8 shows how the plant is sited in relation to inhabited buildings, source of water supply and service road.

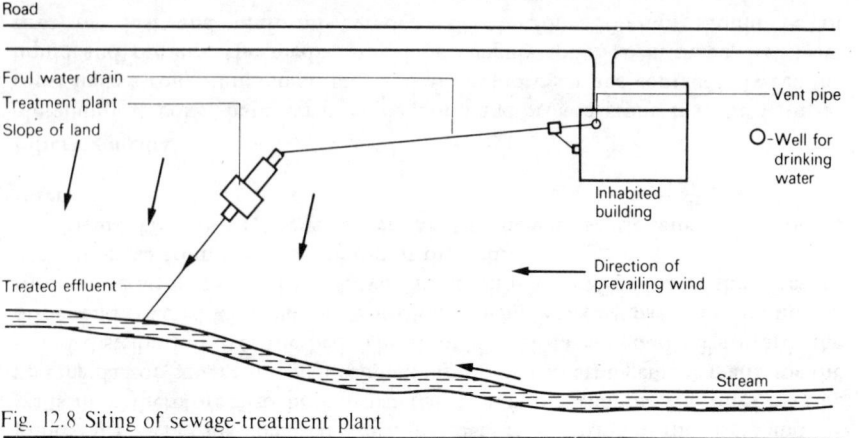

Road
Foul water drain
Treatment plant
Slope of land

Vent pipe

O—Well for drinking water

Inhabited building

Treated effluent

Direction of prevailing wind

Stream

Fig. 12.8 Siting of sewage-treatment plant

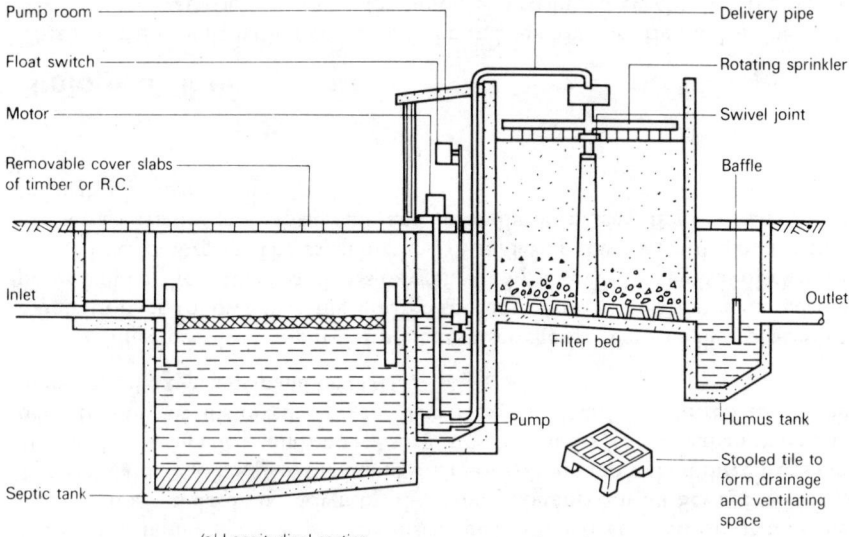

Pump room
Float switch
Motor
Removable cover slabs of timber or R.C.

Inlet

Septic tank

Delivery pipe
Rotating sprinkler
Swivel joint
Baffle

Filter bed

Outlet

Humus tank

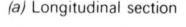

Stooled tile to form drainage and ventilating space

Pump

(a) Longitudinal section

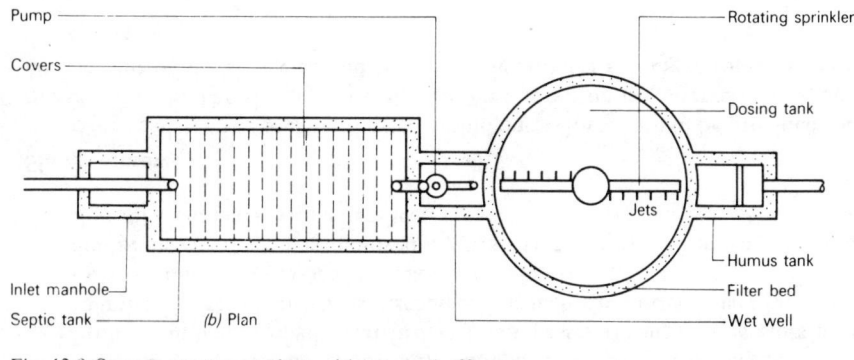

Pump
Covers

Rotating sprinkler

Dosing tank

Jets

Humus tank
Filter bed
Wet well

Inlet manhole
Septic tank   (b) Plan

Fig. 12.9 Sewage-treatment plant with pumped effluent

Table 12.2

Number of persons served by the plant	Minimum distances from inhabited buildings and source of drinking water
20	15.50
50	31.00
100	61.00
200	92.00

Pumped effluent

Wherever possible the effluent from a septic tank should gravitate to the biological filter, but sometimes due to the layout of the site it is necessary to pump the effluent to the filter. Figure 12.9 shows a section and a plan of a plant having a pumping arrangement.

Conservancy systems

(The storage of sewage on site, in a cesspool, chemical closet, or earth closet, for periodic disposal.)

Cesspool: because of the nature of the ground, or the lack of site area, it is sometimes impossible to install a treatment plant. In such cases, a cesspool may be installed which is a watertight chamber built below ground, into which the sewage is discharged from a normal foul water drainage system.

Siting: a cesspool must be sited at least 15—25 m from a dwelling house and at least 18—25 m from any well, spring or stream used for drinking water.

Emptying: some rural councils operate a scheme for emptying cesspools free of charge, but in some areas a charge may be made for the service, or private cesspool-emptying contractors undertake the work. The cesspool-emptying vehicle carry long lengths of hose, so it is not absolutely essential for the cesspool to be close to a road. It is, however, advisable to seek the advice of the local authority or the private contractor before constructing the cesspool, in order to ensure that it can be readily emptied by the vehicle.

Size: this will depend upon the frequency in which the cesspool is emptied and a capacity that will allow 5 weeks between the frequency of emptying is desirable, but more frequent emptying between 3 or 4 weeks is common practice. The Building Regulations 1985 requires a minimum capacity of 18 m³. A capacity based upon 136 litres per head, per day, set against the frequency of emptying, provides a good basis, but the capacity should not normally exceed 45 000 litres.

Construction: cesspools must be absolutely watertight, to prevent the liquids inside from escaping to the soil and also to prevent surface and sub-soil water

entering the cesspool from the outside. A circular chamber is generally considered best for both strength and economy in construction. The normal construction consists of a concrete base, with one-brick thick walls of engineering brick in cement mortar, surrounded with clay paddle on the outside and lined on the inside with either asphalt or water-proof mortar. Alternatively, a cesspool may be constructed of precast concrete pipes surrounded with concrete, or a self-contained glass fibre chamber may be obtained. The inlet drain should be trapped by means of an interceptor and the chamber should be well ventilated by means of a 100 mm bore fresh air inlet and a 100 mm bore gas outlet (see Fig. 12.11).

Building Regulations 1985

Cesspools, septic and settlement tanks

1. Cesspools should hav sufficient capacity to store the foul water from the building until they are emptied; and
2. Cesspools, septic and settlement tanks should be sited so as to be accessible and designed and constructed so as not to contaminate water supplies or be prejudicial to health; and
3. Septic tanks and settlement tanks should have sufficient capacity to break down and settle solid matter in the flow of foul water from the buildings; and
4. Cesspools should have a capacity below the level of the inlet of at least 18 000 litres /18 m^3); and
5. Septic tanks and settlement tanks should have a capacity below the inlet of at least 2700 litres /2.7 m^3); and
6. Septic tanks and settlement tanks, if they are to be desludged using a tanker, should be sited within 30 m of a vehicle access and so that they can be emptied and desludged without the contents being taken through a dwelling or place of work, access may be through an open or covered space.

Design and construction

1. Cesspools and tnaks should be impervious to the contents and to subsoil water. They may be constructed with brick, concrete, glass-reinforced concrete, glass-reinforced plastic or steel. Brickwork should be engineering bricks in cement mortar and at least 200 mm thick. Concrete should be at least 150 mm thick.
2. Cesspools should be covered and ventilated. Tanks should be covered or fenced and, if they are covered, ventilated.
3. Cesspools and tanks if they are covered should be provided with access for emptying, desludging and cleaning. The access should not have any dimension less than 600 mm.
4. The inlet of a cesspool and the inlet of a tank should be provided with access for inspection.
5. Cesspools should have no openings except the inlet and the access.
6. Provision should be made to limit the velocity of flow to a tank. For drains up to 150 mm the velocity may be limited by laying at least 12 m of the incoming drain at a gradient of 1 in 150 or flatter or by providing a dip pipe.

Chemical closets

These may be portable types for use in caravans, camping sites and aircraft, or fixed types incorporating an underground storage tank for use in schools, factories and dwellings in rural areas. The sewage is subjected to the action of a sterilising fluid, which breaks down the solids. Smells are prevented by incorporating a deodorising agent with the chemical, or by an oil film which seals off the surface of the sewage.

Disposal of effluent

In its odourless and inoffensive condition, the treated sewage from a portable closet can be emptied into a trench and covered with soil without the risk of contaminating the ground, causing an offensive odour or attracting flies or vermin. Figure 12.12 shows a detail of a portable chemical closet which is ventilated at the rear; unventilated types may also be obtained.

Figure 12.13 shows the installation of a fixed or permanent type of chemical closet, which operates as follows:

The anti-splash plate and agitator is operated automatically by movement of the seat cover, or lid. When this is lifted, the anti-splash plate is positioned below the pan opening, thus masking the contents of the tank and preventing solids falling directly into the liquid, which prevents splashing. When the seat cover or lid is lowered, the anti-splash plate drops automatically into the tank and is sterilised by the chemical solution. The agitator works in conjunction with the anti-splash plate and is required for the efficient action of the chemical which disintegrates and liquefies the solid matter. The treated sewage is usually discharged through a 100 mm bore drain to a soakaway pit, sited at a suitable distance from the dwelling, or source of water supply.

Earthclosets

In rural areas, where temproary sanitation is required, an earthcloset may be installed.

Fine dry earth is held in a container which falls automatically on to the sewage when the seat is raised. The earth acts as a deodorant and also disintegrates the sewage by the action of bacteria. The contents of the sewage container, after the sewage has been broken down, may be disposed of on the land. Figure 12.14 shows the installation of an earth closet.

Installation

1. The closet should have sufficient opening airbrick for ventilation directly to the external air, situated as near to the ceiling as practicable.
2. The floor of the closet should be of non-absorbent material and should fall towards the entrance door at a gradient of not less than 1 in 25.
3. Thesreceptacle should be of non-absorbent material and placed so that its contents cannot escape by leakage.
4. No part of the receptable or the interior of the closet should have an outlet to a drain.
5. The concrete base of the closet should be not less than 75 mm above the adjoining ground.
6. The arrangement and installation of the closet should not be prejudicial to health.

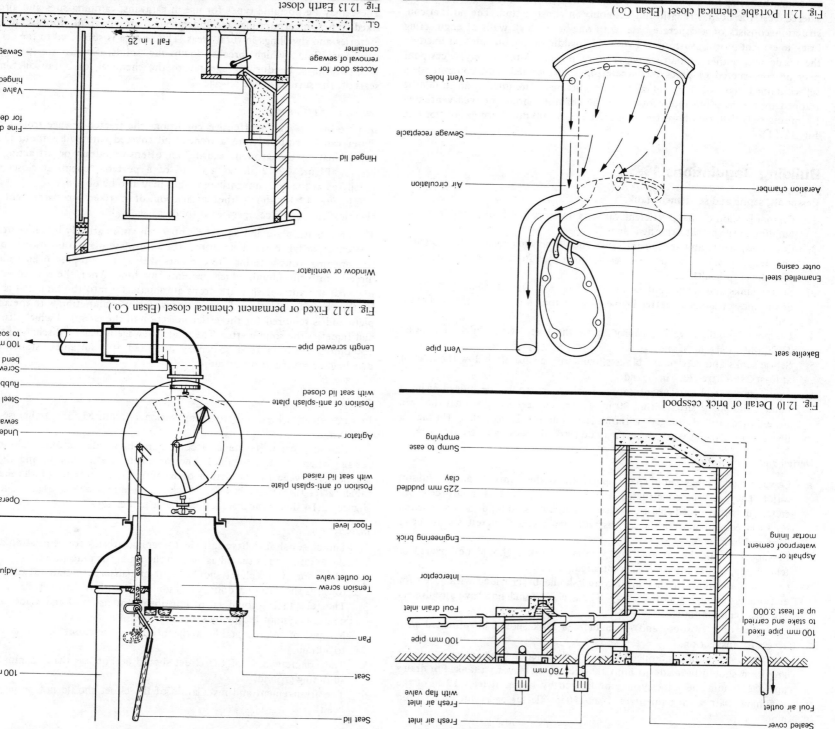

Fig. 12.13 Earth closet

Sewage container

Access door for
removal of sewage
container

Valve operated by
hinged seat

Fine dry earth
for deodorising

Hinged lid

Fall 1 in 25

G.L.

75 mm

Window or ventilator

Fig. 12.11 Portable chemical closet (Elsan Co.)

Vent holes

Sewage receptacle

Air circulation

Aeration chamber

Enamelled steel
outer casing

Bakelite seat

Vent pipe

Fig. 12.12 Fixed or permanent chemical closet (Elsan Co.)

100 mm drain
to soakaway pit

Length screwed pipe

Screwed steel
bend

Rubber seating

Steel plate

Position of anti-splash plate
with seat lid closed

Underground
sewage tank

Agitator

Position of anti-splash plate
with seat lid raised

Operating stem

Floor level

Adjusting sleeve

Tension screw
for outlet valve

Pan

100 mm vent pipe

Seat

Seat lid

Fig. 12.10 Detail of brick cesspool

Sump to ease
emptying

225 mm puddled
clay

Engineering brick

Interceptor

Foul drain inlet

100 mm pipe

Fresh air inlet
with flap valve

Fresh air inlet

Asphalt or
waterproof cement
mortar lining

100 mm pipe fixed
to stake and carried
up at least 3,000

760 mm

Foul air outlet

Sealed cover

148

1. An earth closet should only be entered from the external air and should not open directly into:

    (*a*) a habitable room
    (*b*) a kitchen
    (*c*) a workroom.

2. The closet should be sited so that it will not cause pollution of a water supply which is used, or is likely to be used, for drinking or kitchen purposes.

3. The closet should be sited as far as practicable from any habitable building.

Appendix A
Unvented hot-water storage systems

Definition

The unvented hot-water storage system may be defined as being a secondary curcuit which is not provided with a vent pipe, permanently open to the atmosphere, and is fed with cold water either from the main or a storage cistern. See Figs 2.25 and 2.26.

Building Regulations 1985

To meet the requirements of the Regulations two temperature-activiated devices, operating in sequence are required, as follows:

1. A non re-setting thermal cut-out to BS 3955: *Electrical controls for domestic appliances* Part 3: 1979.

2. A temperature relief valve to BS 6283: *Safety devices for use in hot-water systems* Part 2: 1982, *Specification for temperature relief valves for use at pressures up to and including 10 bar* or Part 3: 1982, *Specification for combined temperature and pressure relief valves for pressures up to and including 10 bar*. The two devices are additional to any thermostatic control which is fitted to maintainsthe temperature of the stored water.

Systems

In a directly heated system, the thermal cut-out should be on the storage system. In an indirectly heated system, the heating coil should only be connected to an energy supply fitted with a temperature-operated energy cut-out. In both directly and indirectly heated systems, the temperature relief valve, as specified in paragraph 2, should be located directly on the storage vessel, preferably within 150 mm of the top but in every case within the top 20 per cent of the volume of water in the vessel.

Installation

The unit or packaged system should be installed by an approved installer and include the installation of a discharge pipe from any safety device releasing hot water. The discharge pipe should be of suitable metal and its size should be the

same as the discharge outlet size on the safety device. The discharge should be via an air break to a tun-dish and the pipe should be laid to a continuous fall and be no longer than 9 metres, unless the bore is increased.

The pipe should discharge in a visible but safe place such as a gully, where there is no risk of contact with the hot water by persons using the building.

Hot-water temperature

The design and installation of unvented hot-water storage systems should be such that the highest water temperature at any time should not exceed 100 °C. *Note:* The system should be in the form of an approved proprietary unit or package.

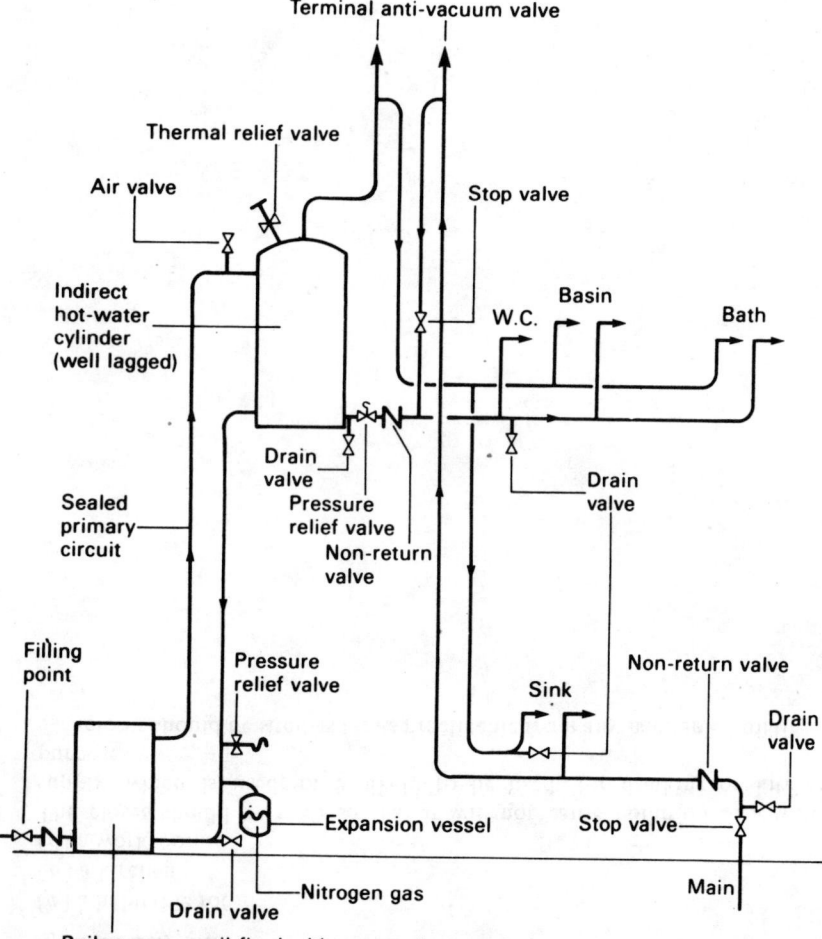

Fig. A.1 Mains-fed unvented hot water storage system

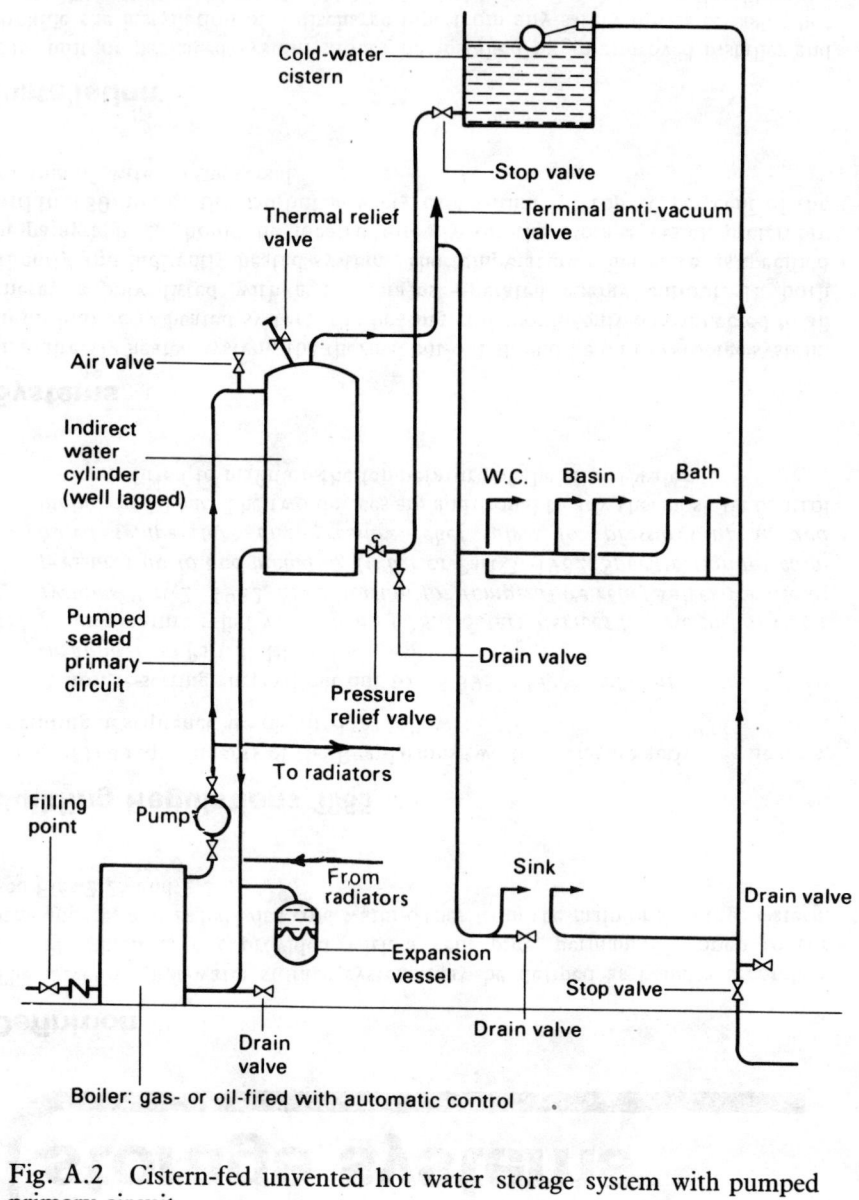

Fig. A.2 Cistern-fed unvented hot water storage system with pumped primary circuit

Bibliography

Relevant BS. British Standards Institution.

Relevant BSCP. British Standards Institution.

Building Regulations 1985, HMSO.

Relevant BRE Digests. HMSO.

C.I.B.S. Guide Book. The Chartered Institution of Building Services.

The Architects Specification. The Architectural Press.

O. Faber and J. R. Kell. *Heating and Air Conditioning of Buildings.* The Architectural Press.

Steam Heating. Spirax Sarco Limited.

R. Chudley. *Construction Technology Volume 3.* Longman Group Limited.

F. Hall. *Plumbing.* Macmillan.

P. Burberry. *Environment and Services.* Batsford Limited.

C. R. Bassett and M. Pritchard. *Heating.* Longman Group Limited.

Sub-Soil Drainage. The Pitch Fibre Association.

Relevant manufacturers' catalogues contained in the Barbour Index and Building Products Index Libraries.

Water Byelaws 1986

Index